AF538441

Design Dimensioning with Computer Graphics Applications

MECHANICAL ENGINEERING

A Series of Textbooks and Reference Books

EDITORS

1. Spring Designer's Handbook, *by Harold Carlson*
2. Computer-Aided Graphics and Design, *by Daniel L. Ryan*
3. Lubrication Fundamentals, *by J. George Wills*
4. Solar Engineering for Domestic Buildings, *by William A. Himmelman*
5. Applied Engineering Mechanics: Statics and Dynamics, *by G. Boothroyd and C. Poli*
6. Centrifugal Pump Clinic, *by Igor J. Karassik*
7. Computer-Aided Kinetics for Machine Design, *by Daniel L. Ryan*
8. Plastics Products Design Handbook, Part A: Materials and Components; Part B: Processes and Design for Processes, *edited by Edward Miller*
9. Turbomachinery: Basic Theory and Applications, *by Earl Logan, Jr.*
10. Vibrations of Shells and Plates, *by Werner Soedel*
11. Flat and Corrugated Diaphragm Design Handbook, *by Mario Di Giovanni*
12. Practical Stress Analysis in Engineering Design, *by Alexander Blake*
13. An Introduction to the Design and Behavior of Bolted Joints, *by John H. Bickford*
14. Optimal Engineering Design: Principles and Applications, *by James N. Siddall*
15. Spring Manufacturing Handbook, *by Harold Carlson*
16. Industrial Noise Control: Fundamentals and Applications, *by Lewis H. Bell*
17. Gears and Their Vibration: A Basic Approach to Understanding Gear Noise, *by J. Derek Smith*
18. Chains for Power Transmission and Material Handling: Design and Applications Handbook, *by the American Chain Association*
19. Corrosion and Corrosion Protection Handbook, *edited by Philip A. Schweitzer*
20. Gear Drive Systems: Design and Application, *by Peter Lynwander*

21. Controlling In-Plant Airborne Contaminants: Systems Design and Calculations, *by John D. Constance*
22. CAD/CAM Systems Planning and Implementation, *by Charles S. Knox*
23. Probabilistic Engineering Design: Principles and Applications, *by James N. Siddall*
24. Traction Drives: Selection and Application, *by Frederick W. Heilich III and Eugene E. Shube*
25. Finite Element Methods: An Introduction, *by Ronald L. Huston and Chris E. Passerello*
26. Mechanical Fastening of Plastics: An Engineering Handbook, *by Brayton Lincoln, Kenneth J. Gomes, and James F. Braden*
27. Lubrication in Practice, Second Edition, *edited by W. S. Robertson*
28. Principles of Automated Drafting, *by Daniel L. Ryan*
29. Practical Seal Design, *by Leonard J. Martini*
30. Engineering Documentation for CAD/CAM Applications, *by Charles S. Knox*
31. Design Dimensioning with Computer Graphics Applications, *by Jerome C. Lange*

OTHER VOLUMES IN PREPARATION

Design Dimensioning with Computer Graphics Applications

JEROME C. LANGE

Milwaukee Area Technical College
Milwaukee, Wisconsin

Marcel Dekker, Inc.   New York and Basel

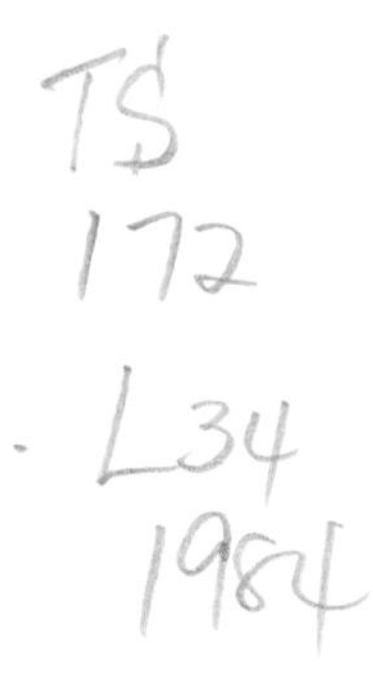

Library of Congress Cataloging in Publication Data

Lange, Jerome C.
Design dimensioning with computer graphics applications.

(Mechanical engineering; 31)
1. Tolerance (Engineering)--Data processing. 2. Design, Industrial--Data processing. 3. Computer graphics. I. Title. II. Series
TS172.L34 1984 620'.00422'0285443 84-4956
ISBN 0-8247-7119-2

MARCEL DEKKER, INC.
270 Madison Avenue, New York, New York 10016

Current printing (last digit):
10 9 8 7 6 5 4 3 2 1

PRINTED IN THE UNITED STATES OF AMERICA

I would like to dedicate this book to my father,
Henry Lange, Sr.

Preface

> One of the most important stages in the manufacture of an interchangeable product is that of design. It is here that the inventor's ideas are translated into tangible form on paper, and here the design is analyzed, perhaps modified; and finally dimensional limits are assigned to each detail component. The designer's duty is to perform the analysis, determine clearances and tolerances which will ensure correct functioning, interchangeability, and economical manufacture and inspection, and to state the acceptance conditions for each component clearly on the drawing.
>
> Even if a logical basis were available for determining the dimensions and assessing the tolerances for components, it is doubtful whether solutions could be stated with complete satisfaction on the drawings, owing to the ill-defined and inadequate vocabulary of drawing language. Tolerances essential for interchangeability have been omitted in the past because there was no ready method of indicating them, and drawings could very often be interpreted in more than one way.

These two quotations are from a paper written by Mr. C. A. Gladman which was published in Great Britain in 1944. Mr. Gladman's paper was discussed at the first America, Britain and Canada (ABC) Conference for the Unification of Engineering Standards in 1945.

Design dimensioning involves the determination and application of tolerances such that individual parts will meet the performance intent of the design and the requirements of interchangeability, therefore satisfying the serviceability, manufacturing, and assembly needs.

Tolerances were first used on engineering drawings at about the turn of the century, although muskets built by Eli Whitney in 1812 for the U.S. Army are among the earliest examples of mass-produced items with interchangeable parts. These parts were built by the use of matched or common fixtures, tools, and gages which exactly duplicated each part (precision) rather than the attainment of exact size relative to a standard measure, such as inches or millimeters (accuracy).

However, as designs became more complex and sophisticated, and parts were beginning to be manufactured at locations different from where they would be assembled, accuracy became the dominant principle of manufacturing interchangeable parts.

With the advent of computer graphics, clearances and interferences between mating parts can be actually determined visually, eliminating many laborious calculations. The fixturing and gaging necessary to mass-produce interchangeable parts and inspect them can be easily designed and drawn using computer graphics.

The intent of this book is to explain the various methods of tolerancing being used for mass production of interchangeable parts. The book also covers the determination of tolerances between mating parts such that the design and manufacturing specifications are satisfied.

The necessary fixturing and gaging used to mass-produce and inspect interchangeable parts are finally investigated with a discussion concerning the converting of inch drawings to metric.

Primarily, the book is intended for more advanced drafting and design students attending technical and engineering colleges who have completed beginning drafting courses. It can also be used as a reference book by drafters, designers, and engineers in industry.

I would like to acknowledge my appreciation to my wife, Judy, who did most of the typesetting for the technical illustrations contained in the book.

Jerome C. Lange

Contents

1

Design Dimensioning

1.1. INTRODUCTION

Proper communication is essential to human existence. Many of the world's problems are due to a lack of effective communication between people, which results in misunderstandings. In the industrial world, poor communication could cause the actual failure of a company.

The mechanical drawing is used extensively as a communication device throughout industry, primarily between design and manufacturing departments, although engineering, marketing, sales, and servicing personnel frequently use it too. A mechanical drawing (Fig. 1.1) is a graphical representation of an object or part with its shape and size shown by orthographic projection methods and the use of dimensions. The dimensions not only specify the size and shape of the object, but they maintain the design integrity (function, assembly, and serviceability) of the part during manufacturing. These dimensions are commonly referred to as design or production dimensions.

Design integrity applied to a part ensures the proper function or performance (motion, strength, fit, etc.) of that part with respect to its mating parts that make up the device, machine, or product. For example, if the impeller shaft bushing and the packing inside the waterpump of your car are worn or incorrectly assembled, coolant from the car's cooling system could leak through this bushing and packing. Losing coolant could cause the car's engine to overheat, resulting in engine damage. Designing the waterpump's bushing and packing to minimize wear (function) and ensure proper assembly (interchangeability) are important aspects of design integrity. If the bushing and

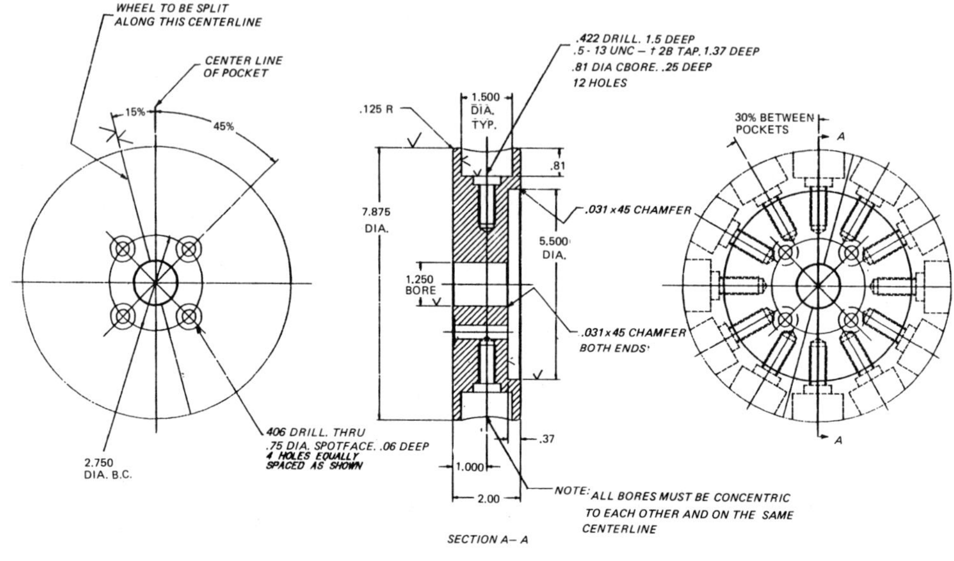

Figure 1.1 A typical mechanical drawing.

packing did fail because of fatigue, and maintenance is required, the ease of replacement (serviceability) would be another important aspect of design integrity. Design integrity of a part is achieved on a mechanical drawing through the use of design dimensions.

1.2. FUNCTION, INTERCHANGEABILITY, AND SERVICEABILITY

Imagine yourself involved in a design project at a company. The project is to calculate the dimensions necessary to place a one inch nominal shaft into a one inch nominal hole with a minimum clearance of 0.0002 and a maximum clearance of 0.0010 of an inch. Also, the particular machine to be used in the manufacturing of this shaft and hole has an accuracy of ± 0.0003 of an inch. The *function* of the shaft and hole is dependent on the clearance requirements and is referred to as an allowance while the manufacturing variance of ± 0.0003 of an inch is known as a *tolerance*. Tolerances are a result of human and machine

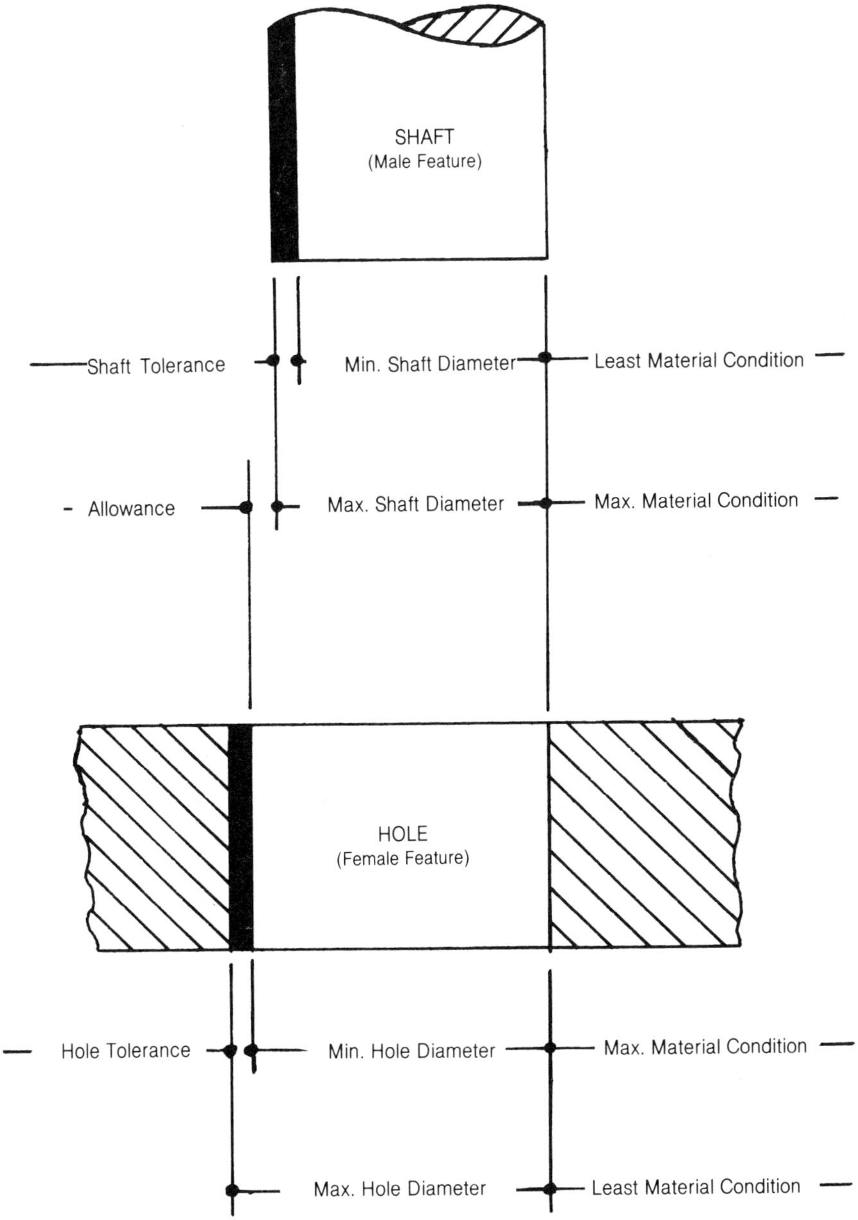

Figure 1.2 Allowance and tolerances applied to male and female features.

error and they must become part of the calculations; otherwise, it would be impossible to mass produce parts and satisfy their design requirements at the same time. A specified *fit* is the result of this calculation for both the shaft and hole. Figure 1.2 shows the relationship between the manufacturing tolerances used in the shop and engineering allowances established by the engineering department resulting in predetermined fits.

Now, imagine if more than one shaft and hole are involved in the previous design and the interrelationship between the various shafts and holes is important. The interrelationship is commonly known as *interchangeability*. As a matter of fact, interchangeability requirements are the most common use for the application of tolerances on a mechanical drawing. These tolerances satisfy two important requirements for interchangeability: (1) manufacturing and assembly needs and (2) universal serviceability.

Duplicating a part during manufacturing is referred to as precision while using dimensions and tolerances for achieving the same result is referred to as accuracy. Manufacturing requires tolerances to meet the accuracy specifications of the part so that it can be assembled correctly and easily with other parts. If for some reason, one or more of these parts fail, tolerances will ensure that the replacement part will fit and function according to previous specifications. This is referred to as universal serviceability.

Correctly assembled parts are not necessarily serviceable. For example, a part could be assembled by using clamps during the drilling operation to ensure the correct hole alignment with its mating part's holes. But if this part ever failed, there would not be any good method of making a replacement part except by using the broken part as a sample and making a duplicate of it. Historically, this has been the case for many companies in the United States.

1.3. BASIC HOLE AND BASIC SHAFT SYSTEMS

When a mechanical designer begins to dimension a part, he or she must continuously remember the influence that the mating part(s) has on the particular part they are dimensioning. With this in mind, each part that mates or fits with another part must either possess a male (basic shaft) or a female (basic hole) feature in assembly. This concept is important, because either the male or female feature must take precedence in the beginning stage of design dimensioning. For instance, the size of a 2.000/2.005-in. diameter hole could be easier to manufacture at a particular company than a 2.000/2.005-in. diameter shaft. The choice of system (basic hole or basic shaft) depends mainly on manufacturing methods and raw material sizes which vary from one company to another.

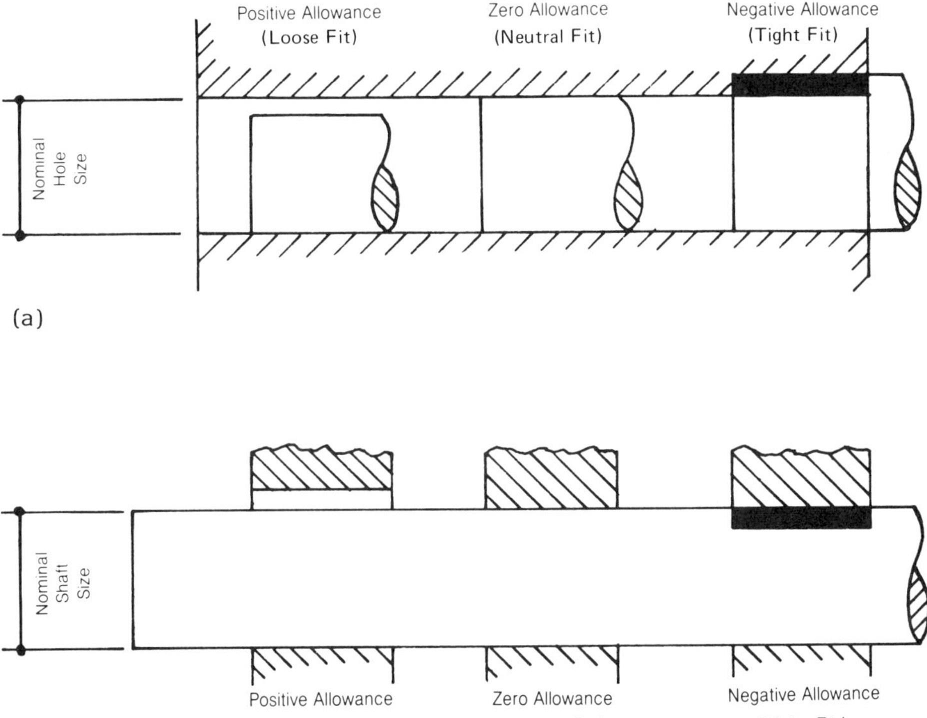

Figure 1.3 The basic (a) hole and (b) shaft systems.

The basic hole and shaft systems can be summarized by the following figures and table. Figure 1.3 shows the relationship between the basic hole and basic shaft systems. This relationship is expanded upon in Figures 1.4 and 1.5 when using the basic hole and basic shaft systems, respectively.

One can see from Figure 1.4 that manufacturing would be interested in the shaft's tolerance of 0.0006 in. and the hole tolerance of 0.0007 in., since these are the tolerance variances that are acceptable when these parts are manufactured. The engineering allowance of −0.0017 is an interference fit or allowance usually calculated by the engineer who is interested in the function and purpose of the fit rather than the actual manufacturing processes. As shown in Figure 1.5, the shaft (0.006) and the hole (0.003) tolerances would be of interest to the shop, and the clearance fit of +0.004 would be of interest to the engineering section of the company. When design dimensioning a drawing,

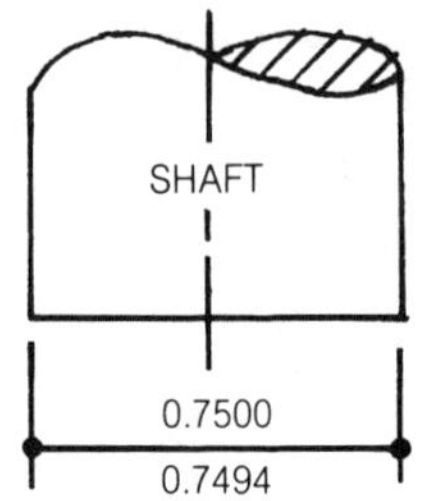

Hole Tolerance = 0.0007 in.

Shaft Tolerance = 0.0006 in.

Allowance = − 0.0017 in.

Type of Fit: Interference

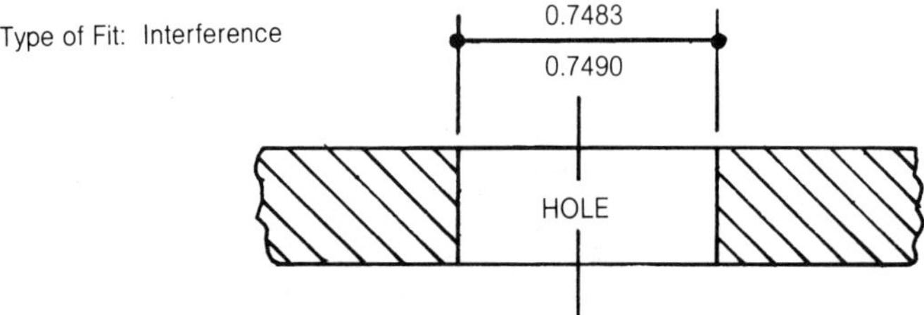

Figure 1.4 Basic shaft system.

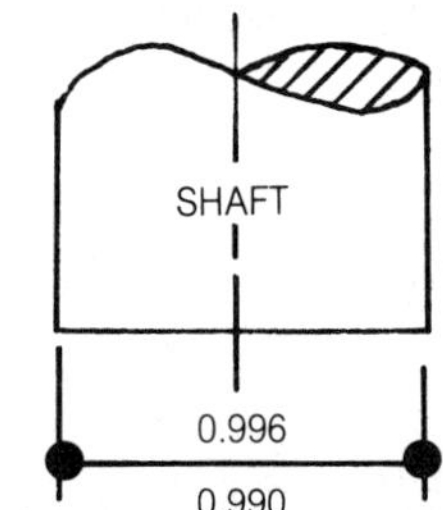

Hole Tolerance = 0.003 in.

Shaft Tolerance = 0.006 in.

Allowance = + 0.004 in.

Type of Fit: Clearance

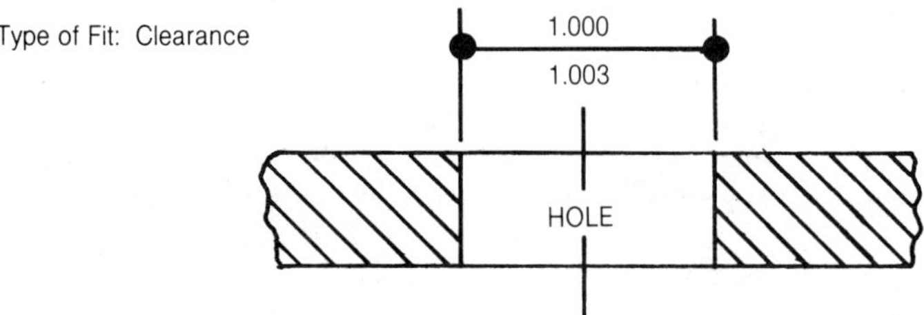

Figure 1.5 Basic hole system.

the designer must satisfy both specified allowances established by the engineer and manufacturing capabilities established by the shop.

To further illustrate the basic hole and basic shaft systems, let's use Figure 1.4 as an example. The *nominal size* or general identification of the shaft would be given as a fraction of an inch in the United States. In other countries this would be a whole number expressed in millimeters. This nominal number generally establishes the system, since 3/4 is a commonly used fraction available for raw stock; thus, enabling the company to obtain the shaft to this dimension. The usage of a decimal (0.7500) conversion of the fraction immediately implies the degree of accuracy required for allowancing and tolerancing. In this particular example, the degree of accuracy is to the fourth decimal place. This decimal is referred to as the *design size*. One must remember to have the same number of decimal places on both the allowances and the upper and lower limits of the tolerances for the hole and the shaft; otherwise, compatibility between tolerances and allowances cannot be met.

Now, if the basic hole system was used instead of the basic shaft system, the hole would be the starting point for our design dimensioning. Table 1.1 is used to compare both systems for this example.

Both systems maintain the same allowance, but the design dimension of 0.7500 moves from the maximum shaft (shaft's maximum material condition) to the minimum hole (hole's maximum material condition) depending on the system being used. Only the tolerance changes to accommodate this move; thus, keeping the allowance the same in both cases.

Table 1.1 Comparing the Shaft System to the Basic Hole System

MANUFACTURING TOLERANCES	BASIC SHAFT	BASIC HOLE
SHAFT		
a. upper limit	.7500	.7517
b. lower limit	.7494	.7511
c. shaft tolerance	.0006	.0006
HOLE		
a. upper limit	.7490	.7507
b. lower limit	.7483	.7500
c. hole tolerance	.0007	.0007
Allowances (maximum material condition)	.7500	.7517
	.7483	.7500
Maximum material condition (MMC) is defined as the minimum clearance or the maximum interference.	– .0017	– .0017

One can clearly understand how these systems of dimensioning achieve correctness of fit between two or more mating parts, and also, have the characteristics of interchangeability as an added feature. This is the reason why mechanical designers should not take the attitude that their job is just to design a part or machine so it will work; and that how it is to be made is someone else's worry. Unless careful consideration is given to tolerances (manufacturing processes) coupled with allowances (engineering design), a mechanical designer's ideas and work are worthless. For a part or machine that is impractical to manufacture and to assemble is as much a poor design as one that will not operate correctly.

1.4. NOMENCLATURE AND DEFINITIONS

A clear understanding of the terms used in design dimensioning is essential, since they are the backbone of this subject. The following definitions are from the USA Standard Limits and Fits for Cylindrical Parts, and are in accord with the American-British-Canadian Conferences.

Actual fit The actual fit between two mating parts is the relation existing between them with respect to the amount of clearance or interference which is present when they are assembled. (Fits are of three general types; clearance, transition and interferences.)

Actual size An actual size is a measured size.

Allowance An allowance is a prescribed difference between the maximum material limits of mating parts. It is a minimum clearance (positive allowance) or maximum interference (negative allowance) between such parts.

Basic hole system A basic hole system is a system of fits in which the design of the hole is the basic size and the allowance, if any, is applied to the shaft.

Basic shaft system A basic shaft system is a system of fits in which the design size of the shaft is the basic size and the allowance, if any, is applied to the hole.

Basic size The basic size is that size from which the limits of size are derived by the application of allowances and tolerances.

Bilateral tolerance A bilateral tolerance is a tolerance in which variation is permitted in both directions from the design size.

Clearance fit A clearance fit is one having limits of size so specified that a clearance always results when mating parts are assembled.

Design size The design size is the basic size with allowance applied, from which the limits of size are derived by the application of tolerances. Where there is no allowance, the design size is the same as the basic size.

Dimension A geometric characteristic such as diameter, length, angle, or center distance.

Fit Fit is the general term used to signify the range of tightness which may result from the application of a specific combination of allowances and tolerances in the design of mating parts.

Interference fit An interference fit is one having limits of size so specified that an interference always results when mating parts are assembled.

Limits of size The limits of size are applicable maximum and minimum sizes.

Maximum material limit A maximum material limit is that limit of size that provides the maximum amount of material for the part. Normally it is the maximum limit of size of an external dimension or the minimum limit of size of an internal dimension.

Minimum material limit A minimum material limit is that limit of size that provides the minimum amount of material for the part. Normally it is the minimum limit of size of an external dimension or the maximum limit of size of an internal dimension.

Nominal size The designation that is used for the purpose of general identification.

Size Size is a designation of magnitude. When a value is assigned to a dimension it is referred to as the size of that dimension.

Tolerance A tolerance is the total permissible variation of a size. The tolerance is the difference between the limits of size.

Tolerance limits A tolerance limit is the variation, positive or negative, by which a size is permitted to depart from the design size.

Transition fit A transition fit is one having limits of size so specified that either a clearance or an interference may result when mating parts are assembled.

Unilateral tolerance A unilateral tolerance is a tolerance in which variation is permitted only in one direction from the design size.

1.5. DETERMINING SIZE DIMENSIONS AND TOLERANCES

Let us now apply the foregoing definitions to the determination of the proper dimensions for two mating features. Consider the projection and slot of Figure 1.6. Design considerations indicate a *nominal* width of 1 = 1/4 inch for both features and a free sliding or clearance fit between them. The precision or accuracy required will be three decimal places; thus, the *basic size* is expressed as 1.250 in. (Fig. 1.7).

The allowance for this fit and specific size is determined next. This is usually done by consulting shop standards or other tables (Tables 1.2—1.6) which have been developed either by experience or experimentation. It is assumed the allowance for this size and fit is found to

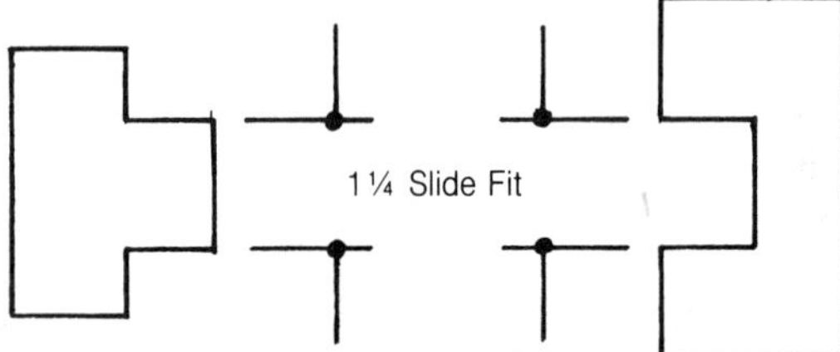

Figure 1.6 Nominal size.

be +0.004 in. Since for a clearance fit, the allowance gives the minimum clearance between the mating features, this 0.004 in. is subtracted from the basic size of the male feature (shaft), which in this case, is the projection. The result is the *design size* of both features as shown in Figure 1.8.

Next, the *tolerance* for each feature is determined, usually, from the same fit tables. Due to manufacturing, the tolerances will be assumed to be ±0.001 in. Since the allowance fixes the minimum clearance, the tolerance must be applied in such a way as to determine the maximum clearance. Thus, the tolerance will be subtracted from the design size of the projection and added to the design size of the slot. These values, together with the two design sizes, form the upper and lower limits of the features as shown in Figure 1.9.

If, however, a bilateral tolerance is required, then 0.001 is added to the design size of the slot and subtracted from the design size of the projection and the bilateral tolerance of ±0.001 is added to both of these values (Fig. 1.10).

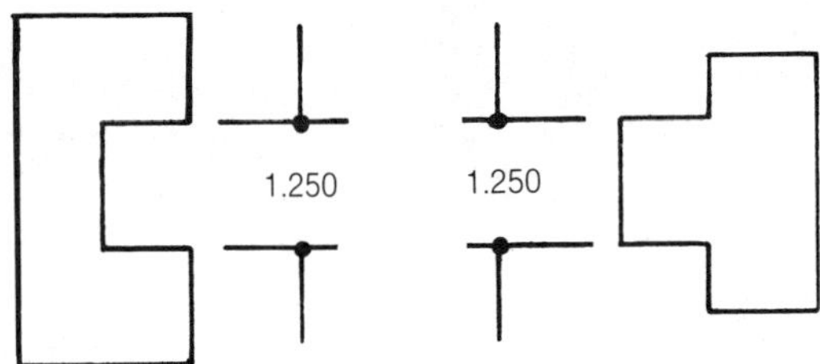

Figure 1.7 Basic size.

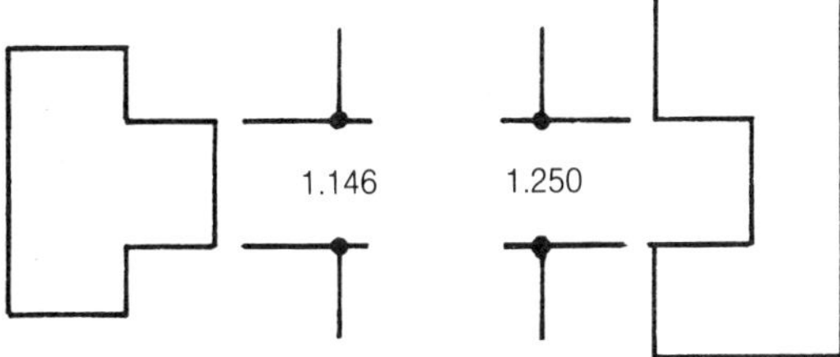

Figure 1.8 Design size.

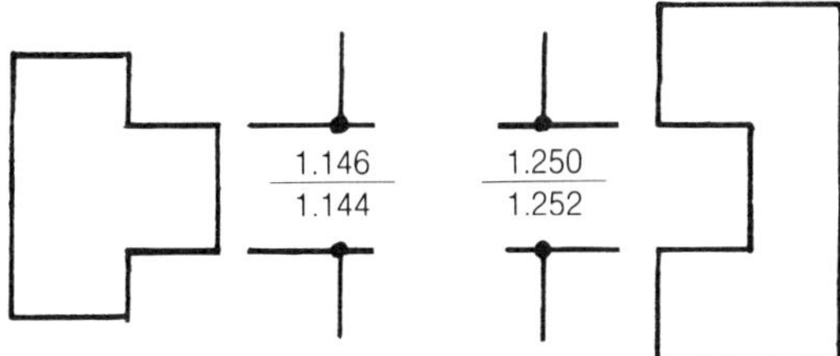

Figure 1.9 Upper and lower limits.

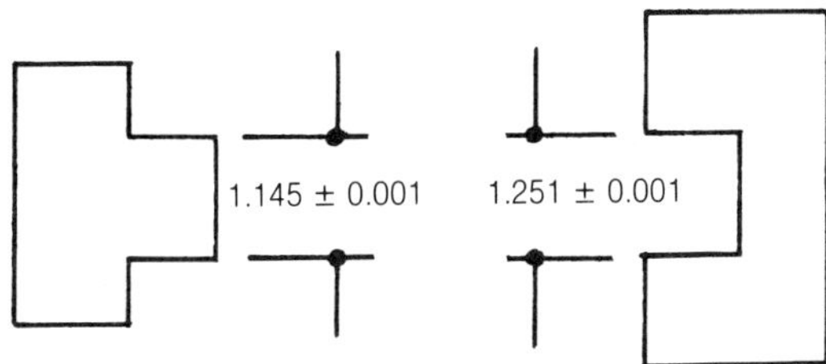

Figure 1.10 Bilateral tolerancing.

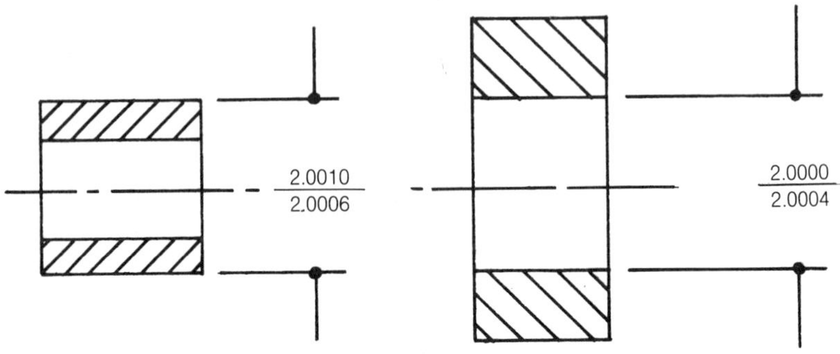

Nominal Size = 2 in.

Basic Size = 2.0000 in.

Allowance = Maximum Interference of 0.0010 in.

Tolerances = ± 0.0002 in.

Figure 1.11 A bushing and hole problem.

Figure 1.11 shows a bushing and a hole into which it fits and is retained permanently. This requires an interference fit; that is, one in which the diameter of the bushing is larger than the hole, so it must be forced into the hole. For this interference fit, the allowance is added to the basic size of the bushing to obtain its design size. Since in this case the allowance determines the maximum interference, the tolerance on the bushing is subtractive while additive for the hole.

It should be noted in both of these cases that the allowance has been applied to the hole (basic hole system). This is usually customary since its surface can be machined more easily and accurately to odd sizes by using various size reamers.

1.6. FITS, CLASSES, AND GRADES

As previously mentioned, an allowance is a prescribed difference in dimensions of mating parts and a tolerance is the total permissible variation in size of these mating parts. When an allowance is combined with its tolerance, a fit is the result. A fit is the relationship existing between two mating parts with respect to the amount of clearance of interference which is present when they are assembled.

All modern systems of fits are based on curves, for example, of the type $c = a\sqrt[3]{d^2}$, in which c is the allowance or tolerance; d is the nominal diameter in inches; and a is a constant whose value depends on the type of clearance fit required (close or loose). In mathematical form, the curves are the results of practical experience.

Fits are predetermined and provide the special relationship mating parts must have if they are to function correctly during their operation and satisfy interchangeability requirements. There are five standard classes of fits used in the United States with several grades under each class. The fits are:

RC — Running or sliding clearance fit
LC — Locational clearance fit
LT — Transition clearance or interference fit
LN — Locational interference fit
FN — Force or shrink fit

It is important to make the distinction between the class of a fit and the grade of a fit. The class of a fit for a part is determined by its allowance, the minimum permissible clearance or the maximum permissible interference, as the case may be. Accordingly, the term class may be applied also to a shaft, to indicate the class of fit it gives with a basic hole. Similarly, in a basic shaft system, a hole may be considered as being of the same class as the fit given by that hole with a basic shaft. The grade of a fit for a part is determined by the permissible variation in its relative looseness or tightness. This variation depends on the tolerance given on the mating parts, that is, on the grade of workmanship.

The concepts of "class" and "grade" are illustrated in Figures 1.12 — 1.15. The five tolerances shown at the left in Figure 1.6 apply to basic holes of the same basic diameter and vary from 0.0010 to 0.0001 inches. Since all the holes are basic, they are of the same class. However, their grades are different, improving from left to right. A similar series of tolerances is shown for basic shafts of the same class but different grades.

Figure 1.13 shows five fits having the same allowance of 0.0002 inches. Therefore, these fits are of the same class. They also have the same maximum clearance (0.0012 in.) and, therefore, the same range of variation in clearance (maximum clearance minus minimum clearance). This range is represented for each fit on the right side by the length of a heavy line. Since the five ranges are equal, the five fits are of the same grade.

The five fits in Figure 1.14 have the same allowance of 0.002 in., and, are of the same class. Their maximum clearances are different (0.0011, 0.0010, 0.0008, 0.0006 and 0.0005 in., respectively) and, therefore, also their range of variations in clearance, as indicated

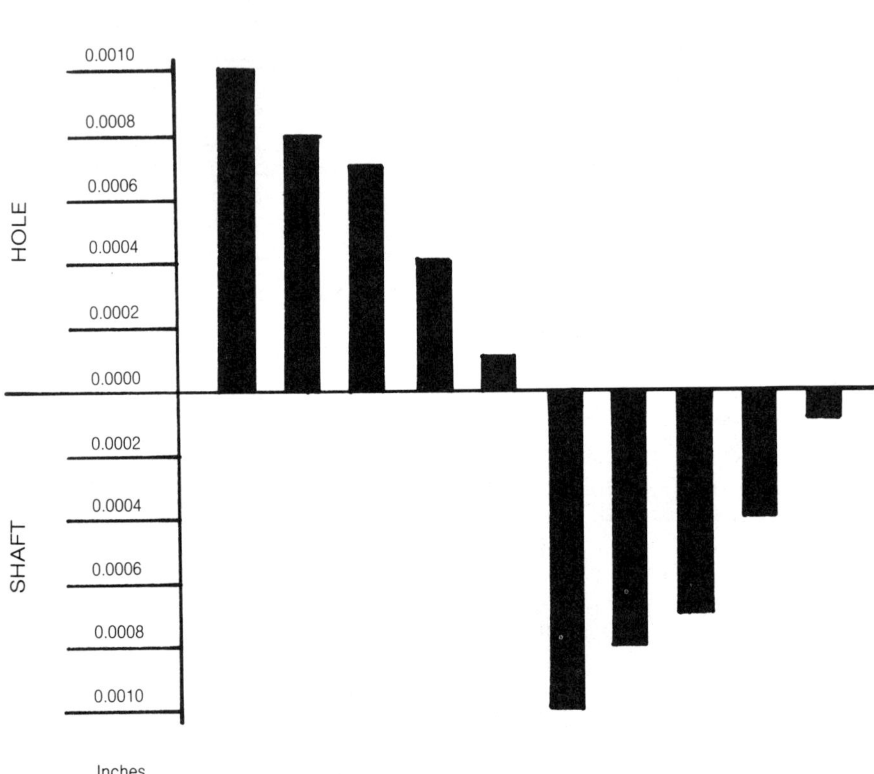

Figure 1.12 Different grades of holes and shafts (0.0010, 0.0008, 0.0007, 0.0004, and 0.0001).

again by the five heavy lines on the right. These show that the grade of the fits improves from left to right.

Finally, Figure 1.15 represents five fits, all of which are different in regard to class as well as to grade. From left to right the allowance decreases from 0.0006 to 0.0000 inches.

The classification of the standard fits appears in Tables 1.2—1.6. Running and sliding fits, listed in Table 1.2, are numbered from RC 1 through RC 9. Class RC 1 is a close-clearance fit intended for accurate location of parts which must be assembled without much play. Class 7 is a free-running fit used where accuracy is not essential and there is likely to be a large temperature change.

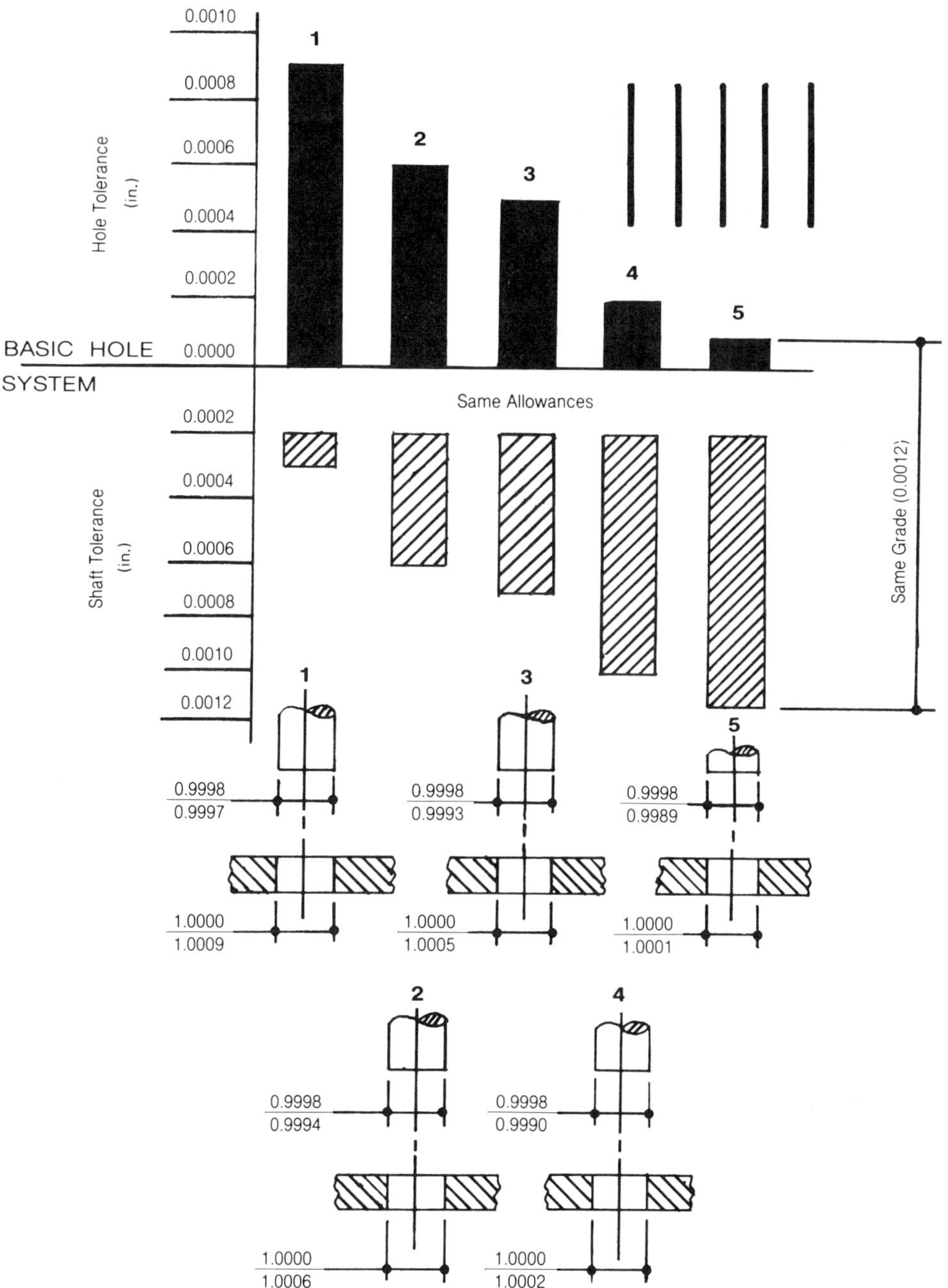

Figure 1.13 Five fits with the same class or allowance and the same grades.

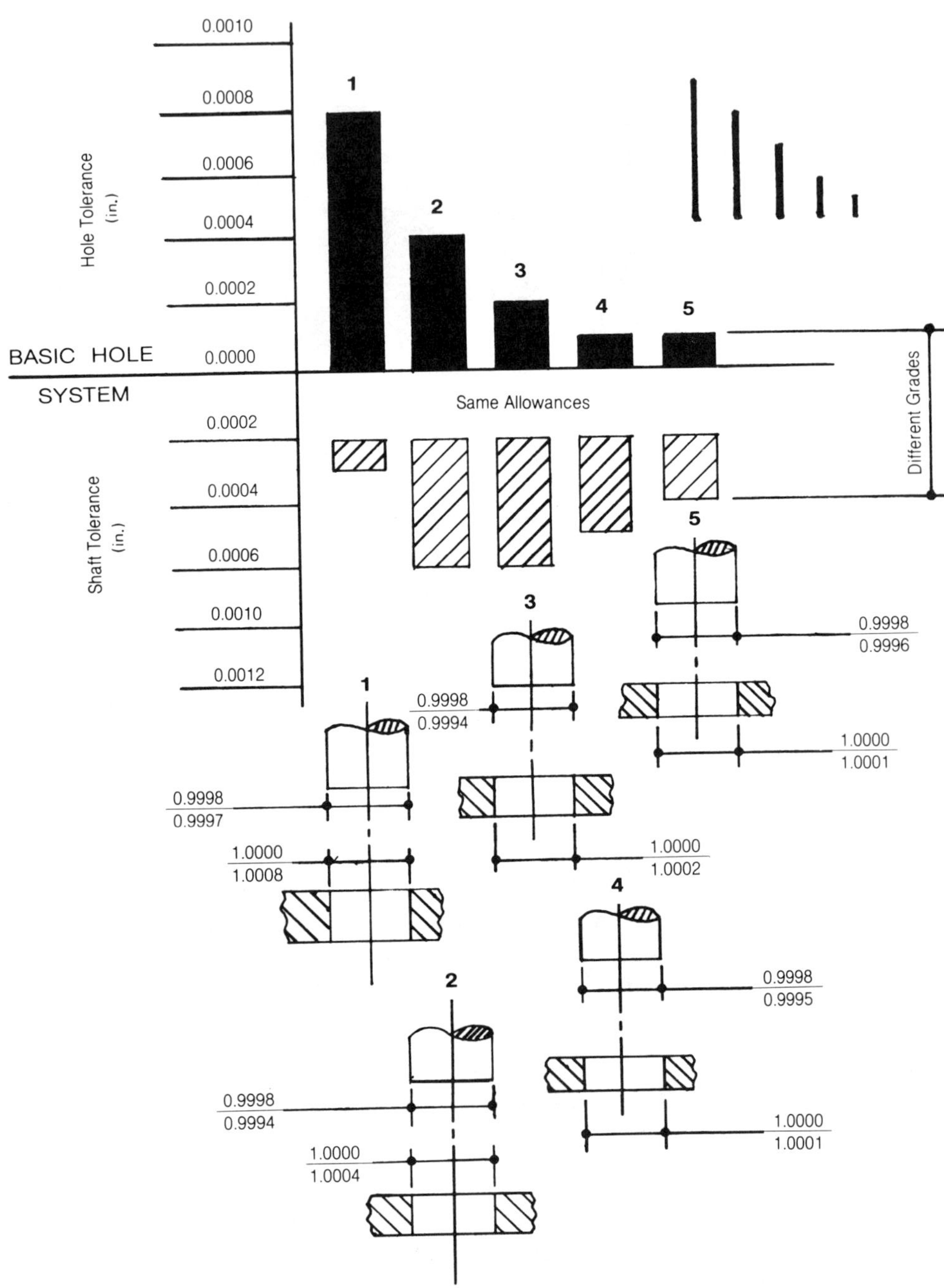

Figure 1.14 Five fits with the same class or allowance and different grades.

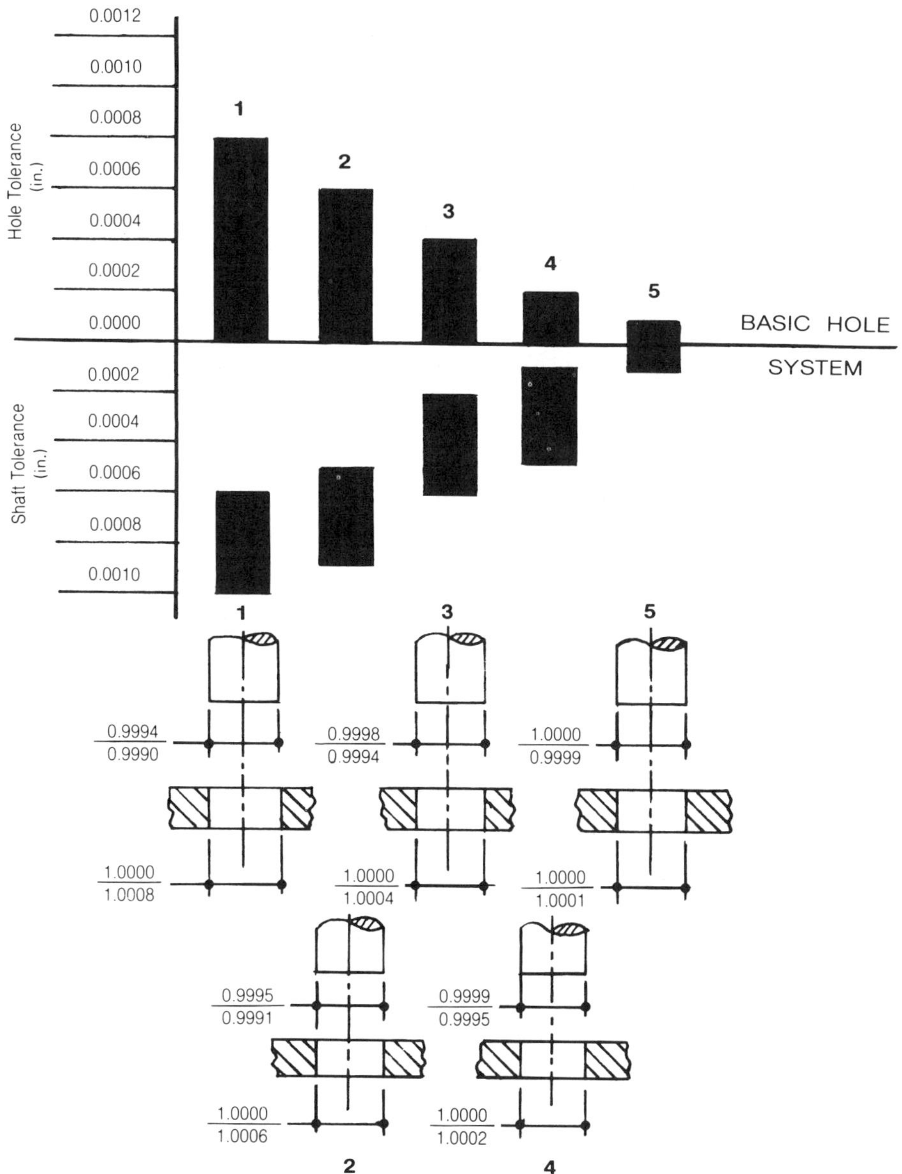

Figure 1.15 Five fits with different classes or allowances and different grades.

Table 1.2 Running and Sliding Fits

Nominal Size Range, Inches	Class RC 1			Class RC 2			Class RC 3			Class RC 4		
	Clearance*	Standard Tolerance Limits		Clearance*	Standard Tolerance Limits		Clearance*	Standard Tolerance Limits		Clearance*	Standard Tolerance Limits	
Over To		Hole H5	Shaft g4		Hole H6	Shaft g5		Hole H7	Shaft f6		Hole H8	Shaft f7
	Values shown below are in thousandths of an inch											
0– 0.12	0.1 0.45	+0.2 0	−0.1 −0.25	0.1 0.55	+0.25 0	−0.1 −0.3	0.3 0.95	+0.4 0	−0.3 −0.55	0.3 1.3	+0.6 0	−0.3 −0.7
0.12– 0.24	0.15 0.5	+0.2 0	−0.15 −0.3	0.15 0.65	+0.3 0	−0.15 −0.35	0.4 1.12	+0.5 0	−0.4 −0.7	0.4 1.6	+0.7 0	−0.4 −0.9
0.24– 0.40	0.2 0.6	+0.25 0	−0.2 −0.35	0.2 0.85	+0.4 0	−0.2 −0.45	0.5 1.5	+0.6 0	−0.5 −0.9	0.5 2.0	+0.9 0	−0.5 −1.1
0.40– 0.71	0.25 0.75	+0.3 0	−0.25 −0.45	0.25 0.95	+0.4 0	−0.25 −0.55	0.6 1.7	+0.7 0	−0.6 −1.0	0.6 2.3	+1.0 0	−0.6 −1.3
0.71– 1.19	0.3 0.95	+0.4 0	−0.3 −0.55	0.3 1.2	+0.5 0	−0.3 −0.7	0.8 2.1	+0.8 0	−0.8 −1.3	0.8 2.8	+1.2 0	−0.8 −1.6
1.19– 1.97	0.4 1.1	+0.4 0	−0.4 −0.7	0.4 1.4	+0.6 0	−0.4 −0.8	1.0 2.6	+1.0 0	−1.0 −1.6	1.0 3.6	+1.6 0	−1.0 −2.0
1.97– 3.15	0.4 1.2	+0.5 0	−0.4 −0.7	0.4 1.6	+0.7 0	−0.4 −0.9	1.2 3.1	+1.2 0	−1.2 −1.9	1.2 4.2	+1.8 0	−1.2 −2.4
3.15– 4.73	0.5 1.5	+0.6 0	−0.5 −0.9	0.5 2.0	+0.9 0	−0.5 −1.1	1.4 3.7	+1.4 0	−1.4 −2.3	1.4 5.0	+2.2 0	−1.4 −2.8
4.73– 7.09	0.6 1.8	+0.7 0	−0.6 −1.1	0.6 2.3	+1.0 0	−0.6 −1.3	1.6 4.2	+1.6 0	−1.6 −2.6	1.6 5.7	+2.5 0	−1.6 −3.2
7.09– 9.85	0.6 2.0	+0.8 0	−0.6 −1.2	0.6 2.6	+1.2 0	−0.6 −1.4	2.0 5.0	+1.8 0	−2.0 −3.2	2.0 6.6	+2.8 0	−2.0 −3.8
9.85–12.41	0.8 2.3	+0.9 0	−0.8 −1.4	0.8 2.9	+1.2 0	−0.8 −1.7	2.5 5.7	+2.0 0	−2.5 −3.7	2.5 7.5	+3.0 0	−2.5 −4.5
12.41–15.75	1.0 2.7	+1.0 0	−1.0 −1.7	1.0 3.4	+1.4 0	−1.0 −2.0	3.0 6.6	+2.2 0	−3.0 −4.4	3.0 8.7	+3.5 0	−3.0 −5.2
15.75–19.69	1.2 3.0	+1.0 0	−1.2 −2.0	1.2 3.8	+1.6 0	−1.2 −2.2	4.0 8.1	+2.5 0	−4.0 −5.6	4.0 10.5	+4.0 0	−4.0 −6.5

See footnotes at end of table.

Table 1.2 (Continued)

Nominal Size Range, Inches	Class RC 5			Class RC 6			Class RC 7			Class RC 8			Class RC 9		
	Clearance*	Standard Tolerance Limits		Clearance*	Standard Tolerance Limits		Clearance*	Standard Tolerance Limits		Clearance*	Standard Tolerance Limits		Clearance*	Standard Tolerance Limits	
Over To		Hole H8	Shaft e7		Hole H9	Shaft e8		Hole H9	Shaft d8		Hole H10	Shaft c9		Hole H11	Shaft
	Values shown below are in thousandths of an inch														
0 – 0.12	0.6 1.6	+0.6 0	− 0.6 − 1.0	0.6 2.2	+1.0 0	− 0.6 − 1.2	1.0 2.6	+1.0 0	− 1.0 − 1.6	2.5 5.1	+ 1.6 0	− 2.5 − 3.5	4.0 8.1	+ 2.5 0	− 4.0 − 5.6
0.12 – 0.24	0.8 2.0	+0.7 0	− 0.8 − 1.3	0.8 2.7	+1.2 0	− 0.8 − 1.5	1.2 3.1	+1.2 0	− 1.2 − 1.9	2.8 5.8	+ 1.8 0	− 2.8 − 4.0	4.5 9.0	+ 3.0 0	− 4.5 − 6.0
0.24 – 0.40	1.0 2.5	+0.9 0	− 1.0 − 1.6	1.0 3.3	+1.4 0	− 1.0 − 1.9	1.6 3.9	+1.4 0	− 1.6 − 2.5	3.0 6.6	+ 2.2 0	− 3.0 − 4.4	5.0 10.7	+ 3.5 0	− 5.0 − 7.2
0.40 – 0.71	1.2 2.9	+1.0 0	− 1.2 − 1.9	1.2 3.8	+1.6 0	− 1.2 − 2.2	2.0 4.6	+1.6 0	− 2.0 − 3.0	3.5 7.9	+ 2.8 0	− 3.5 − 5.1	6.0 12.8	+ 4.0 0	− 6.0 − 8.8
0.71 – 1.19	1.6 3.6	+1.2 0	− 1.6 − 2.4	1.6 4.8	+2.0 0	− 1.6 − 2.8	2.5 5.7	+2.0 0	− 2.5 − 3.7	4.5 10.0	+ 3.5 0	− 4.5 − 6.5	7.0 15.5	+ 5.0 0	− 7.0 −10.5
1.19 – 1.97	2.0 4.6	+1.6 0	− 2.0 − 3.0	2.0 6.1	+2.5 0	− 2.0 − 3.6	3.0 7.1	+2.5 0	− 3.0 − 4.6	5.0 11.5	+ 4.0 0	− 5.0 − 7.5	8.0 18.0	+ 6.0 0	− 8.0 −12.0
1.97 – 3.15	2.5 5.5	+1.8 0	− 2.5 − 3.7	2.5 7.3	+3.0 0	− 2.5 − 4.3	4.0 8.8	+3.0 0	− 4.0 − 5.8	6.0 13.5	+ 4.5 0	− 6.0 − 9.0	9.0 20.5	+ 7.0 0	− 9.0 −13.5
3.15 – 4.73	3.0 6.6	+2.2 0	− 3.0 − 4.4	3.0 8.7	+3.5 0	− 3.0 − 5.2	5.0 10.7	+3.5 0	− 5.0 − 7.2	7.0 15.5	+ 5.0 0	− 7.0 −10.5	10.0 24.0	+ 9.0 0	−10.0 −15.0
4.73 – 7.09	3.5 7.6	+2.5 0	− 3.5 − 5.1	3.5 10.0	+4.0 0	− 3.5 − 6.0	6.0 12.5	+4.0 0	− 6.0 − 8.5	8.0 18.0	+ 6.0 0	− 8.0 −12.0	12.0 28.0	+10.0 0	−12.0 −18.0
7.09 – 9.85	4.0 8.6	+2.8 0	− 4.0 − 5.8	4.0 11.3	+4.5 0	− 4.0 − 6.8	7.0 14.3	+4.5 0	− 7.0 − 9.8	10.0 21.5	+ 7.0 0	−10.0 −14.5	15.0 34.0	+12.0 0	−15.0 −22.0
9.85 – 12.41	5.0 10.0	+3.0 0	− 5.0 − 7.0	5.0 13.0	+5.0 0	− 5.0 − 8.0	8.0 16.0	+5.0 0	− 8.0 −11.0	12.0 25.0	+ 8.0 0	−12.0 −17.0	18.0 38.0	+12.0 0	−18.0 −26.0
12.41 – 15.75	6.0 11.7	+3.5 0	− 6.0 − 8.2	6.0 15.5	+6.0 0	− 6.0 − 9.5	10.0 19.5	+6.0 0	−10.0 −13.5	14.0 29.0	+ 9.0 0	−14.0 −20.0	22.0 45.0	+14.0 0	−22.0 −31.0
15.75 – 19.69	8.0 14.5	+4.0 0	− 8.0 −10.5	8.0 18.0	+6.0 0	− 8.0 −12.0	12.0 22.0	+6.0 0	−12.0 −16.0	16.0 32.0	+10.0 0	−16.0 −22.0	25.0 51.0	+16.0 0	−25.0 −35.0

All data above heavy lines are in accord with ABC agreements. Symbols H5, g4, etc. are hole and shaft designations in ABC system. Limits for sizes above 19.69 inches are also given in the USA Standard.

* Pairs of values shown represent minimum and maximum amounts of clearance resulting from application of standard tolerance limits.

Source: H. L. Horton, P. B. Schubert, and G. Garratt, eds., *Machinery's Handbook*, Industrial Press, New York, 1982.

Table 1.3 Clearance Locational Fits

Nominal Size Range, Inches	Class LC 1			Class LC 2			Class LC 3			Class LC 4			Class LC 5		
	Clearance*	Standard Tolerance Limits		Clearance*	Standard Tolerance Limits		Clearance*	Standard Tolerance Limits		Clearance*	Standard Tolerance Limits		Clearance*	Standard Tolerance Limits	
Over To		Hole H6	Shaft h5		Hole H7	Shaft h6		Hole H8	Shaft h7		Hole H10	Shaft h9		Hole H7	Shaft g6
	Values shown below are in thousandths of an inch														
0 – 0.12	0 0.45	+0.25 0	0 −0.2	0 0.65	+0.4 0	0 −0.25	0 1	+0.6 0	0 −0.4	0 2.6	+1.6 0	0 −1.0	0.1 0.75	+0.4 0	−0.1 −0.35
0.12 – 0.24	0 0.5	+0.3 0	0 −0.2	0 0.8	+0.5 0	0 −0.3	0 1.2	+0.7 0	0 −0.5	0 3.0	+1.8 0	0 −1.2	0.15 0.95	+0.5 0	−0.15 −0.45
0.24 – 0.40	0 0.65	+0.4 0	0 −0.25	0 1.0	+0.6 0	0 −0.4	0 1.5	+0.9 0	0 −0.6	0 3.6	+2.2 0	0 −1.4	0.2 1.2	+0.6 0	−0.2 −0.6
0.40 – 0.71	0 0.7	+0.4 0	0 −0.3	0 1.1	+0.7 0	0 −0.4	0 1.7	+1.0 0	0 −0.7	0 4.4	+2.8 0	0 −1.6	0.25 1.35	+0.7 0	−0.25 −0.65
0.71 – 1.19	0 0.9	+0.5 0	0 −0.4	0 1.3	+0.8 0	0 −0.5	0 2	+1.2 0	0 −0.8	0 5.5	+3.5 0	0 −2.0	0.3 1.6	+0.8 0	−0.3 −0.8
1.19 – 1.97	0 1.0	+0.6 0	0 −0.4	0 1.6	+1.0 0	0 −0.6	0 2.6	+1.6 0	0 −1	0 6.5	+4.0 0	0 −2.5	0.4 2.0	+1.0 0	−0.4 −1.0
1.97 – 3.15	0 1.2	+0.7 0	0 −0.5	0 1.9	+1.2 0	0 −0.7	0 3	+1.8 0	0 −1.2	0 7.5	+4.5 0	0 −3	0.4 2.3	+1.2 0	−0.4 −1.1
3.15 – 4.73	0 1.5	+0.9 0	0 −0.6	0 2.3	+1.4 0	0 −0.9	0 3.6	+2.2 0	0 −1.4	0 8.5	+5.0 0	0 −3.5	0.5 2.8	+1.4 0	−0.5 −1.4
4.73 – 7.09	0 1.7	+1.0 0	0 −0.7	0 2.6	+1.6 0	0 −1.0	0 4.1	+2.5 0	0 −1.6	0 10.0	+6.0 0	0 −4	0.6 3.2	+1.6 0	−0.6 −1.6
7.09 – 9.85	0 2.0	+1.2 0	0 −0.8	0 3.0	+1.8 0	0 −1.2	0 4.6	+2.8 0	0 −1.8	0 11.5	+7.0 0	0 −4.5	0.6 3.6	+1.8 0	−0.6 −1.8
9.85 – 12.41	0 2.1	+1.2 0	0 −0.9	0 3.2	+2.0 0	0 −1.2	0 5	+3.0 0	0 −2.0	0 13.0	+8.0 0	0 −5	0.7 3.9	+2.0 0	−0.7 −1.9
12.41 – 15.75	0 2.4	+1.4 0	0 −1.0	0 3.6	+2.2 0	0 −1.4	0 5.7	+3.5 0	0 −2.2	0 15.0	+9.0 0	0 −6	0.7 4.3	+2.2 0	−0.7 −2.1
15.75 – 19.69	0 2.6	+1.6 0	0 −1.0	0 4.1	+2.5 0	0 −1.6	0 6.5	+4 0	0 −2.5	0 16.0	+10.0 0	0 −6	0.8 4.9	+2.5 0	−0.8 −2.4

See footnotes at end of table.

Table 1.3 (Continued)

Nominal Size Range, Inches	Class LC 6			Class LC 7			Class LC 8			Class LC 9			Class LC 10			Class LC 11		
	Clearance*	Standard Tolerance Limits		Clearance*	Standard Tolerance Limits		Clearance*	Standard Tolerance Limits		Clearance*	Standard Tolerance Limits		Clearance*	Standard Tolerance Limits		Clearance*	Standard Tolerance Limits	
Over To		Hole H9	Shaft f8		Hole H10	Shaft e9		Hole H10	Shaft d9		Hole H11	Shaft c10		Hole H12	Shaft		Hole H13	Shaft
	Values shown below are in thousandths of an inch																	
0– 0.12	0.3 1.9	+1.0 0	−0.3 −0.9	0.6 3.2	+1.6 0	− 0.6 − 1.6	1.0 2.0	+ 1.6 0	− 1.0 − 2.0	2.5 6.6	+ 2.5 0	− 2.5 − 4.1	4 12	+ 4 0	− 4 − 8	5 17	+ 6 0	− 5 −11
0.12– 0.24	0.4 2.3	+1.2 0	−0.4 −1.1	0.8 3.8	+1.8 0	− 0.8 − 2.0	1.2 4.2	+ 1.8 0	− 1.2 − 2.4	2.8 7.6	+ 3.0 0	− 2.8 − 4.6	4.5 14.5	+ 5 0	− 4.5 − 9.5	6 20	+ 7 0	− 6 −13
0.24– 0.40	0.5 2.8	+1.4 0	−0.5 −1.4	1.0 4.6	+2.2 0	− 1.0 − 2.4	1.6 5.2	+ 2.2 0	− 1.6 − 3.0	3.0 8.7	+ 3.5 0	− 3.0 − 5.2	5 17	+ 6 0	− 5 −11	7 25	+ 9 0	− 7 −16
0.40– 0.71	0.6 3.2	+1.6 0	−0.6 −1.6	1.2 5.6	+2.8 0	− 1.2 − 2.8	2.0 6.4	+ 2.8 0	− 2.0 − 3.6	3.5 10.3	+ 4.0 0	− 3.5 − 6.3	6 20	+ 7 0	− 6 −13	8 28	+10 0	− 8 −18
0.71– 1.19	0.8 4.0	+2.0 0	−0.8 −2.0	1.6 7.1	+3.5 0	− 1.6 − 3.6	2.5 8.0	+ 3.5 0	− 2.5 − 4.5	4.5 13.0	+ 5.0 0	− 4.5 − 8.0	7 23	+ 8 0	− 7 −15	10 34	+12 0	−10 −22
1.19– 1.97	1.0 5.1	+2.5 0	−1.0 −2.6	2.0 8.5	+4.0 0	− 2.0 − 4.5	3.6 9.5	+ 4.0 0	− 3.0 − 5.5	5.0 15.0	+ 6 0	− 5.0 − 9.0	8 28	+10 0	− 8 −18	12 44	+16 0	−12 −28
1.97– 3.15	1.2 6.0	+3.0 0	−1.0 −3.0	2.5 10.0	+4.5 0	− 2.5 − 5.5	4.0 11.5	+ 4.5 0	− 4.0 − 7.0	6.0 17.5	+ 7 0	− 6.0 −10.5	10 34	+12 0	−10 −22	14 50	+18 0	−14 −32
3.15– 4.73	1.4 7.1	+3.5 0	−1.4 −3.6	3.0 11.5	+5.0 0	− 3.0 − 6.5	5.0 13.5	+ 5.0 0	− 5.0 − 8.5	7 21	+ 9 0	− 7 −12	11 39	+14 0	−11 −25	16 60	+22 0	−16 −38
4.73– 7.09	1.6 8.1	+4.0 0	−1.6 −4.1	3.5 13.5	+6.0 0	− 3.5 − 7.5	6 16	+ 6 0	− 6 −10	8 24	+10 0	− 8 −14	12 44	+16 0	−12 −28	18 68	+25 0	−18 −43
7.09– 9.85	2.0 9.3	+4.5 0	−2.0 −4.8	4.0 15.5	+7.0 0	− 4.0 − 8.5	7 18.5	+ 7 0	− 7 −11.5	10 29	+12 0	−10 −17	16 52	+18 0	−16 −34	22 78	+28 0	−22 −50
9.85–12.41	2.2 10.2	+5.0 0	−2.2 −5.2	4.5 17.5	+8.0 0	− 4.5 − 9.5	7 20	+ 8 0	− 7 −12	12 32	+12 0	−12 −20	20 60	+20 0	−20 −40	28 88	+30 0	−28 −58
12.41–15.75	2.5 12.0	+6.0 0	−2.5 −6.0	5.0 20.0	+9.0 0	− 5 −11	8 23	+ 9 0	− 8 −14	14 37	+14 0	−14 −23	22 66	+22 0	−22 −44	30 100	+35 0	−30 −65
15.75–19.69	2.8 12.8	+6.0 0	−2.8 −6.8	5.0 21.0	+10.0 0	− 5 −11	9 25	+10 0	− 9 −15	16 42	+16 0	−16 −26	25 75	+25 0	−25 −50	35 115	+40 0	−35 −75

All data above heavy lines are in accordance with American-British-Canadian (ABC) agreements. Symbols H6, H7, s6, etc. are hole and shaft designations in ABC system. Limits for sizes above 19.69 inches are not covered by ABC agreements but are given in the USA standard.

* Pairs of values shown represent minimum and maximum amounts of interference resulting from application of standard tolerance limits.

Source: H. L. Horton, P. B. Schubert, and G. Garratt, eds., *Machinery's Handbook*, Industrial Press, New York, 1982.

Table 1.4 Transition Locational Fits

Nominal Size Range, Inches	Class LT 1			Class LT 2			Class LT 3			Class LT 4			Class LT 5			Class LT 6		
	Fit*	Standard Tolerance Limits		Fit*	Standard Tolerance Limits		Fit*	Standard Tolerance Limits		Fit*	Standard Tolerance Limits		Fit*	Standard Tolerance Limits		Fit*	Standard Tolerance Limits	
Over To		Hole H7	Shaft js6		Hole H8	Shaft js7		Hole H7	Shaft k6		Hole H8	Shaft k7		Hole H7	Shaft n6		Hole H7	Shaft n7
	Values shown below are in thousandths of an inch																	
0 – 0.12	−0.12 +0.52	+0.4 0	+0.12 −0.12	−0.2 +0.8	+0.6 0	+0.2 −0.2							−0.5 +0.15	+0.4 0	+0.5 +0.25	−0.65 +0.15	+0.4 0	+0.65 +0.25
0.12– 0.24	−0.15 +0.65	+0.5 0	+0.15 −0.15	−0.25 +0.95	+0.7 0	+0.25 −0.25							−0.6 +0.2	+0.5 0	+0.6 +0.3	−0.8 +0.2	+0.5 0	+0.8 +0.3
0.24– 0.40	−0.2 +0.8	+0.6 0	+0.2 −0.2	−0.3 +1.2	+0.9 0	+0.3 −0.3	−0.5 +0.5	+0.6 0	+0.5 +0.1	−0.7 +0.8	+0.9 0	+0.7 +0.1	−0.8 +0.2	+0.6 0	+0.8 +0.4	−1.0 +0.2	+0.6 0	+1.0 +0.4
0.40– 0.71	−0.2 +0.9	+0.7 0	+0.2 −0.2	−0.35 +1.35	+1.0 0	+0.35 −0.35	−0.5 +0.6	+0.7 0	+0.5 +0.1	−0.8 +0.9	+1.0 0	+0.8 +0.1	−0.9 +0.2	+0.7 0	+0.9 +0.5	−1.2 +0.2	+0.7 0	+1.2 +0.5
0.71– 1.19	−0.25 +1.05	+0.8 0	+0.25 −0.25	−0.4 +1.6	+1.2 0	+0.4 −0.4	−0.6 +0.7	+0.8 0	+0.6 +0.1	−0.9 +1.1	+1.2 0	+0.9 +0.1	−1.1 +0.2	+0.8 0	+1.1 +0.6	−1.4 +0.2	+0.8 0	+1.4 +0.6
1.19– 1.97	−0.3 +1.3	+1.0 0	+0.3 −0.3	−0.5 +2.1	+1.6 0	+0.5 −0.5	−0.7 +0.9	+1.0 0	+0.7 +0.1	−1.1 +1.5	+1.6 0	+1.1 +0.1	−1.3 +0.3	+1.0 0	+1.3 +0.7	−1.7 +0.3	+1.0 0	+1.7 +0.7
1.97– 3.15	−0.3 +1.5	+1.2 0	+0.3 −0.3	−0.6 +2.4	+1.8 0	+0.6 −0.6	−0.8 +1.1	+1.2 0	+0.8 +0.1	−1.3 +1.7	+1.8 0	+1.3 +0.1	−1.5 +0.4	+1.2 0	+1.5 +0.8	−2.0 +0.4	+1.2 0	+2.0 +0.8
3.15– 4.73	−0.4 +1.8	+1.4 0	+0.4 −0.4	−0.7 +2.9	+2.2 0	+0.7 −0.7	−1.0 +1.3	+1.4 0	+1.0 +0.1	−1.5 +2.1	+2.2 0	+1.5 +0.1	−1.9 +0.4	+1.4 0	+1.9 +1.0	−2.4 +0.4	+1.4 0	+2.4 +1.0
4.73– 7.09	−0.5 +2.1	+1.6 0	+0.5 −0.5	−0.8 +3.3	+2.5 0	+0.8 −0.8	−1.1 +1.5	+1.6 0	+1.1 +0.1	−1.7 +2.4	+2.5 0	+1.7 +0.1	−2.2 +0.4	+1.6 0	+2.2 +1.2	−2.8 +0.4	+1.6 0	+2.8 +1.2
7.09– 9.85	−0.6 +2.4	+1.8 0	+0.6 −0.6	−0.9 +3.7	+2.8 0	+0.9 −0.9	−1.4 +1.6	+1.8 0	+1.4 +0.2	−2.0 +2.6	+2.8 0	+2.0 +0.2	−2.6 +0.4	+1.8 0	+2.6 +1.4	−3.2 +0.4	+1.8 0	+3.2 +1.4
9.85–12.41	−0.6 +2.6	+2.0 0	+0.6 −0.6	−1.0 +4.0	+3.0 0	+1.0 −1.0	−1.4 +1.8	+2.0 0	+1.4 +0.2	−2.2 +2.8	+3.0 0	+2.2 +0.2	−2.6 +0.6	+2.0 0	+2.6 +1.4	−3.4 +0.6	+2.0 0	+3.4 +1.4
12.41–15.75	−0.7 +2.9	+2.2 0	+0.7 −0.7	−1.0 +4.5	+3.5 0	+1.0 −1.0	−1.6 +2.0	+2.2 0	+1.6 +0.2	−2.4 +3.3	+3.5 0	+2.4 +0.2	−3.0 +0.6	+2.2 0	+3.0 +1.6	−3.8 +0.6	+2.2 0	+3.8 +1.6
15.75–19.69	−0.8 +3.3	+2.5 0	+0.8 −0.8	−1.2 +5.2	+4.0 0	+1.2 −1.2	−1.8 +2.3	+2.5 0	+1.8 +0.2	−2.7 +3.8	+4.0 0	+2.7 +0.2	−3.4 +0.7	+2.5 0	+3.4 +1.8	−4.3 +0.7	+2.5 0	+4.3 +1.8

All data above heavy lines are in accord with ABC agreements. Symbols H7, js6, etc. are hole and shaft designations in ABC system.
* Pairs of values shown represent maximum amount of interference (−) and maximum amount of clearance (+) resulting from application of standard tolerance limits.

Source: H. L. Horton, P. B. Schubert, and G. Garratt, eds., *Machinery's Handbook*, Industrial Press, New York, 1982.

Table 1.5 Interference Locational Fits

Nominal Size Range, Inches	Class LN 1			Class LN 2			Class LN 3		
	Limits of Interference	Standard Limits		Limits of Interference	Standard Limits		Limits of Interference	Standard Limits	
Over To		Hole H6	Shaft n5		Hole H7	Shaft p6		Hole H7	Shaft r6
	Values shown below are given in thousandths of an inch								
0 – 0.12	0 0.45	+0.25 0	+0.45 +0.25	0 0.65	+0.4 0	+0.65 +0.4	0.1 0.75	+0.4 0	+0.75 +0.5
0.12– 0.24	0 0.5	+0.3 0	+0.5 +0.3	0 0.8	+0.5 0	+0.8 +0.5	0.1 0.9	+0.5 0	+0.9 +0.6
0.24– 0.40	0 0.65	+0.4 0	+0.65 +0.4	0 1.0	+0.6 0	+1.0 +0.6	0.2 1.2	+0.6 0	+1.2 +0.8
0.40– 0.71	0 0.8	+0.4 0	+0.8 +0.4	0 1.1	+0.7 0	+1.1 +0.7	0.3 1.4	+0.7 0	+1.4 +1.0
0.71– 1.19	0 1.0	+0.5 0	+1.0 +0.5	0 1.3	+0.8 0	+1.3 +0.8	0.4 1.7	+0.8 0	+1.7 +1.2
1.19– 1.97	0 1.1	+0.6 0	+1.1 +0.6	0 1.6	+1.0 0	+1.6 +1.0	0.4 2.0	+1.0 0	+2.0 +1.4
1.97– 3.15	0.1 1.3	+0.7 0	+1.3 +0.7	0.2 2.1	+1.2 0	+2.1 +1.4	0.4 2.3	+1.2 0	+2.3 +1.6
3.15– 4.73	0.1 1.6	+0.9 0	+1.6 +1.0	0.2 2.5	+1.4 0	+2.5 +1.6	0.6 2.9	+1.4 0	+2.9 +2.0
4.73– 7.09	0.2 1.9	+1.0 0	+1.9 +1.2	0.2 2.8	+1.6 0	+2.8 +1.8	0.9 3.5	+1.6 0	+3.5 +2.5
7.09– 9.85	0.2 2.2	+1.2 0	+2.2 +1.4	0.2 3.2	+1.8 0	+3.2 +2.0	1.2 4.2	+1.8 0	+4.2 +3.0
9.85–12.41	0.2 2.3	+1.2 0	+2.3 +1.4	0.2 3.4	+2.0 0	+3.4 +2.2	1.5 4.7	+2.0 0	+4.7 +3.5
12.41–15.75	0.2 2.6	+1.4 0	+2.6 +1.6	0.3 3.9	+2.2 0	+3.9 +2.5	2.3 5.9	+2.2 0	+5.9 +4.5
15.75–19.69	0.2 2.8	+1.6 0	+2.8 +1.8	0.3 4.4	+2.5 0	+4.4 +2.8	2.5 6.6	+2.5 0	+6.6 +5.0

All data in this table are in accordance with American-British-Canadian (ABC) agreements.

Limits for sizes above 19.69 inches are not covered by ABC agreements but are given in the USA Standard.

Symbols H7, p6, etc. are hole and shaft designations in ABC system.

*Pairs of values shown represent minimum and maximum amounts of interference resulting from application of standard tolerance limits.

Source: H. L. Horton, P. B. Schubert, and G. Garratt, eds., *Machinery's Handbook,* Industrial Press, New York, 1982.

Table 1.6 Force and Shrink Fits

Nominal Size Range, Inches	Class FN 1			Class FN 2			Class FN 3			Class FN 4			Class FN 5		
	Interference*	Standard Tolerance Limits		Interference*	Standard Tolerance Limits		Interference*	Standard Tolerance Limits		Interference*	Standard Tolerance Limits		Interference*	Standard Tolerance Limits	
Over To		Hole H6	Shaft		Hole H7	Shaft s6		Hole H7	Shaft t6		Hole H7	Shaft u6		Hole H8	Shaft x7
	Values shown below are in thousandths of an inch														
0–0.12	0.05 0.5	+0.25 0	+0.5 +0.3	0.2 0.85	+0.4 0	+0.85 +0.6				0.3 0.95	+0.4 0	+0.95 +0.7	0.3 1.3	+0.6 0	+ 1.3 + 0.9
0.12–0.24	0.1 0.6	+0.3 0	+0.6 +0.4	0.2 1.0	+0.5 0	+1.0 +0.7				0.4 1.2	+0.5 0	+1.2 +0.9	0.5 1.7	+0.7 0	+ 1.7 + 1.2
0.24–0.40	0.1 0.75	+0.4 0	+0.75 +0.5	0.4 1.4	+0.6 0	+1.4 +1.0				0.6 1.6	+0.6 0	+1.6 +1.2	0.5 2.0	+0.9 0	+ 2.0 + 1.4
0.40–0.56	0.1 0.8	+0.4 0	+0.8 +0.5	0.5 1.6	+0.7 0	+1.6 +1.2				0.7 1.8	+0.7 0	+1.8 +1.4	0.6 2.3	+1.0 0	+ 2.3 + 1.6
0.56–0.71	0.2 0.9	+0.4 0	+0.9 +0.6	0.5 1.6	+0.7 0	+1.6 +1.2				0.7 1.8	+0.7 0	+1.8 +1.4	0.8 2.5	+1.0 0	+ 2.5 + 1.8
0.71–0.95	0.2 1.1	+0.5 0	+1.1 +0.7	0.6 1.9	+0.8 0	+1.9 +1.4				0.8 2.1	+0.8 0	+2.1 +1.6	1.0 3.0	+1.2 0	+ 3.0 + 2.2
0.95–1.19	0.3 1.2	+0.5 0	+1.2 +0.8	0.6 1.9	+0.8 0	+1.9 +1.4	0.8 2.1	+0.8 0	+2.1 +1.6	1.0 2.3	+0.8 0	+2.3 +1.8	1.3 3.3	+1.2 0	+ 3.3 + 2.5
1.19–1.58	0.3 1.3	+0.6 0	+1.3 +0.9	0.8 2.4	+1.0 0	+2.4 +1.8	1.0 2.6	+1.0 0	+2.6 +2.0	1.5 3.1	+1.0 0	+3.1 +2.5	1.4 4.0	+1.6 0	+ 4.0 + 3.0
1.58–1.97	0.4 1.4	+0.6 0	+1.4 +1.0	0.8 2.4	+1.0 0	+2.4 +1.8	1.2 2.8	+1.0 0	+2.8 +2.2	1.8 3.4	+1.0 0	+3.4 +2.8	2.4 5.0	+1.6 0	+ 5.0 + 4.0
1.97–2.56	0.6 1.8	+0.7 0	+1.8 +1.3	0.8 2.7	+1.2 0	+2.7 +2.0	1.3 3.2	+1.2 0	+3.2 +2.5	2.3 4.2	+1.2 0	+4.2 +3.5	3.2 6.2	+1.8 0	+ 6.2 + 5.0
2.56–3.15	0.7 1.9	+0.7 0	+1.9 +1.4	1.0 2.9	+1.2 0	+2.9 +2.2	1.8 3.7	+1.2 0	+3.7 +3.0	2.8 4.7	+1.2 0	+4.7 +4.0	4.2 7.2	+1.8 0	+ 7.2 + 6.0
3.15–3.94	0.9 2.4	+0.9 0	+2.4 +1.8	1.4 3.7	+1.4 0	+3.7 +2.8	2.1 4.4	+1.4 0	+4.4 +3.5	3.6 5.9	+1.4 0	+5.9 +5.0	4.8 8.4	+2.2 0	+ 8.4 + 7.0
3.94–4.73	1.1 2.6	+0.9 0	+2.6 +2.0	1.6 3.9	+1.4 0	+3.9 +3.0	2.6 4.9	+1.4 0	+4.9 +4.0	4.6 6.9	+1.4 0	+6.9 +6.0	5.8 9.4	+2.2 0	+ 9.4 + 8.0

See footnotes at end of table.

Table 1.6 (Continued)

Nominal Size Range, Inches	Class FN 1			Class FN 2			Class FN 3			Class FN 4			Class FN 5		
	Interference*	Standard Tolerance Limits		Interference*	Standard Tolerance Limits		Interference*	Standard Tolerance Limits		Interference*	Standard Tolerance Limits		Interference*	Standard Tolerance Limits	
Over To		Hole H6	Shaft		Hole H7	Shaft s6		Hole H7	Shaft t6		Hole H7	Shaft u6		Hole H8	Shaft x7
	Values shown below are in thousandths of an inch														
4.73–5.52	1.2 2.9	+1.0 0	+2.9 +2.2	1.9 4.5	+1.6 0	+ 4.5 + 3.5	3.4 6.0	+1.6 0	+ 6.0 + 5.0	5.4 8.0	+1.6 0	+ 8.0 + 7.0	7.5 11.6	+2.5 0	+11.6 +10.0
5.52– 6.30	1.5 3.2	+1.0 0	+3.2 +2.5	2.4 5.0	+1.6 0	+ 5.0 + 4.0	3.4 6.0	+1.6 0	+ 6.0 + 5.0	5.4 8.0	+1.6 0	+ 8.0 + 7.0	9.5 13.6	+2.5 0	+13.6 +12.0
6.30– 7.09	1.8 3.5	+1.0 0	+3.5 +2.8	2.9 5.5	+1.6 0	+ 5.5 + 4.5	4.4 7.0	+1.6 0	+ 7.0 + 6.0	6.4 9.0	+1.6 0	+ 9.0 + 8.0	9.5 13.6	+2.5 0	+13.6 +12.0
7.09– 7.88	1.8 3.8	+1.2 0	+3.8 +3.0	3.2 6.2	+1.8 0	+ 6.2 + 5.0	5.2 8.2	+1.8 0	+ 8.2 + 7.0	7.2 10.2	+1.8 0	+10.2 + 9.0	11.2 15.8	+2.8 0	+15.8 +14.0
7.88– 8.86	2.3 4.3	+1.2 0	+4.3 +3.5	3.2 6.2	+1.8 0	+ 6.2 + 5.0	5.2 8.2	+1.8 0	+ 8.2 + 7.0	8.2 11.2	+1.8 0	+11.2 +10.0	13.2 17.8	+2.8 0	+17.8 +16.0
8.86– 9.85	2.3 4.3	+1.2 0	+4.3 +3.5	4.2 7.2	+1.8 0	+ 7.2 + 6.0	6.2 9.2	+1.8 0	+ 9.2 + 8.0	10.2 13.2	+1.8 0	+13.2 +12.0	13.2 17.8	+2.8 0	+17.8 +16.0
9.85–11.03	2.8 4.9	+1.2 0	+4.9 +4.0	4.0 7.2	+2.0 0	+ 7.2 + 6.0	7.0 10.2	+2.0 0	+10.2 + 9.0	10.0 13.2	+2.0 0	+13.2 +12.0	15.0 20.0	+3.0 0	+20.0 +18.0
11.03–12.41	2.8 4.9	+1.2 0	+4.9 +4.0	5.0 8.2	+2.0 0	+ 8.2 + 7.0	7.0 10.2	+2.0 c	+10.2 + 9.0	12.0 15.2	+2.0 0	+15.2 +14.0	17.0 22.0	+3.0 0	+22.0 +20.0
12.41–13.98	3.1 5.5	+1.4 0	+5.5 +4.5	5.8 9.4	+2.2 0	+ 9.4 + 8.0	7.8 11.4	+2.2 0	+11.4 +10.0	13.8 17.4	+2.2 0	+17.4 +16.0	18.5 24.2	+3.5 0	+24.2 +22.0
13.98–15.75	3.6 6.1	+1.4 0	+6.1 +5.0	5.8 9.4	+2.2 0	+ 9.4 + 8.0	9.8 13.4	+2.2 0	+13.4 +12.0	15.8 19.4	+2.2 0	+19.4 +18.0	21.5 27.2	+3.5 0	+27.2 +25.0
15.75–17.72	4.4 7.0	+1.6 0	+7.0 +6.0	6.5 10.6	+2.5 0	+10.6 + 9.0	9.5 13.6	+2.5 0	+13.6 +12.0	17.5 21.6	+2.5 0	+21.6 +20.0	24.0 30.5	+4.0 0	+30.5 +28.0
17.72–19.69	4.4 7.0	+1.6 0	+7.0 +6.0	7.5 11.6	+2.5 0	+11.6 +10.0	11.5 15.6	+2.5 0	+15.6 +14.0	19.5 23.6	+2.5 0	+23.6 +22.0	26.0 32.5	+4.0 0	+32.5 +30.0

All data above heavy lines are in accordance with American-British-Canadian (ABC) agreements. Symbols H6, H7, s6, etc. are hole and shaft designations in ABC system. Limits for sizes above 19.69 inches are not covered by ABC agreements but are given in the USA standard.

* Pairs of values shown represent minimum and maximum amounts of interference resulting from application of standard tolerance limits.

Source: H. L. Horton, P. B. Schubert, and G. Garratt, eds., *Machinery's Handbook*, Industrial Press, New York, 1982.

Clearance locational fits, listed in Table 1.3, are intended for parts which are usually stationary but can be assembled or dissassembled easily. They are numbered from LC 1 through LC 11.

Transition locational fits, listed in Table 1.4, are used where location accuracy is important and a small degree of clearance or interference is permissible. Classification numbers are from LT 1 through LT 6.

Interference locational fits, listed in Table 1.5, are used where location accuracy is essential and slight metal interference is necessary to get the accuracy. There are only three classes: LN 1, LN 2, and LN 3.

Force and shrink fits, listed in Table 1.6, are used for permanent assembly where a radial pressure between the shaft and the hole is needed to prevent slippage. They are numbered from FN 1 through FN 5.

The previous tables guide the mechanical designer in the design dimensioning of parts that mate with each other. For example, a clearance fit of RC 3 is required between a 0.8750-in. diameter shaft and hole on a particular assembly. Table 1.2 is used to find the necessary values to be added or subtracted to or from the design size. The design size of 0.8750 falls in the nominal size range of 0.71—1.19 inches. Moving along this row horizontally until intersecting the column of class RC 3, yields +0.0008 and +0.0000 inches as the values to be added to the hole and −0.0008 and −0.0013 as the values to be subtracted from the shaft. The hole's maximum and minimum limits would be 0.8750 + 0.0008 or 0.8758 and 0.8750 + 0.0000 or 0.8750, respectively. The shaft's maximum and minimum limits are 0.8750 − 0.0008 or 0.8742 and 0.8750 − 0.0013 or 0.8737.

These tables of fits are based on the basic hole system rather than the basic shaft system. Different tables are required if the basic shaft system is to be used.

1.7. SPECIFYING DESIGN DIMENSIONS ON MECHANICAL DRAWINGS

In Figure 1.16, a mechanical drawing of a flange is shown with its associated dimensions. Extension and dimension lines are used with these dimensions to help clarify what feature or features are being dimensioned. The numerical value of the dimension is the design size of the feature, and tolerances are added and subtracted from this value to obtain the high and low values (upper and lower limits), respectively. The difference between the upper and lower limits of a design size is known as the tolerance band associated with this design size. The minimum clearance or maximum interference resulting between a shaft and a hole is commonly known as allowance or the maximum material condition (MMC).

There are two systems of placing tolerances on a mechanical drawing besides using limits. (1) The unilateral system, in which the tolerance

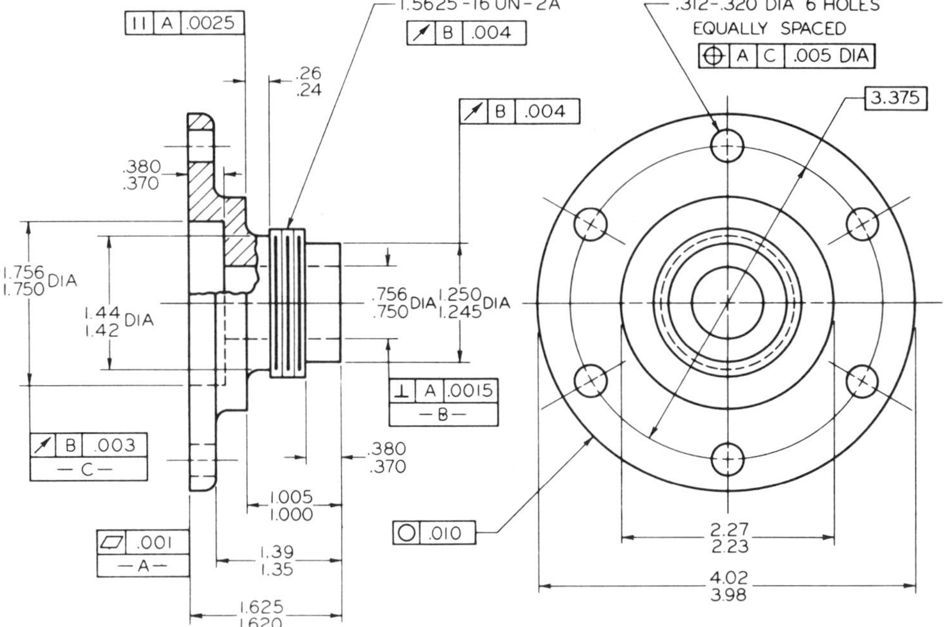

Figure 1.16 A mechanical drawing of a flange.

on each part (shaft and hole) is disposed in only one direction from the design size, plus for the hole and minus for the shaft. (2) The bilateral system, in which the tolerance extends in both directions from the design size.

In the unilateral system, variation from design size is in only one direction and may be either plus or minus, depending on whether the tolerance applies to a hole or a shaft. In this system the minimum (low) limit of the hole and the maximum (high) limit of the shaft are considered the design sizes of the respective parts, from which sizes the tolerances are applied.

Examples of unilateral and bilateral tolerances are shown in Figure 1.17 for a 1.0000-in. basic shaft with an allowance of 0.0002 of an inch. The left side of Figure 1.17 shows holes H1 and H2 with tolerances applied in the plus direction from the design size, and shafts S1 and S2 with tolerances applied in the minus direction. The fit H1/H2 is a coarser grade than H1/S1, but of a finer grade than H2/S2. Also, the fit H2/S1 is of a finer grade than H2/S2. All four combinations are of the same class and vary only in grade. The values of maximum clearance are summarized in Table 1.7.

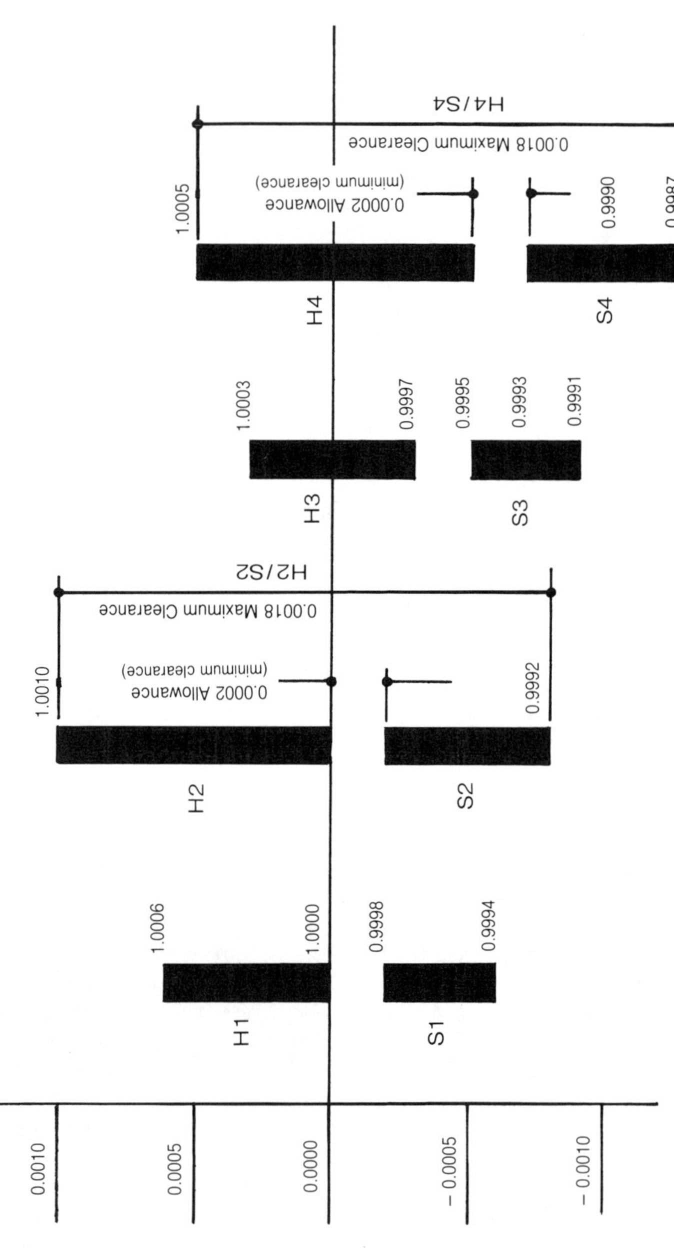

Figure 1.17 Unilateral tolerances (left) and bilateral tolerances (right).

Table 1.7 Values of Unilateral Clearances

DESCRIPTION	COMBINATION H1/S1	H1/S2	H2/S1	H2/S2
Allowance (minimum clearance)	0.0002	0.0002	0.0002	0.0002
Hole Tolerance	0.0006	0.0006	0.0010	0.0010
Shaft Tolerance	0.0004	0.0006	0.0004	0.0006
Maximum Clearance	0.0012	0.0014	0.0016	0.0018

The unilateral system permits changing tolerances on holes or shafts without impairing the essential character of the respective fits. This is an advantage in designing a new product. It may be desirable on initial tooling to adopt close tolerances H1 and S1 for the mating parts. However, after a manufacturing technique has been developed and the product proved in the field, it may be found possible to increase the tolerances of the mating parts to H2 and S2 if the increased maximum looseness is not objectionable to good performance. The adoption of larger manufacturing tolerances will realize, in most cases, lower manufacturing costs.

The right side of Figure 1.17 shows the same two fits shown in the left side, but the tolerances are specified in a different manner. The tolerance is applied in both directions from design holes H3 and H4 together with shafts S3 and S4. This system is known as the bilateral system of tolerances. It is not possible to change the tolerances on holes and shafts of this set of examples and retain the same class of fit as shown on the right. The values of maximum and minimum clearance are summarized in Table 1.8.

The combinations H3/S4 and H4/S3 result in allowances of 0.0004 and 0.0000 in., respectively, constituting two different classes of fits. Fits H3/S3 and H4/S4 each have the desired allowance 0.0002. It is not possible in a bilateral system of tolerance to change smoothly from one grade of work to another, as it is with the unilateral system.

Table 1.8 Values of Bilateral Clearances

DESCRIPTION	COMBINATION H3/S3	H3/S4	H4/S3	H4/S4
Allowance (minimum clearance)	0.0002	0.0004	0.0000	0.0002
Hole Tolerance	0.0006	0.0006	0.0010	0.0010
Shaft Tolerance	0.0004	0.0006	0.0004	0.0006
Maximum Clearance	0.0012	0.0016	0.0014	0.0018

1.8. MACHINING PROCESSES VERSUS TOLERANCE GRADES

Most shops contain lathes, jig borers, milling machines, and other metal removal machines. Each machine is designed to remove metal to a specified accuracy depending upon the size of the work, the condition of the equipment, and the skill of the machinist or operator.

To have some idea of the machining process that will be expected to produce work within a given grade, Figure 1.18 is used. Table 1.9 shows the tolerance band that goes with each grade according to the size range of the shaft or hole.

The shaft shown in Figure 1.19, for example, has a tolerance band of 0.0012 of an inch. According to Table 1.9, this is grade 7 machine work. From Figure 1.18, we would expect to have to ream the hole but the shaft could probably be turned.

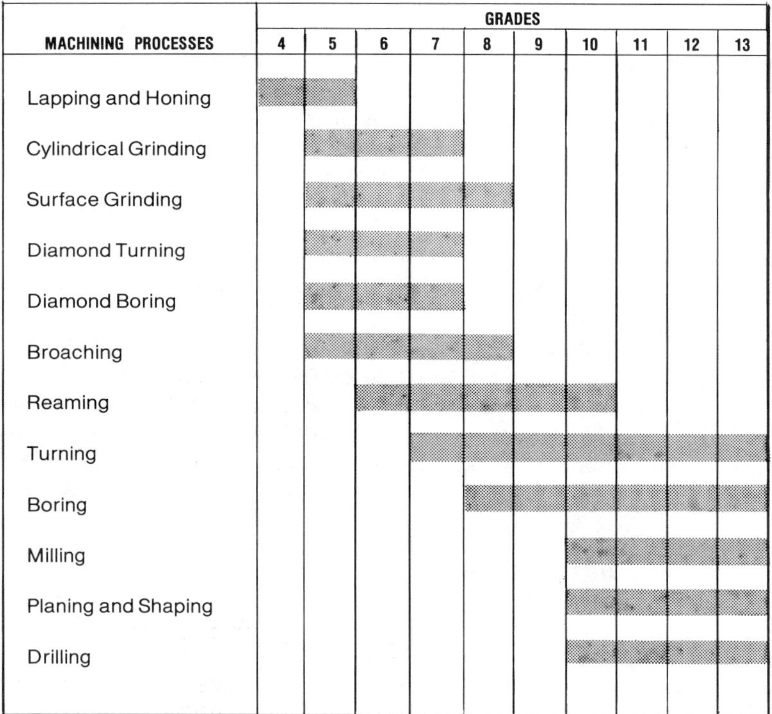

Figure 1.18 Machining processes expected to produce work within a given tolerance.

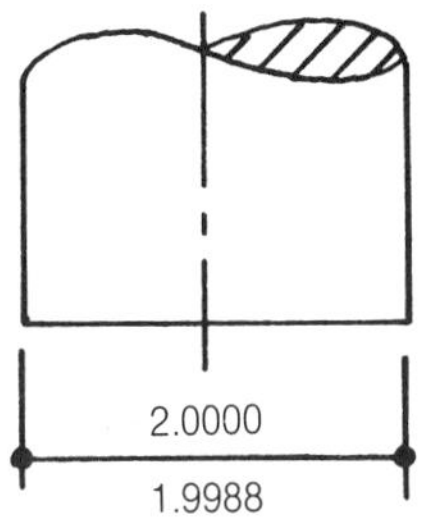

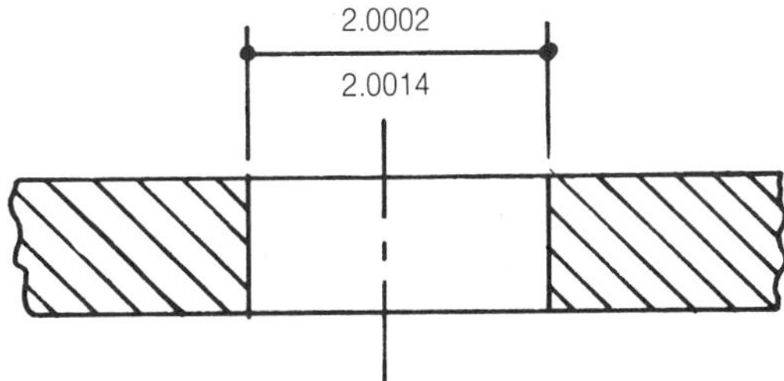

Figure 1.19 Determining the correct manufacturing processes by using grades.

Table 1.9 Standard Tolerances Arranged According to Size of Part and Grade of Machine Work

Nominal Size Range Inches Over To	Grade 4	Grade 5	Grade 6	Grade 7	Grade 8	Grade 9	Grade 10	Grade 11	Grade 12	Grade 13
0.04– 0.12	0.15	0.20	0.25	0.4	0.6	1.0	1.6	2.5	4	6
0.12– 0.24	0.15	0.20	0.3	0.5	0.7	1.2	1.8	3.0	5	7
0.24– 0.40	0.15	0.25	0.4	0.6	0.9	1.4	2.2	3.5	6	9
0.40– 0.71	0.2	0.3	0.4	0.7	1.0	1.6	2.8	4.0	7	10
0.71– 1.19	0.25	0.4	0.5	0.8	1.2	2.0	3.5	5.0	8	12
1.19– 1.97	0.3	0.4	0.6	1.0	1.6	2.5	4.0	6	10	16
1.97– 3.15	0.3	0.5	0.7	1.2	1.8	3.0	4.5	7	12	18
3.15– 4.73	0.4	0.6	0.9	1.4	2.2	3.5	5	9	14	22
4.73– 7.09	0.5	0.7	1.0	1.6	2.5	4.0	6	10	16	25
7.09– 9.85	0.6	0.8	1.2	1.8	2.8	4.5	7	12	18	28
9.85– 12.41	0.6	0.9	1.2	2.0	3.0	5.0	8	12	20	30
12.41– 15.75	0.7	1.0	1.4	2.2	3.5	6	9	14	22	35
15.75– 19.69	0.8	1.0	1.6	2.5	4	6	10	16	25	40
19.69– 30.09	0.9	1.2	2.0	3	5	8	12	20	30	50
30.09– 41.49	1.0	1.6	2.5	4	6	10	16	25	40	60
41.49– 56.19	1.2	2.0	3	5	8	12	20	30	50	80
56.19– 76.39	1.6	2.5	4	6	10	16	25	40	60	100
76.39–100.9	2.0	3	5	8	12	20	30	50	80	125
100.9– 131.9	2.5	4	6	10	16	25	40	60	100	160
131.9– 171.9	3	5	8	12	20	30	50	80	125	200
171.9– 200	4	6	10	16	25	40	60	100	160	250

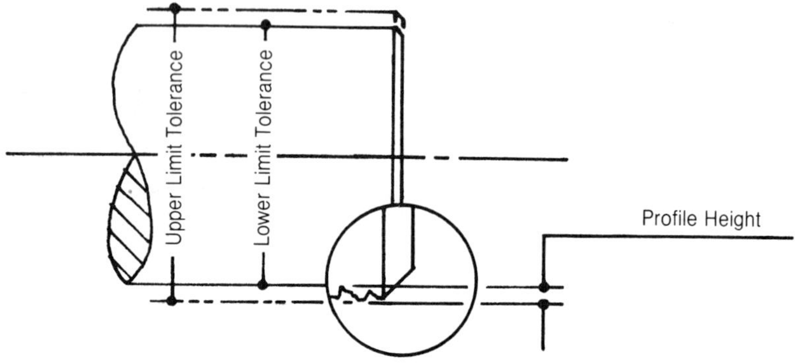

Figure 1.20 Uncertainty in size of measurement.

1.9. SURFACE QUALITY

Surface finish is an important part of measuring and determining tolerances. There is a direct relationship between the dimensional tolerance on a part and the permissible surface roughness. If one is to measure a dimension accurately, it is evident that the variations introduced by surface roughness will influence measurements taken at various locations along the surface. This uncertainty in size of measurement is shown in Figure 1.20.

It is a general rule that the variations introduced by surface roughness should not exceed the tolerances placed on the dimension. The surface finish designation is an average or a root-mean-square average of peak-to-valley heights along a specified length of the surface. Figure 1.21 shows a highly magnified surface and how a root-mean-square average is calculated.

The roughness of various manufacturing processes are shown in Figure 1.22. The increase in refinement or surface quality means an increase in manufacturing costs.

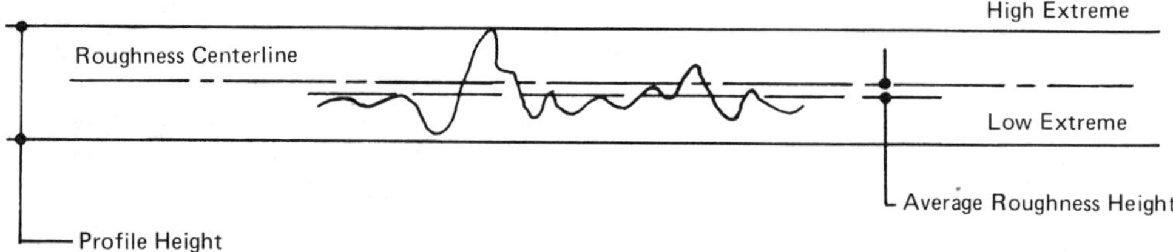

Figure 1.21 Highly magnified surface showing surface irregularities.

SURFACE ROUGHNESS: APPROXIMATE VALUES-WILL VARY WITH MATERIAL AND EQUIPMENT USED-FOR SPECIFIC VALUES CHECK WITH FABRICATION ENGR.

LEFT OF HEAVY LINE : PRACTICAL FINISHES AT COMMERCIAL COSTS

RIGHT OF HEAVY LINE : OBTAINABLE FINISHES AT INCREASED COSTS

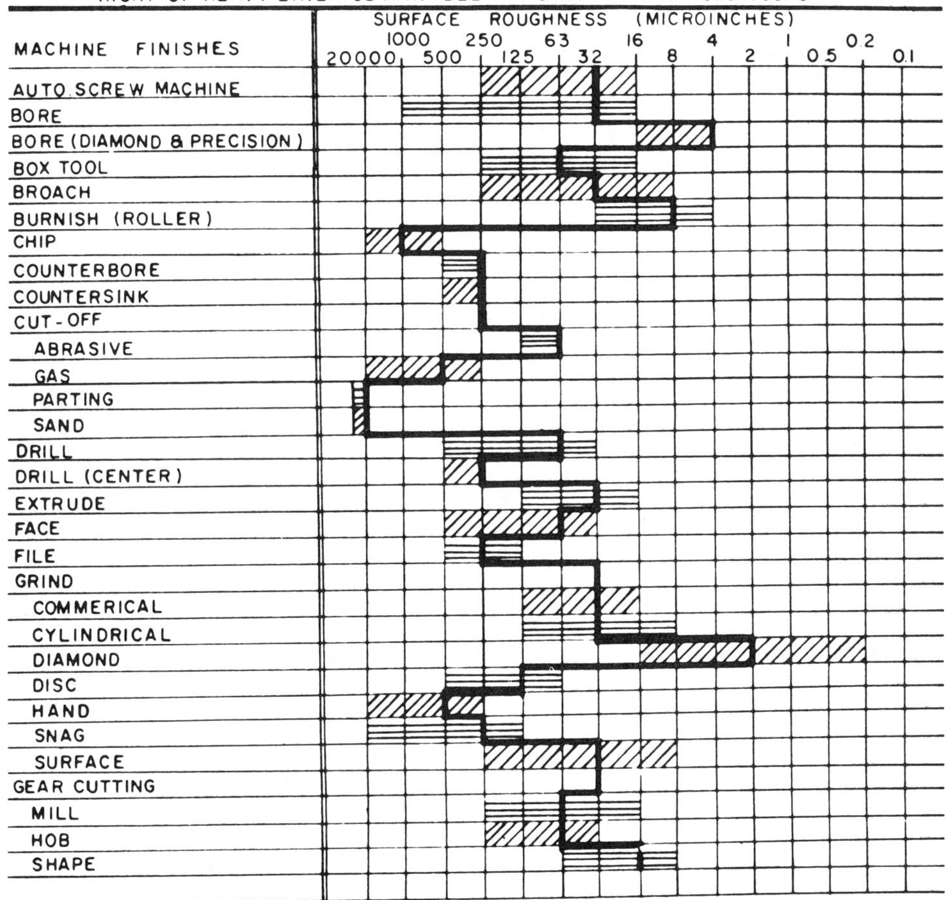

Figure 1.22 Surface texture versus process.

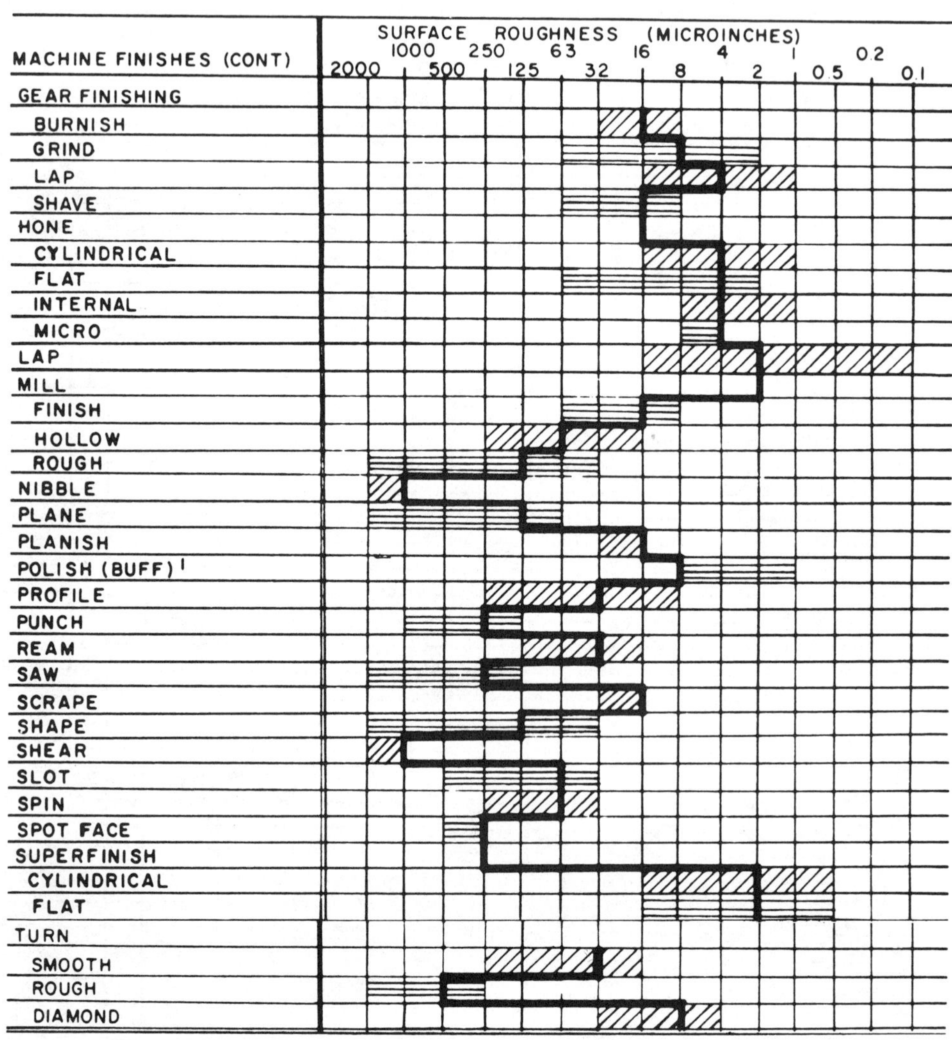

Figure 1.22 (Continued)

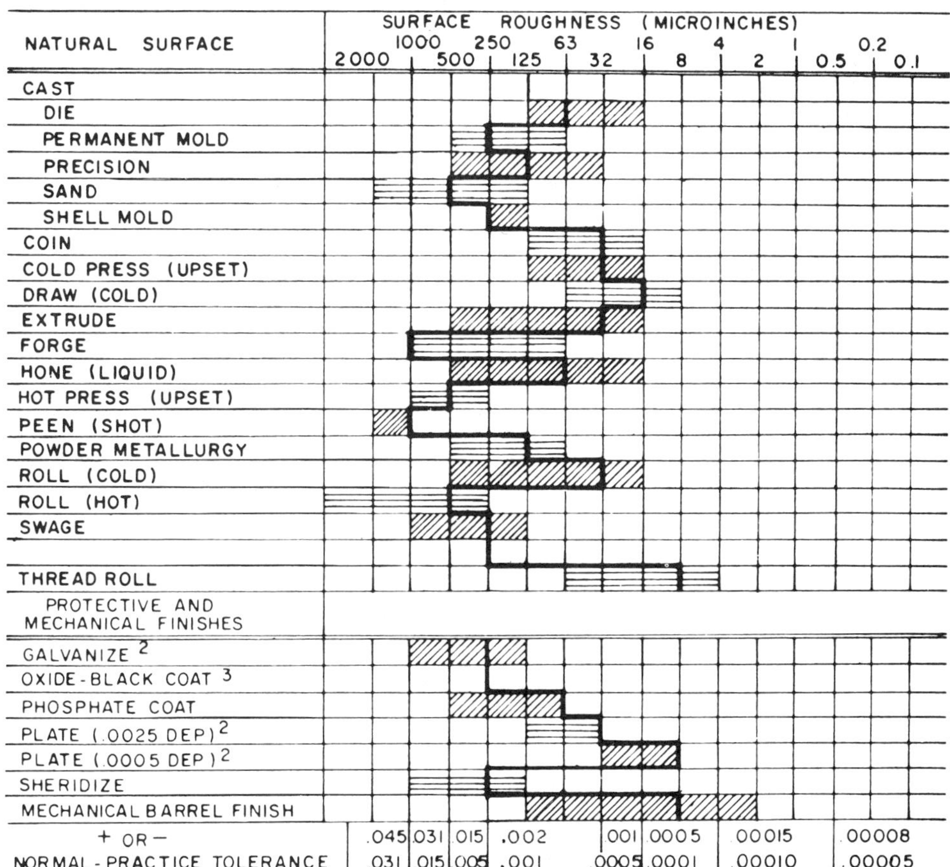

1. DEPENDENT ON PREVIOUS FINISHES, GRIT AND GRADE OF ABRASIVE.
2. ROUGHNESS INCREASES WITH THICKNESS OF DEPOSIT.
3. SURFACE ON WHICH APPLIED DOES NOT CHANGE.

Figure 1.22 (Continued)

1.10. USING DATUMS IN DESIGN DIMENSIONING

The relationship between a hole and a shaft can be specified through the use of allowances and tolerances, but frequently, there are more than one hole and shaft involved. When this is the case, the designer must ensure the proper interrelationships between the various holes and shafts regardless of allowances, tolerances, and surface quality. This task is accomplished by the use of datums.

We live in a three-dimensional world where all objects are made up of points, lines, and planes, and all mating parts of a machine, for example, either come in contact at points, along lines, or against surfaces (planes). One can understand why these geometric elements (points, lines, and planes) were chosen as datums or starting locations in our effort to describe parts on drawings and a method of manufacturing directions in the shop.

The three mutually perpendicular plane system, commonly known as the cartesian coordinate system among mathematicians, is used as the

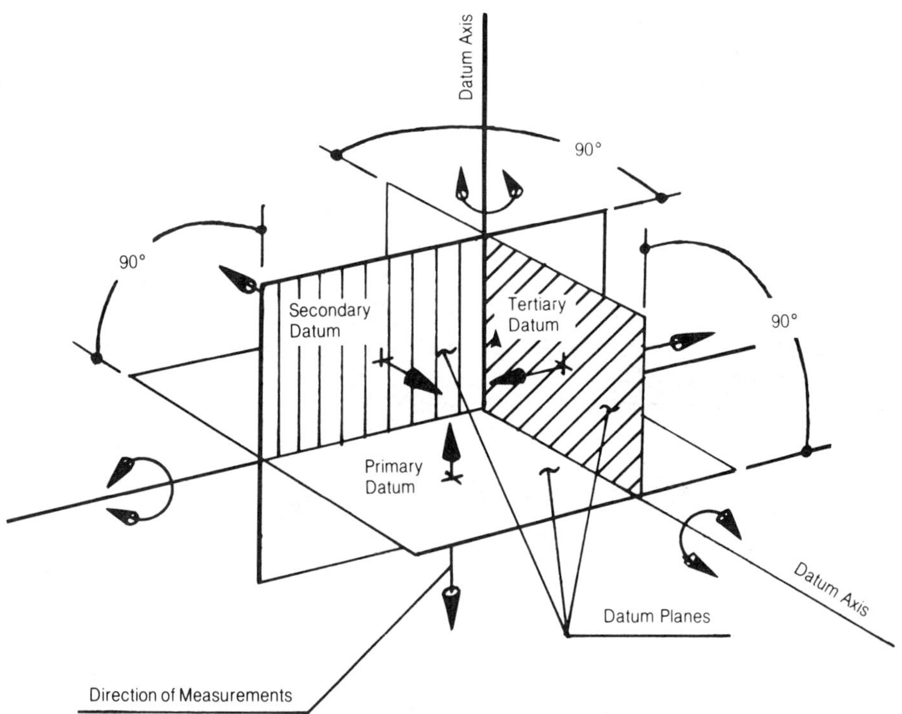

Figure 1.23 The cartesian coordinate system.

datum reference frame for dimensioning objects. These three datum planes are designated as primary, secondary, and tertiary datums (Fig. 1.23).

A point is the simplest geometric element in space. It is defined in space using the x, y, and z directions along the lines of intersection (axes) of the three mutually intersecting planes. This point is referred to as the tertiary datum. Two points define a straight line in space, referred to as the secondary datum. A primary datum consists of three points or two intersecting lines. Two intersecting lines define a plane. Therefore, a primary datum is a plane, a secondary datum is a line, and a tertiary datum is a point.

Most parts are not of the simple rectangular shape; therefore, locating the datums on irregular-shaped parts could lead to more work than is anticipated. An experienced designer can pick out these important datums quickly, since most datums are duplicated on mating parts. For instance, the centerline of a shaft, a bushing, and a hole in the frame would all share the same datum. The centerline of the shaft is the same datum as the centerline of the bushing and the centerline of the hole in the frame.

This knowledge of datum selection can only be fully understood and appreciated in real application problems. Here the designer must choose the datums on all the mating parts of a machine and dimension these parts in such a way that they will function according to design in final assembly. The applications of datums will be discussed further in the following chapters.

1.11. WHAT IS COMPUTER GRAPHICS?

A new era of enhanced creativity and productivity is possible with the application of computer graphics to the areas of mechanical drafting, design, and engineering. Computer graphics is a partnership between the computer and the human being, combining the best abilities of each to form a powerful drafting, design, and engineering team (Fig. 1.24). This interaction between people and the computers is accomplished through the use of cathode-ray tube (CRT) displays (Fig. 1.25), alphanumeric keyboards (Fig 1.26), digitizing tablets and pens (Fig. 1.27), and special computer languages.

A person is able to communicate or interact with the computer and receive a direct response from it. This two-way conversation may be graphical or pictorial in nature with the response time being only seconds. For example, the designer may generate a picture on the cathod-ray tube display by using a digitizing tablet and a pen. The computer understands the picture due to previous programming, makes calculations based on it, and presents answers or a revised picture to the designer within a few seconds. The process can be repeated, simu-

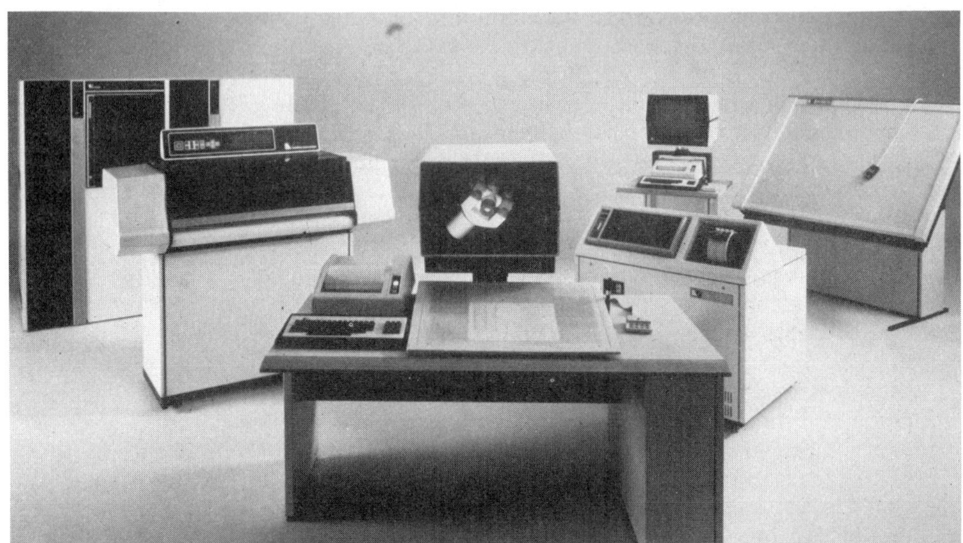

Figure 1.24 An interactive computer graphics system (ICGS).

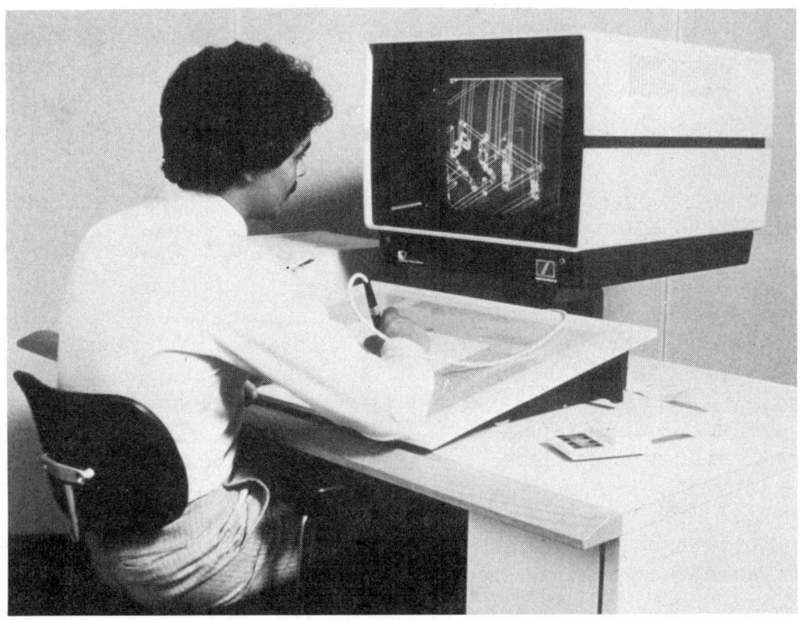

Figure 1.25 An ICGS cathode-ray tube display. (Courtesy of Computervision Corp., Bedford, Massachusetts.)

Figure 1.26 An alphanumeric keyboard. (Courtesy of Computervision Corp., Bedford, Massachusetts.)

Figure 1.27 A digitizing tablet and pen. (Courtesy of Computervision Corp., Bedford, Massachusetts.)

lating the same process that is performed on a conventional drafting board, but at a faster rate.

Most computer graphics systems (CGS) provide the same capabilities as the drawing instruments used by the drafter and designer, so that the user can construct a drawing on a CRT display in a similar manner as one would draw on a drafting board.

A part is drawn on a CGS by using a graphical vocabulary for constructing points, lines, circles, arcs, etc. The operator can call up the graphical vocabulary by use of a menu or by means of a pushbutton panel which enables the operator to use the following graphical constructions. These following graphical constructions represent only, a small portion that are available on most computer graphics systems.

A point may be placed by using any of these options:

1. Locating the point by means of a digitizing pen and a tracking cross.
2. Inputting the coordinates via the keyboard.
3. Designating the intersection of lines and/or arcs.

A line may be drawn by using any of these options:

1. Drawing it between two existing points.
2. Locating the line's end points via the digitizing pen.
3. Inputting the coordinates of the line's end points via the keyboard.
4. Drawing it through a point, tangent to a circle.
5. Drawing it parallel to an existing line at a specified distance from it.
6. Drawing it through a point at a specified angle to an existing line.
7. Drawing it tangent to two circles.

A circle may be drawn by using any of these options:

1. Drawing it with a specified center and radius.
2. Drawing it with a specified center, tangent to a line.
3. Drawing it with a specified center, tangent to another circle.

A circular arc may be drawn by using any of these options:

1. Drawing it tangent to two lines, with a specified radius.
2. Drawing it tangent to a line and an arc, with a specified radius.
3. Drawing it tangent to two arcs, with a specified radius.

In these constructions, the digitizing pen is used to select points, lines, and circles as required. In addition to being able to use the basic construction features, operators can change the scale and center of the picture at will, establish whatever local coordinate systems they wish and erase selected elements.

Figure 1.28 shows a drawing of a part which is to be copied precisely into the computer, using the CRT display and a digitizing pen. Note

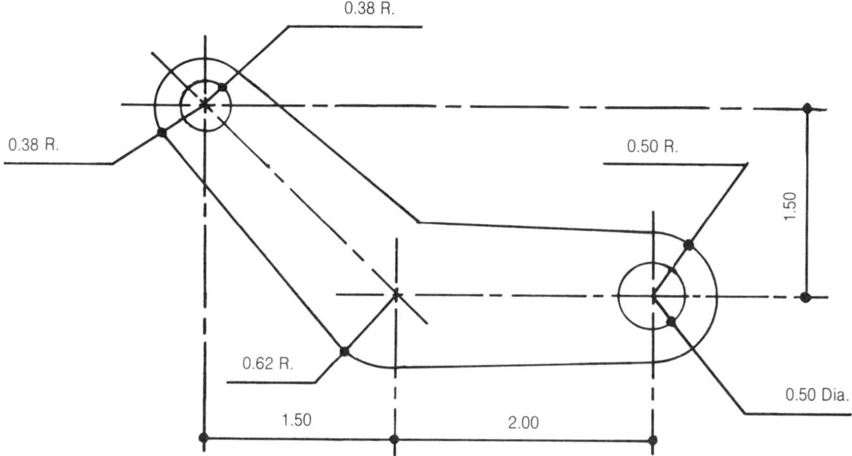

Figure 1.28 A part to be copied precisely by using computer graphics.

that the drawing gives various dimensions. As a first step, certain reference parts of the figure are entered precisely, using the keyboard to input coordinates or using the digitizing pen and numeric pick functions from a digitizing tablet. As shown in Figure 1.29, initially five circles are drawn by specifying their centers and radii. The circles then appear on the CRT display; they don't have to be drawn freehand. In the next step, tangent lines are drawn between the circles. This occurs after the user selects the "tangent" option by pushing the required button or selecting the required command from the menu and indicates which tangent line is wanted. This is done by using the digitizing pen and its tracking cross, which appears on the CRT display (Fig. 1.25). Since there are four possible tangent lines between the four circles, the computer automatically draws the tangent which is closest to the portion of the circle indicated by the tracking cross. Figure 1.30 shows the results of the procedure.

Now the three circles are erased and the arcs of the small radii and tangent to the appropriate lines are added. The result is shown in Figure 1.31. The graphics vocabulary used in this example is an early restricted type, used in the beginning of computer graphics and, because of this, some awkward sequences to perform the operations were used which would be quite simple on a drafting board. For instance, the fillets on the part would have been placed directly on the part since there is a fillet command available on most computer graphics systems today.

Computer graphics enables the designer to draw a three-dimensional (3-D) mathematical model of an object in space. All points, lines, and

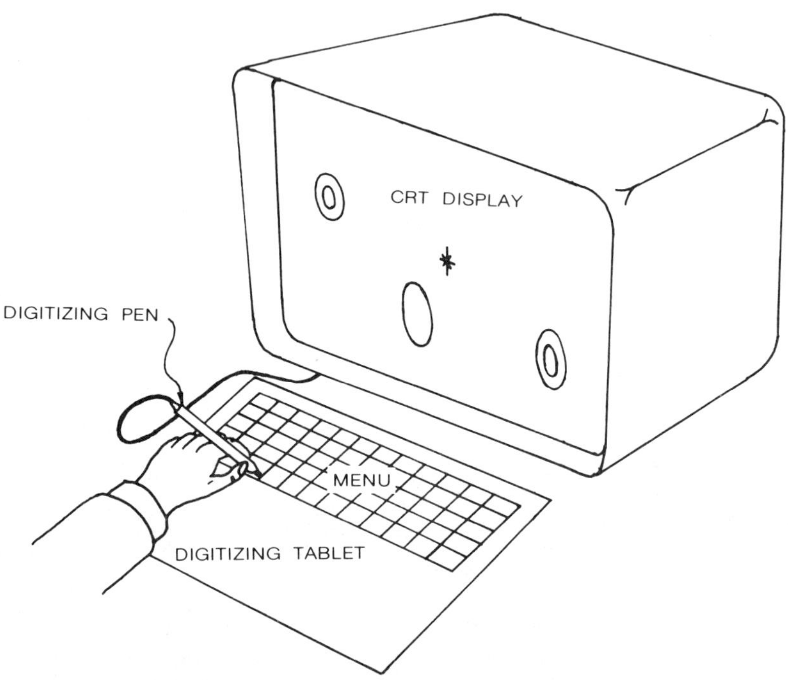

Figure 1.29 The five circles are drawn by specifying their centers and radii.

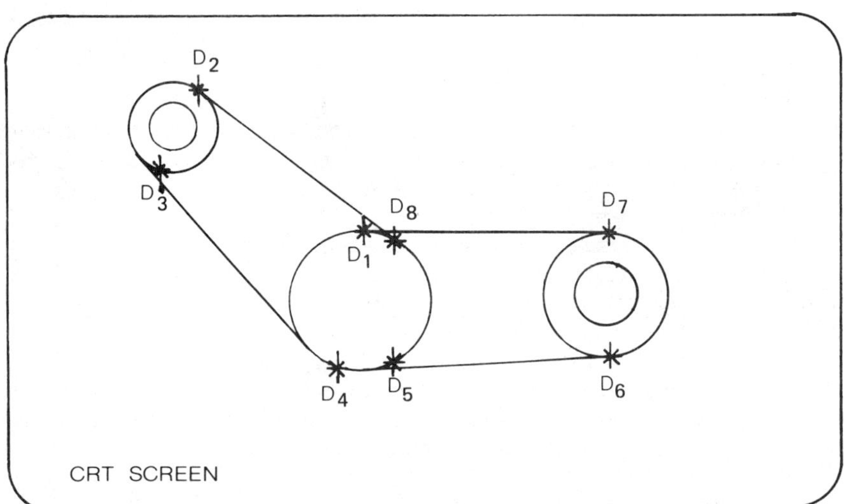

Figure 1.30 Drawing tangents to circles.

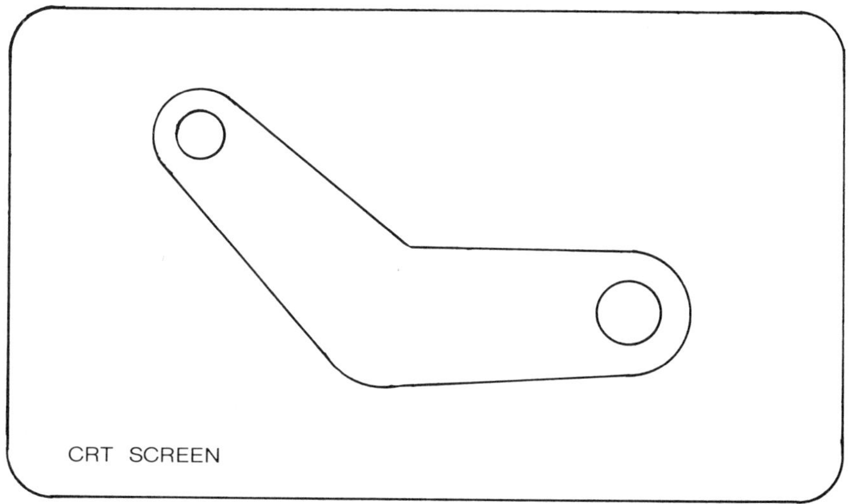

Figure 1.31 Erasing the three circles and adding the small radii.

planes making up this 3-D model are defined mathematically in space. Therefore, the designer has a powerful tool available for calculating distances, angles, lengths, areas, perimeters, volumes, and so on of geometric elements in space with extreme quickness and accuracy. Points, lines, and planes making up the object can be manipulated (rotated, moved, copied, stretched, etc.) and stored in files with ease using computer graphics. If the design needs to be altered or improved for any reason, the time for carrying out this redesign is minimal on a computer graphics system. The more design cycles that can be carried out within the limitations of time or budget the better the results.

The benefits of a computer graphics system are summarized below:

1. The user can quickly see and correct any mistakes on his or her drawing.
2. A three-dimensional view can be constructed and shown on the computer graphics cathode-ray tube (CRT) display. This three-dimensional view is a spatial model and is very beneficial to the designer and engineer during the design process and for engineering analysis work.
3. Repetitive tasks can be preprogrammed as part of the computer's software. The computer performs the monotonous tasks, so that the operator can spend more time in creative design and problem solving.
4. The data generated by a computer graphics system can be used by other computers easily. This feature is very important when

it comes to analysis, manufacturing, robotics, forecasting, marketing, scheduling, etc.

Many companies are using computer graphics for design dimensioning exclusively. The various parts comprising a machine or product are drawn on the CRT screen in their respective locations. Clearances and/or interferences between mating parts can be measured on the CRT screen to the millionth of an inch. This degree of accuracy cannot be easily obtained by using conventional drafting board techniques.

Parts can be moved easily through their operating motion by using computer graphics. This feature enables the designer to locate troublesome areas: parts conflicting with each other during their cycle of motion.

The application of computer graphics to such areas as coordinate dimensioning, geometric-position dimensioning, tolerance stackups and inch-to-metric conversion will be investigated. The questions and problems following each chapter may be used to reinforce the chapter's material.

2

Rectangular Coordinate Dimensioning and Tolerancing

2.1. GENERAL RULES, TERMS, AND PHILOSOPHY

Rectangular coordinating dimensioning is a method of placing numerical values (dimensions) in the x, y, and z directions, in order to describe a part's shape explicitly. The tolerances (manufacturing variations) are placed directly to these coordinate dimensions or to applicable tolerances specified in a general note.

In Figure 2.1, the location of a single hole is shown using coordinate dimensions and tolerances. These indicate the location of the hole's axis and result in a rectangular or wedge-type zone within which the axis of the hole must lie.

A square tolerance zone (Fig. 2.2) results if the two coordinate tolerances are equal, while unequal tolerances form a rectangular tolerance zone (Fig. 2.3). When polar dimensioning is used and one of the locating dimensions is a radius, it will yield a circular section of a tolerance zone (Fig. 2.4).

These tolerance zones extend for the full depth of the hole, that is, the whole length of the axis. Most views show this axis as a point and the tolerance zone as an area. The reader must remember that these zones are three-dimensional and extend through the part (Figs. 2.2—2.4).

The maximum allowable variation of the feature axis may be anywhere within the rectangular tolerance zone. For a square tolerance zone, the maximum allowabe variation of the axis occurs along a 45 degree line starting from the center of this tolerance zone (Fig. 2.3). This line is the diagonal of the tolerance zone. For rectangular tolerance

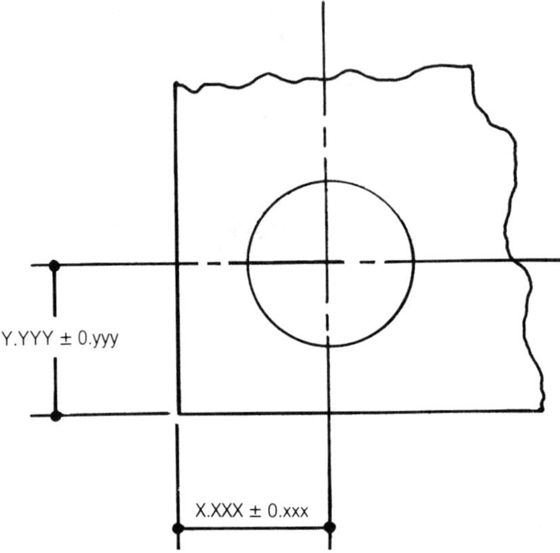

Figure 2.1 Locating a single hole.

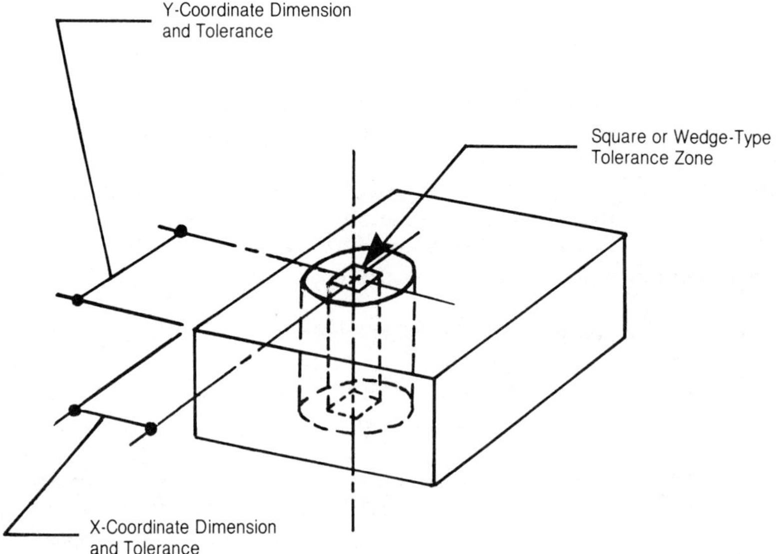

Figure 2.2 A square tolerance zone.

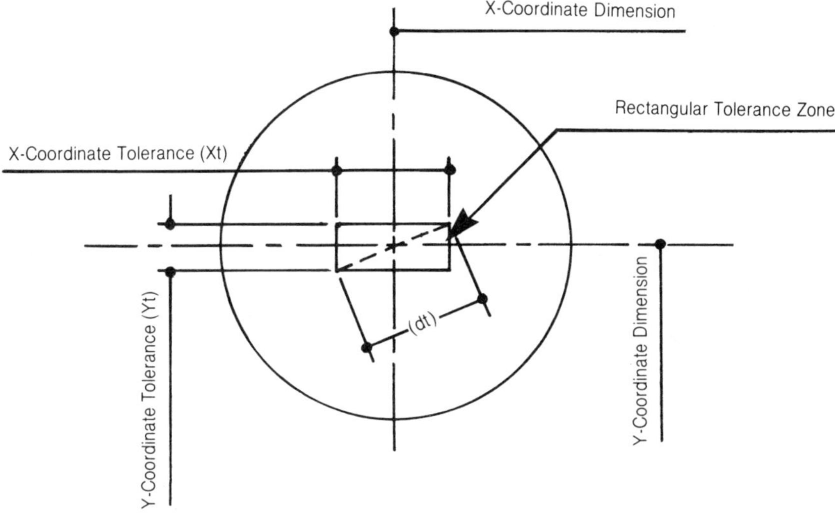

Figure 2.3 A rectangular tolerance zone resulting from unequal tolerances.

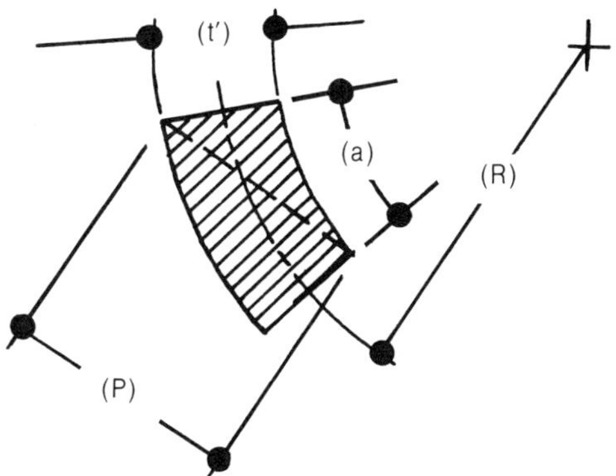

Figure 2.4 A tolerance zone from polar dimensioning.

zones, this maximum tolerance (length of this diagonal) is the square root of the sum of the squares of the individual tolerances, or expressed mathematically as $dt = \sqrt{(xt)^2 + (yt)^2}$ where dt = mazimum tolerance variation, xt = tolerance along the x axis and yt = tolerance along the y axis. For polar coordinates, extreme variation is shown in Figure 2.4. The extreme variation is given by $p = \sqrt{(c)^2 + (t)^2}$ where p = maximum tolerance variation, c = 2 r tan a/2, a = angular tolerance, t = t' cos a/2, t' = tolerance on radius and R = mean radius.

In determining the deviation of part surfaces from perfect form as implied by the drawings, the range of variation is measured. Gaging or measuring this variation is usually done by the use of indicating gages, which play an important part in open set-up inspection (inspection of a part using a surface plate, a right-angle plate, and an indicating gage).

There are many types of such indicating gages available. They may be mechanical, optical, electrical, or pneumatic in operation. Standard mechanical dial indicators (Fig. 2.5) would be used to measure the total surface variation, more commonly known as total indicator reading (TIR). Since the indicator measures the surface variation, the difference in readings between the highest and the lowest points will be the total variation or total indicator reading as shown in Figure 2.6.

Rectangular coordinate tolerancing is simple and easily understood; therefore, it is a method commonly used in the United States. Also,

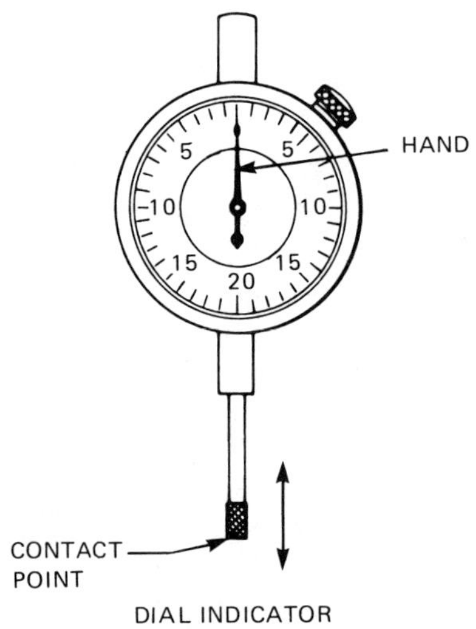

Figure 2.5 A standard dial indicator.

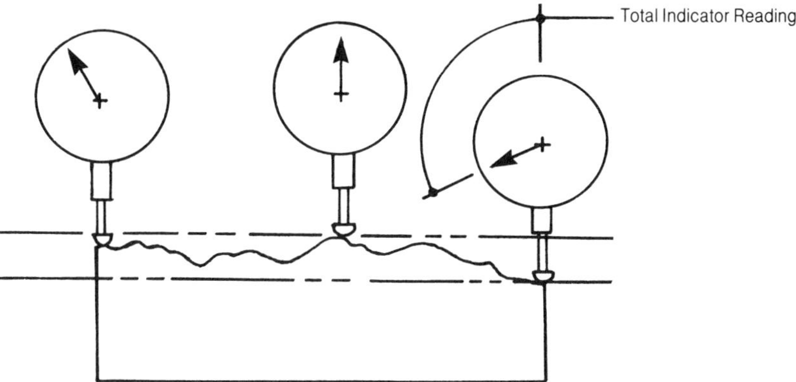

Figure 2.6 Total indicator reading using a dial indicator.

this method permits direct measurements to be made with standard instruments and does not require the use of special purpose functional gages, or other calculations. The methods used when measuring (inspection techniques) these surface variations will be covered in Chapter 3, Sec. 3.4.

2.2. DEFINING THE FORM OF INDIVIDUAL FEATURES WITH RESPECT TO NO DATUMS

Although the sizes of a part may be indicated quite precisely by the dimension figures, these often do not define the true geometric shape of the part with sufficient accuracy. Consider, for example, a simple cylindrical part (Fig. 2.7), dimensioned with considerable exactness. With dimensions shown, this part might be manufactured like any of the shapes shown in Figures 2.8—2.11 and it would have to be accepted as conforming to the specifications of the drawing. Any of these shapes might be acceptable as far as the function of the part was concerned. However, if it is necessary to have the part conform more closely to the exact geometric shape of Figure 2.7 (interchangeability purposes), then this requirement must be indicated together with the exact amount by which it may vary from the desired form. Such expressions are known as tolerances of form, or geometric tolerances. They may refer to the flatness, straightness, roundness, cylindricity, perpendicularity, angularity, parallelism, concentricity, or symmetry of the part.

When tolerances of form are not specified, as in Figure 2.7, it is understood that the part is acceptable if it is within the sizes given regardless of form, as in Figures 2.8—2.11. If however, it is neces-

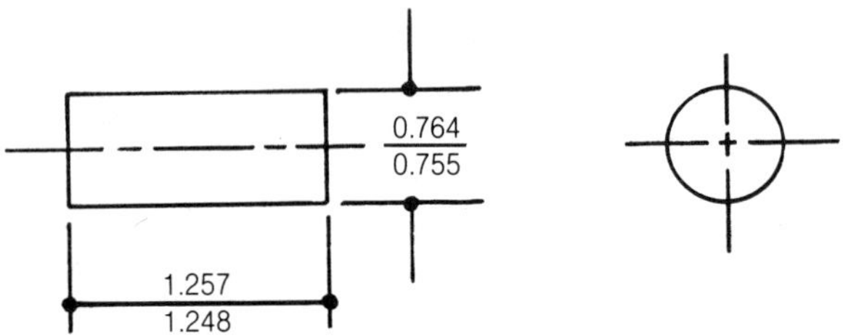

Figure 2.7 A dimensioned and toleranced shaft.

sary for the part to be straighter than the 0.005 in. tolerance shown in Figure 2.7, then an additional tolerance of form must be supplied. Thus, a form tolerance does not need to be stated unless it is smaller than the dimensional tolerances involved. Furthermore, when tolerances of form are stated, they must be observed regardless of the actual finished size of the plate.

Form tolerances fall into two categories: (1) those that need no datums and (2) those that refer to datums. This section discusses those form tolerances that do not need datums, which include straightness and roundness.

Not too many years ago, form tolerances were placed on drawings in the form of a note, with a leader pointing to the feature that was to be toleranced. The note consisted of three items: (1) the geometric relationship involved (straightness, perpendicularity, etc.), (2) the datum surface or axis about which the relationship applies (used for datum-oriented form tolerances), and (3) the amount of the tolerance. Many companies still use this method of indicating form tolerancing; therefore, the following examples show these methods of form tolerancing and they have been included as a part of rectangular coordinate dimensioning and tolerancing.

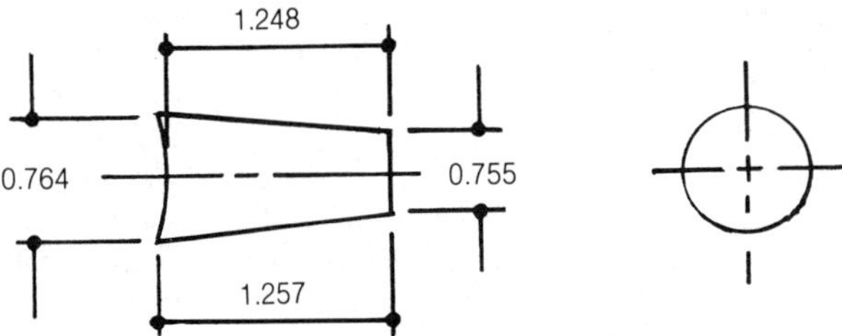

Figure 2.8 A first interpretation of the shaft.

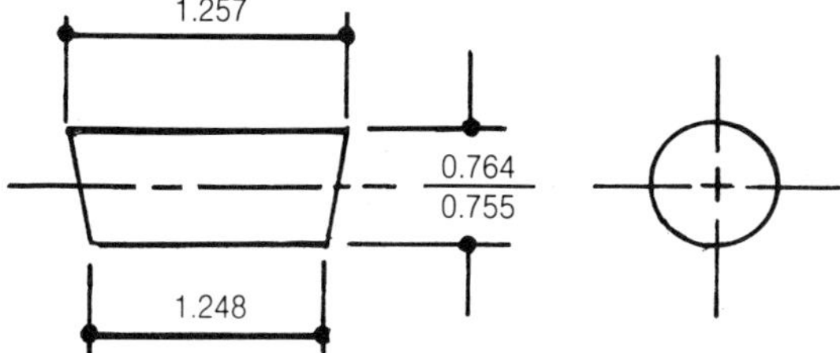

Figure 2.9 A second interpretation of the shaft.

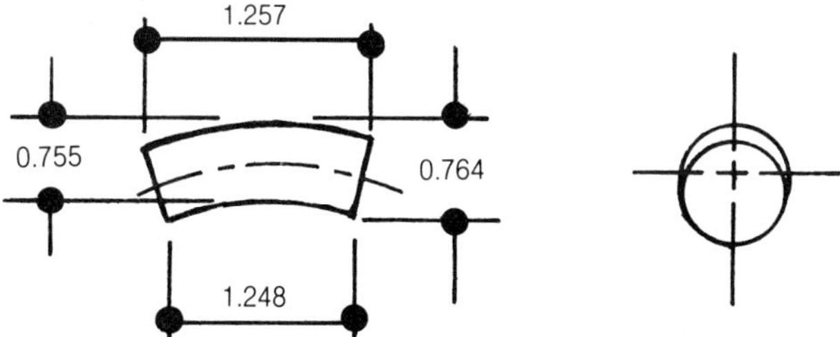

Figure 2.10 A third interpretation of the shaft.

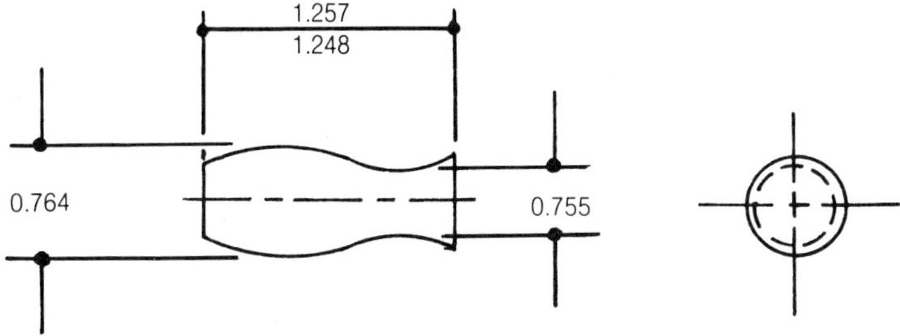

Figure 2.11 A fourth interpretation of the shaft.

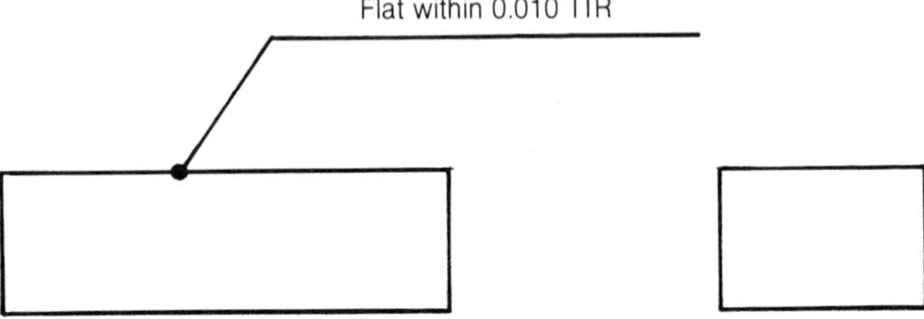

Figure 2.12 A flatness callout.

2.2.1. Flatness

Flatness is where a surface has all of its elements in one plane. A flatness callout appears in Figure 2.12.

In Figure 2.13 the toleranced surface must be between two planes 0.010 in. apart.

2.2.2. Straightness

Straightness is that condition where no element of a surface deviates from a straight line. A straightness callout appears in Figure 2.14.

In Figure 2.15, no element may deviate more than 0.005 inches.

2.2.3. Roundness

Roundness is that condition on a surface of revolution, such as a cylinder or cone, where all points on the plane normal to the axis are equidistant from the center. A roundness callout appears in Figure 2.16.

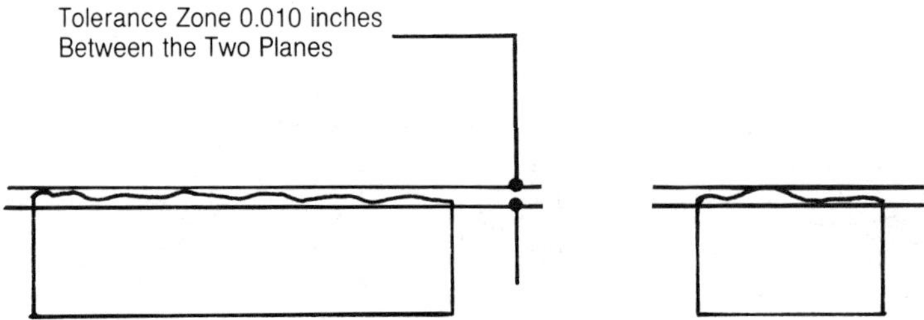

Figure 2.13 Flatness interpretation.

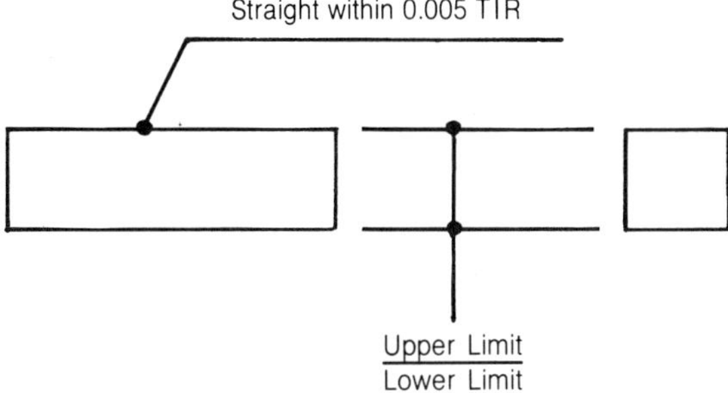

Figure 2.14 A straightness callout.

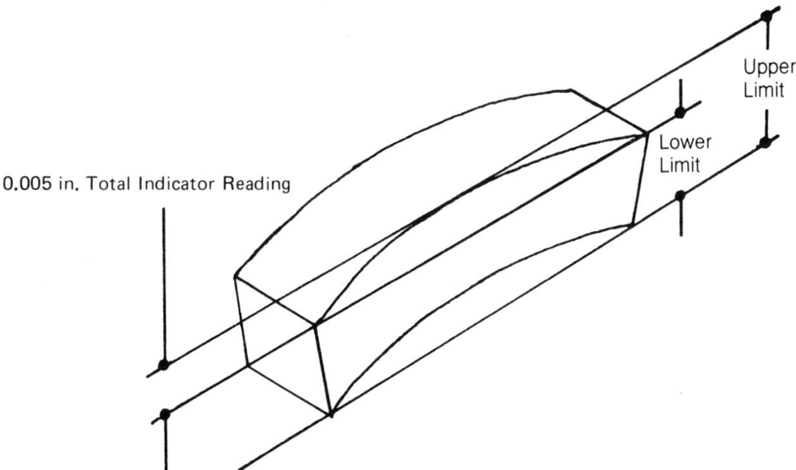

Figure 2.15 Straightness interpretation.

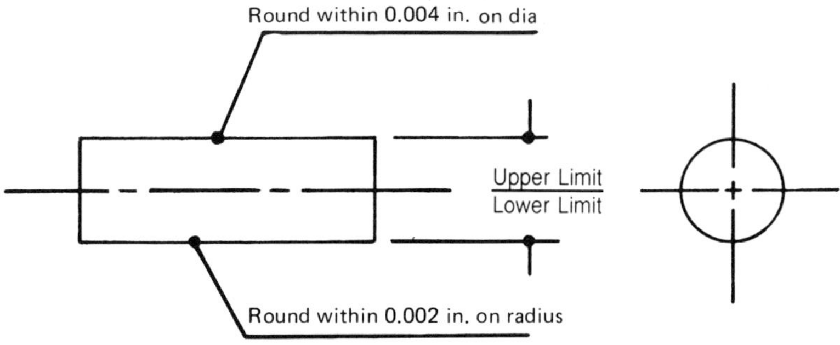

Figure 2.16 A roundness callout.

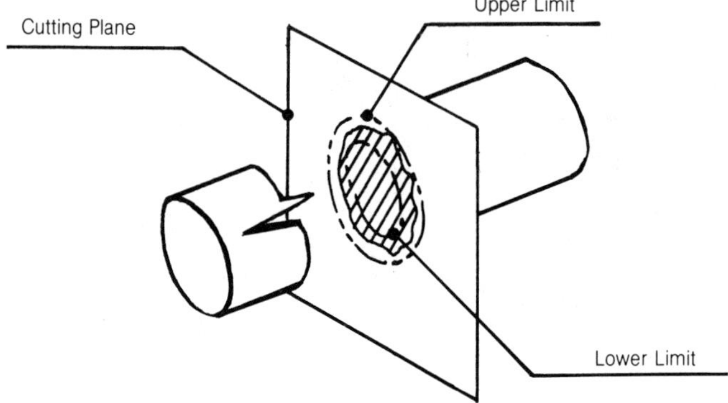

Figure 2.17 Roundness interpretation.

In Figure 2.17, the surface must lie between two concentric circles 0.002 in. apart.

2.3. DEFINING THE FORM OF RELATED FEATURES WITH RESPECT TO DATUMS

The form tolerances of perpendicularity, angularity, and parallelism need a datum or datums to be associated with them. A datum is considered a perfect geometric element which measurements can be taken from for the location of the part's various features.

2.3.1. Perpendicularity

Perpendicularity is the condition where surfaces or lines are at right angles to each other. Perpendicularity callouts appear in Figures 2.18 and 2.19.

In Figure 2.20, the toleranced surface must be between two planes 0.004 in. apart and perpendicular to the datum surface noted as A. In Figure 2.21, the axis of the hole must be within a cylinder of 0.004 in. in diameter and perpendicular to the datum axis noted as B.

2.3.2. Angularity

Angularity is when surfaces and lines are at an angle other than 90 degrees with each other. An angularity callout appears in Figure 2.22.

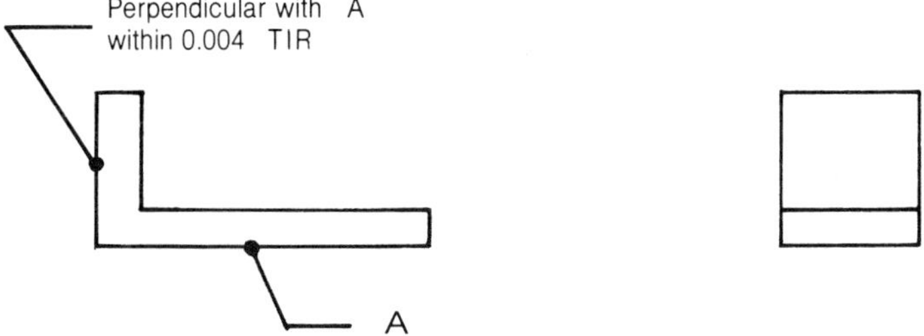

Figure 2.18 A perpendicular callout (surface).

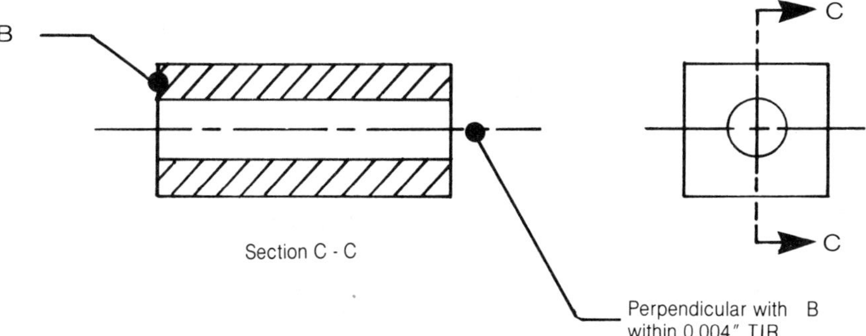

Figure 2.19 Perpendicularity interpretation (surface).

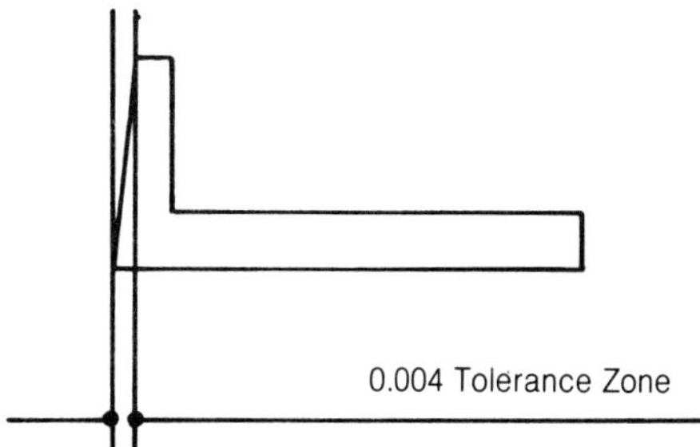

Figure 2.20 A perpendicular callout (axis).

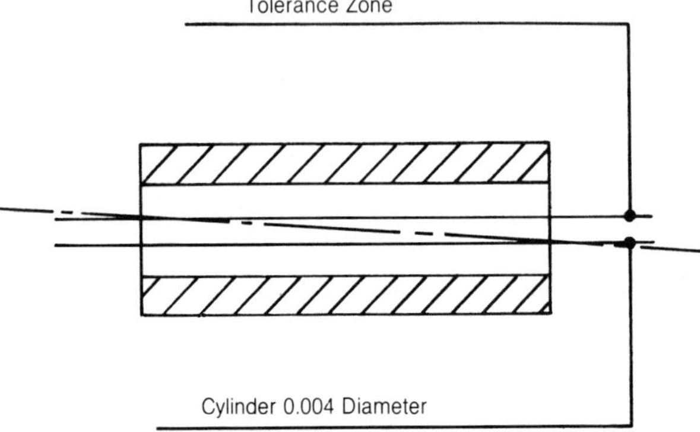

Figure 2.21 Perpendicularity interpretation (axis).

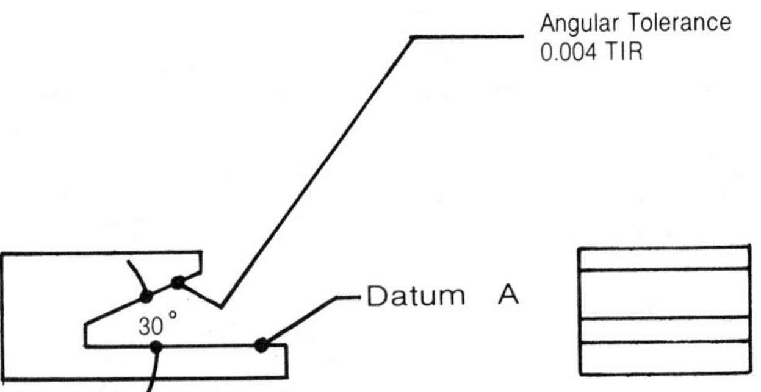

Figure 2.22 An angularity callout.

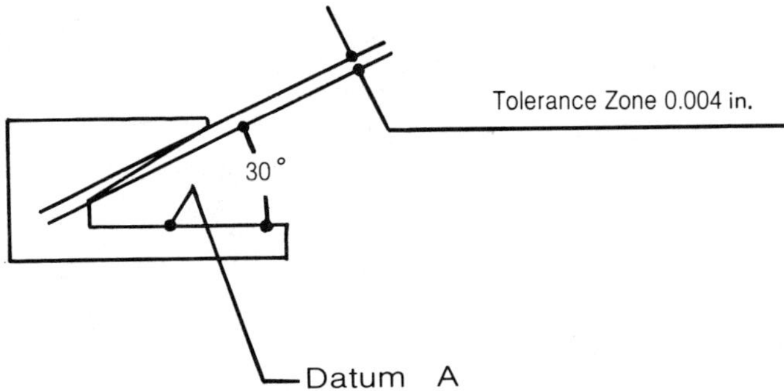

Figure 2.23 Angularity interpretation.

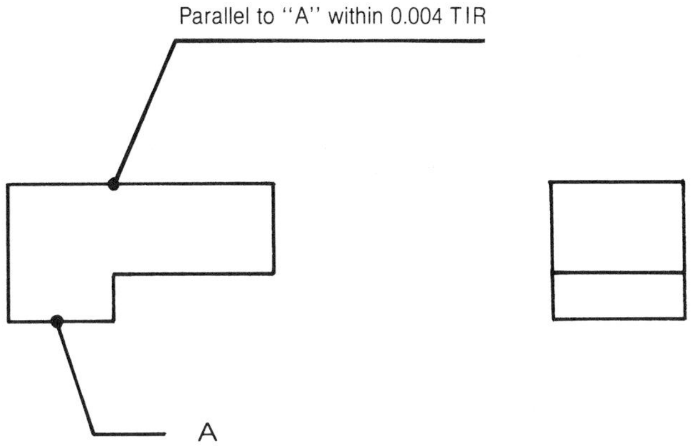

Figure 2.24 A parallelism callout (surface).

In Figure 2.23, the toleranced angular surface must lie between two planes 0.004 in. apart and making the specified angle with datum plane A.

2.3.3. Parallelism

Parallelism is where two planes or straight lines are equidistant at all points. Parallelism callouts appear in Figures 2.24 and 2.25.

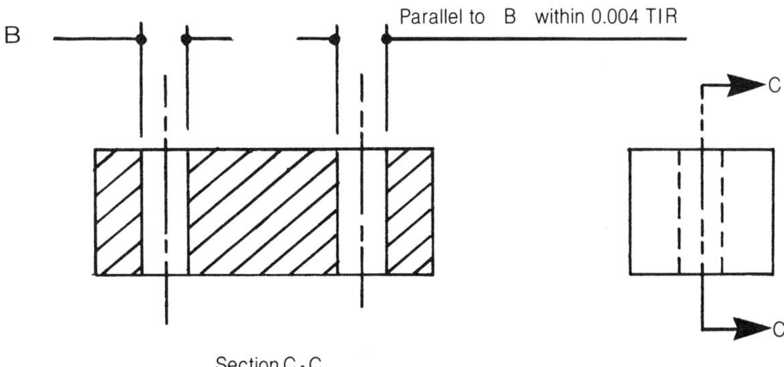

Figure 2.25 Parallelism interpretation (surface).

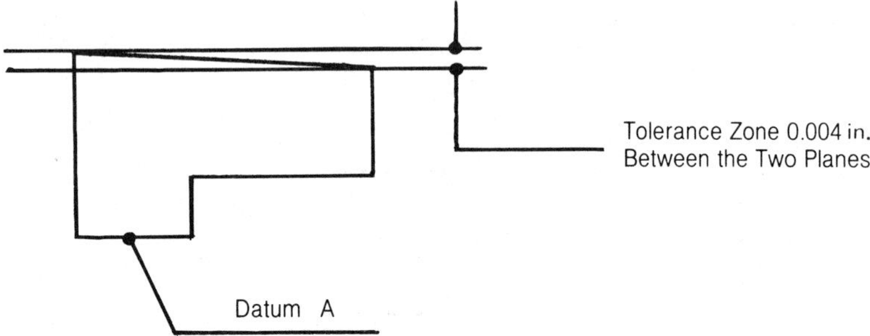

Figure 2.26 A parallelism callout (axis).

In Figure 2.26, the toleranced surface must be between two planes 0.004 in. apart and parallel to datum A. In Figure 2.27, the axis of the hole must lie within a cylinder of 0.003 in. in diameter and parallel to datum B.

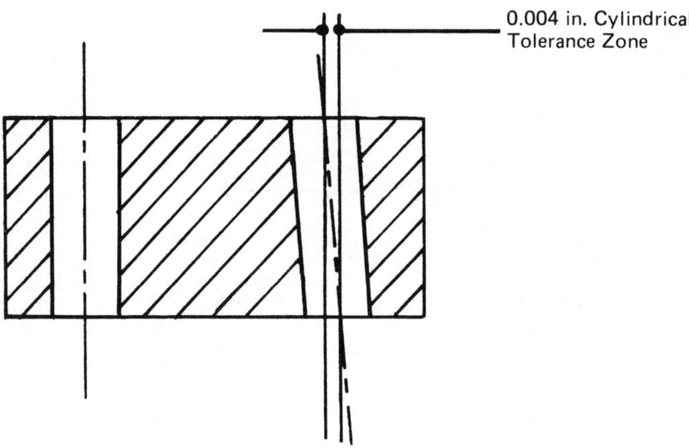

Figure 2.27 Parallelism interpretation (axis).

2.4. DEFINING LOCATION TOLERANCES

So far, the discussion has centered around the control of the shape or form of a feature, such as a hole or surface. One must remember that the location of these features are just as important as their shape. Location control of features is an important aspect of interchangeability, which was discussed in the previous chapter.

Features that mate with other features must be located in such a manner as to satisfy the intended design specifications. Using the rectangular coordinate method of dimensioning for locating these features supposedly ensures this goal. Features will share a rectangular type of tolerance zone and the size of this rectangular zone will depend greatly on the dimensioning scheme used. Examples of such dimensioning schemes are covered in the following section.

2.4.1. Fundamentals of Locational Tolerances

Figures 2.28—2.31 show various rectangular dimensioning schemes for the location of nine holes in a plate.

In Figure 2.28 the dimensions and their respective ±0.005 tolerances are taken from either of the two sides of the plate. The resulting tolerance zone of each hole is independent of each other. The hole in the upper right-hand corner of the plate (Fig. 2.29) has a tolerance

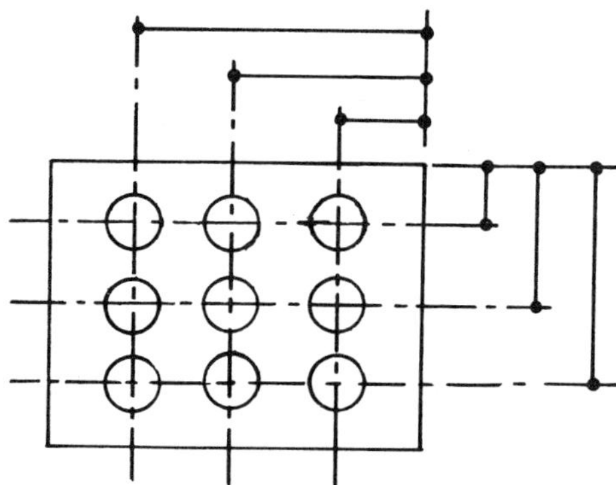

Figure 2.28 Dimensioning using surfaces as datums.

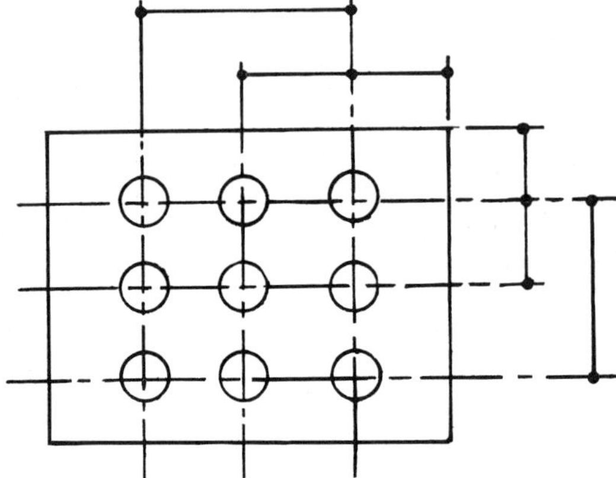

Figure 2.29 Dimensioning using a hole as a datum.

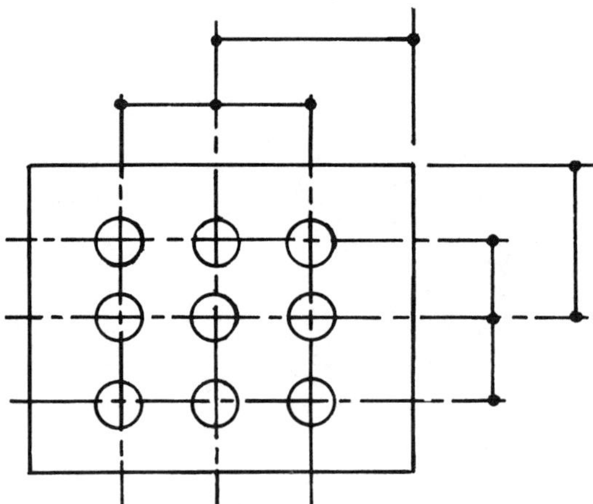

Figure 2.30 Dimensioning using the center of the hole pattern as a datum.

±0.003 zone associated with it. This zone is also shared by the remaining eight holes, resulting in their respective ±0.008 zones. In Figure 2.30, the hole located in the center of the hole pattern has a ±0.003 tolerance associated with it. The remaining eight holes are located from this hole, and therefore, share a ±0.008 tolerance zone. Last but not least, Figure 2.31 illustrates a dimensioning scheme where each hole is located from the previous hole. The resulting tolerance zone for each hole is an addition of ±0.005 from the previous hole.

It appears that Figure 2.28 is the best dimensioning scheme, since the tolerance zone associated with each hole yields the smallest value of locational variation. Therefore, this dimensioning scheme would tend to satisfy the condition of interchangeability a lot better than the other three dimensioning schemes.

From the previous discussion one can conclude that how dimensions are placed on a part is *very* important. Since in Figure 2.28, the locational dimensions for the nine holes start from a common geometric element (surface, line, or point) on the part, this geometric element is very important to the designer and the machinist in the shop. These common geometric elements are referred to as datums.

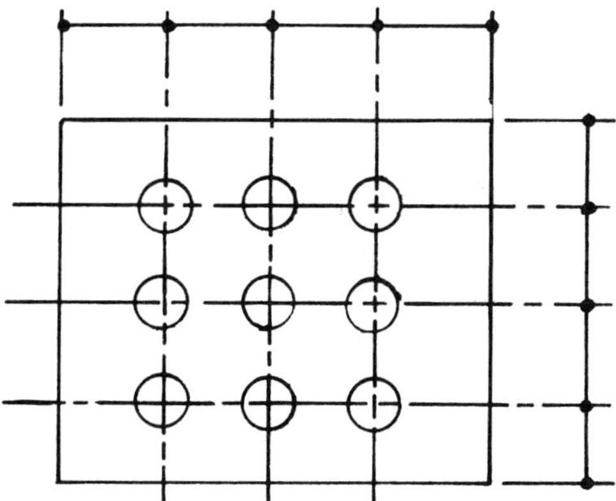

Figure 2.31 Chain dimensioning.

2.4.2. Implied and Specified Datums

Datums are considered simple geometric elements (points, lines, or surfaces) from which features are located. Frequently, these datums are the same datums used in the shop from which measurements are taken to locate the feature on the part during the manufacturing stage. In the shop, these datums are sometimes referred to as setups.

Implied datums are those datums suggested by the placement of the dimensions on the drawing. Figure 2.32 shows a drawing with several implied datums due to the dimensioning scheme.

It is advisable to specify the datums on a drawing to avoid misunderstandings. These types of datums are known as specified datums. Figure 2.33 shows a drawing with several specified datums and their respective dimensions.

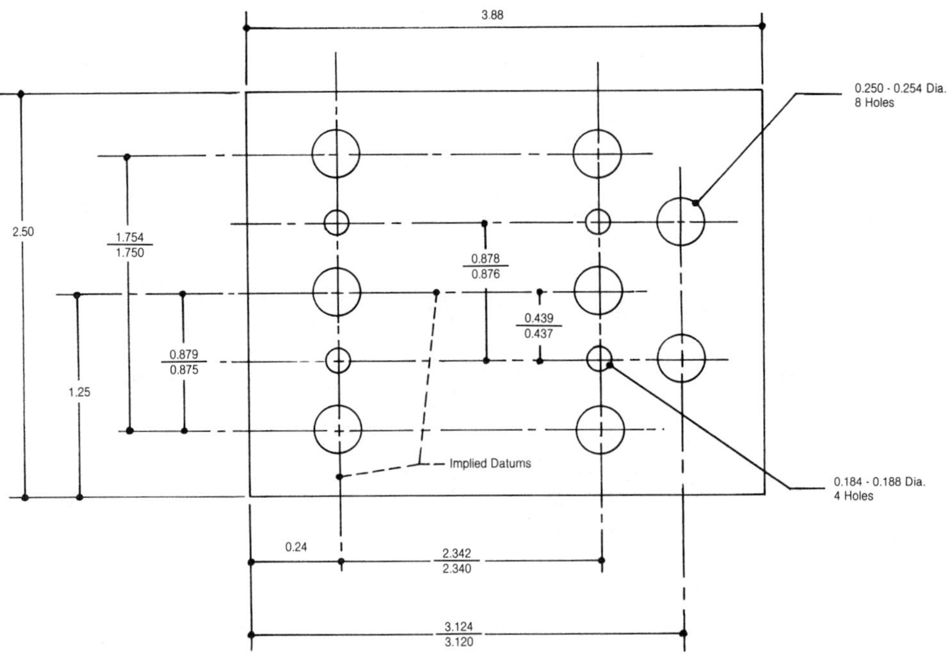

Figure 2.32 Implied datums.

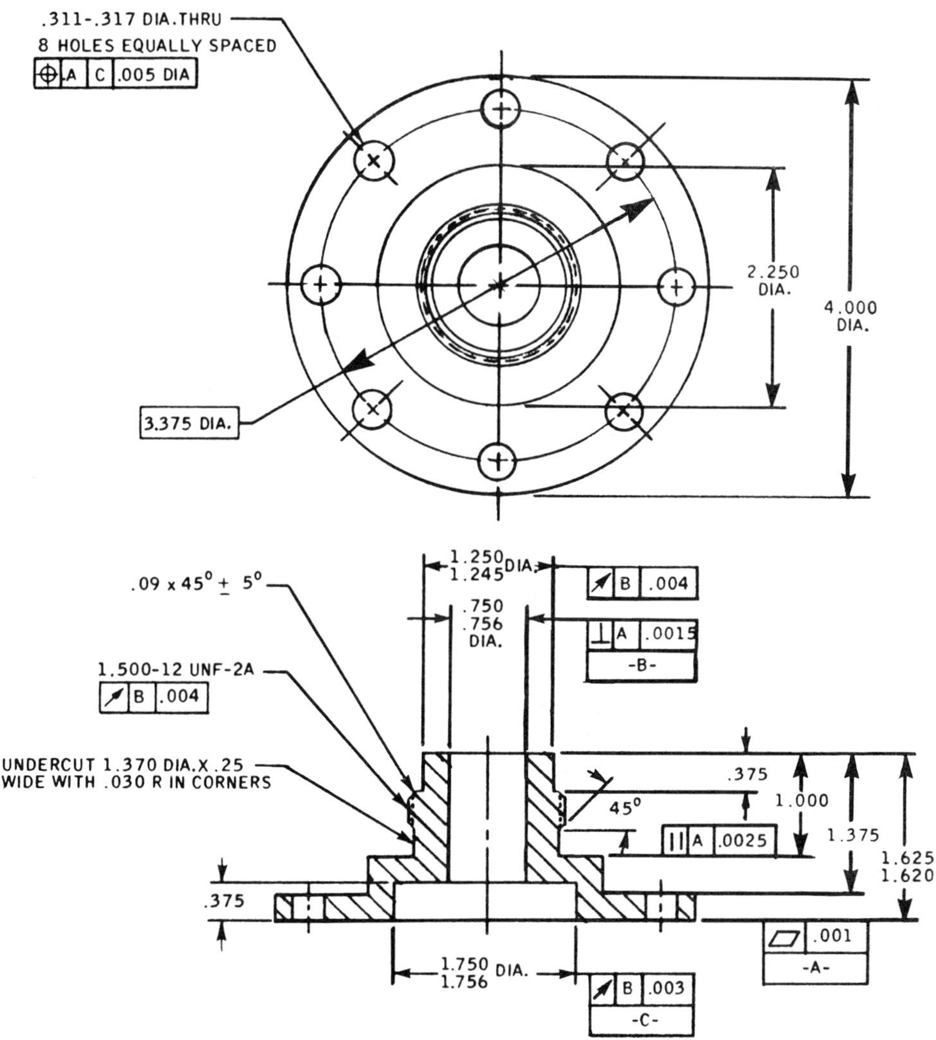

Figure 2.33 Specified datums.

2.5. SELECTION OF DATUMS

The first step in the dimensioning of a working drawing is to determine those points, centerlines, or finish surfaces which:

1. Will specifically control the location of those features of the given part that mate or fit with others.
2. Can be readily established by the machinists and used as points from which their measurements can be made.

These are called datum points, datum lines, or datum surfaces. They are obtained by first studying the machine or product assembly to determine the principle points, centerlines, or surfaces about which the various parts fit together or mate. These become the primary datums and must be determined (primary, secondary, and tertiary).

For example, inspection of the idler bracket assembly of Figure 2.34 indicates that the pulley's hole shares the same centerline that is shared by the shaft and the bracket's hole. This centerline is considered the primary datum (-A-) in the x direction, and in the y direction it is considered the secondary datum (-B-). The face of the idler bracket is the tertiary datum (-C-). These datums are shown in Figure 2.35.

It should be noted that datum points themselves often must be located on the part before they can be used as a base for the location of other features. Frequently, such locations must be made from rough or inaccurate surfaces. Once the datums are fixed, however, all of the other features are located from them, and not with respect to one another or the original rough surfaces.

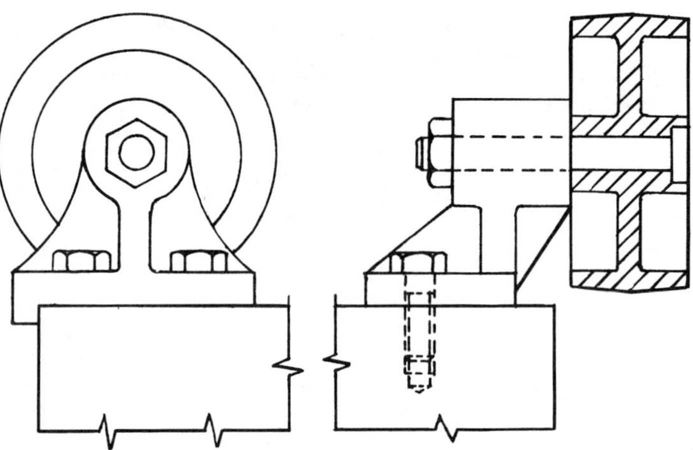

Figure 2.34 Idler bracket assembly.

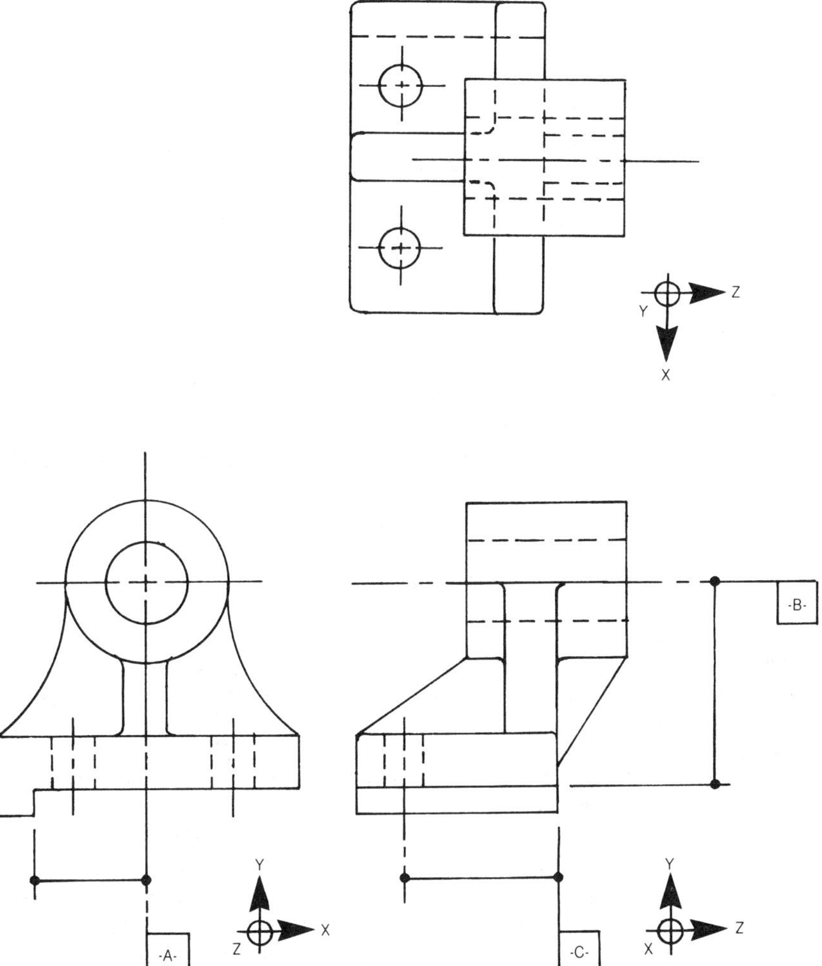

Figure 2.35 Placing datums on the idler bracket drawing.

While centerlines are often used as datum lines, surfaces are to be preferred, as they present an established face from which the machinist can easily make a measurement. Surfaces used as datums, however, must be accurately flat and smooth to insure the accuracy of the measurement made from them. Finished surfaces on castings or machined surfaces on steel parts may qualify as such surfaces.

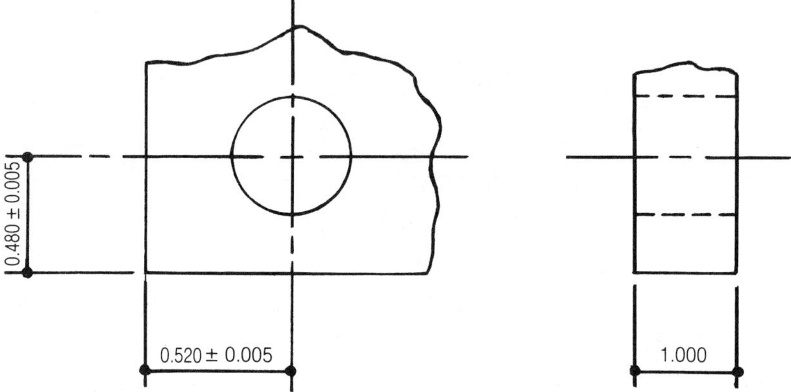

Figure 2.36 Locating a hole using bilateral coordinate tolerance dimensioning.

2.5.1. Using Floating Fasteners

In Figure 2.36, a hole is located by using a bilateral coordinate toleranced dimension. The coordinate tolerance of ±0.005 in. establishes the boundary of the zone within which the hole center must lie. Since the tolerances are equal in both the x and y directions, the resulting bilateral tolerance zone would be square in shape with sides equal to 0.010 of an inch. This 0.010 inch square zone extends down into the part a distance equal to the depth of the hole which , in this case, would be 1.000 in. in depth. The tolerance zone formed by these bilateral coordinate tolerances dimensions is shown in Figure 2.37.

The parts shown in Figures 2.38 and 2.39 are to be assembled with a bolt and a nut. This assembly is shown in Figure 2.40. The location of the clearance holes are dimensioned toleranced; however, the clearance hole size has been omitted in Figures 2.38 and 2.39. If these two parts are assumed to have perfect form, then the tolerances zones will coincide when the parts are assembled. Figure 2.41 illustrates these two tolerance zones. A top view of these two tolerance zones is shown in Figure 2.42 and the worst condition of assembly will occur when the clearance hole centers are farthest apart. This condition will occur when the clearance hole centers are at diagonally opposite corners of their respective tolerance zones. Shown in Figure 2.43, is the minimum clearance hole in one of the parts. The hole center is at the upper right-hand corner of the tolerance zone. The radius of a minimum clearance hole of the other part is superimposed onto this tolerance zone with its hole center at the lower left-hand corner of the tol-

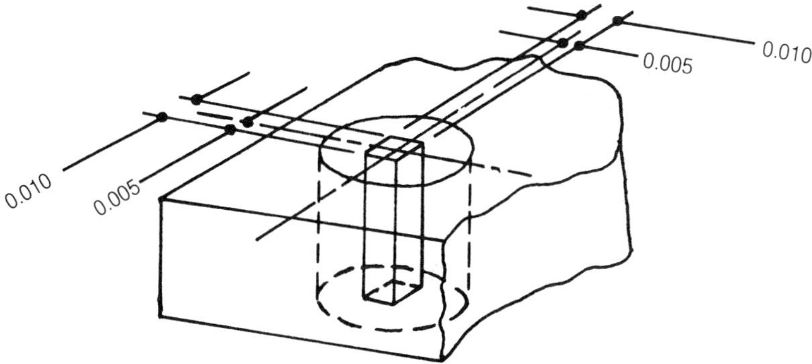

Figure 2.37 The tolerance zone created by bilateral tolerancing of hole.

erance zone (Fig. 2.44). The minimum clearance hole radius of this hole is also H/2. Since the clearance hole centers are located at diagonally opposite corners of the tolerance zone; this represents the worst condition of assembly.

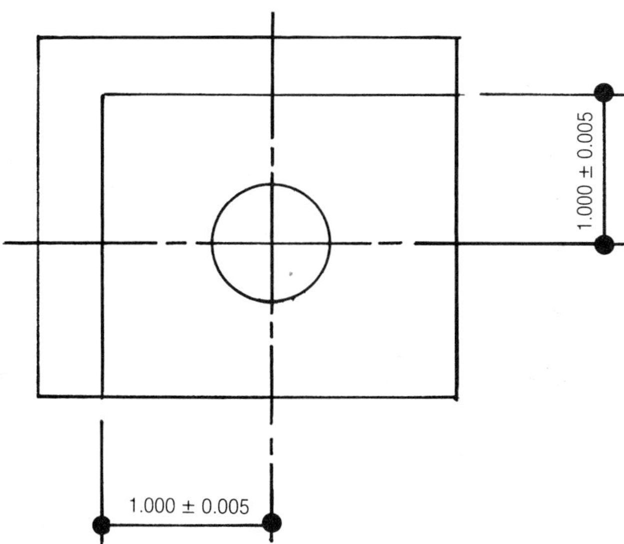

Figure 2.38 Part with one hole.

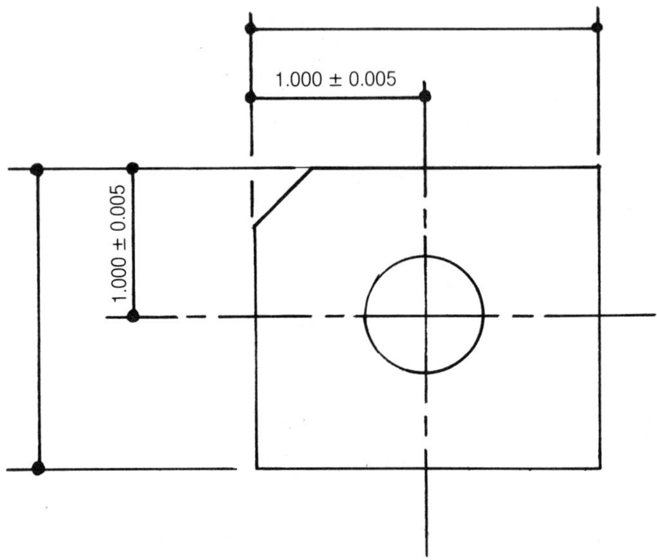

Figure 2.39 Mating part with the other hole.

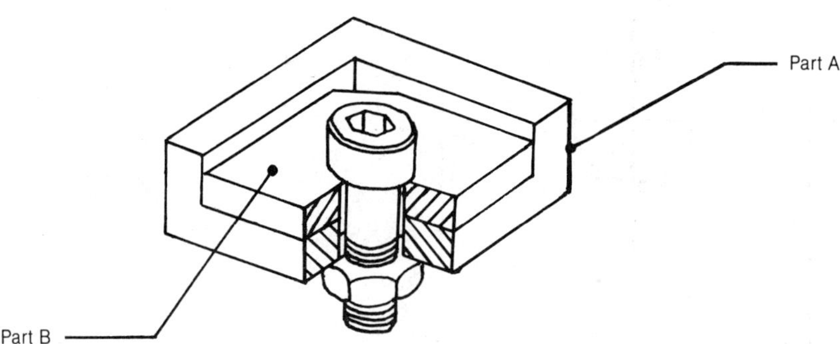

Figure 2.40 Assembly of the two parts with a bolt and nut fastening them together.

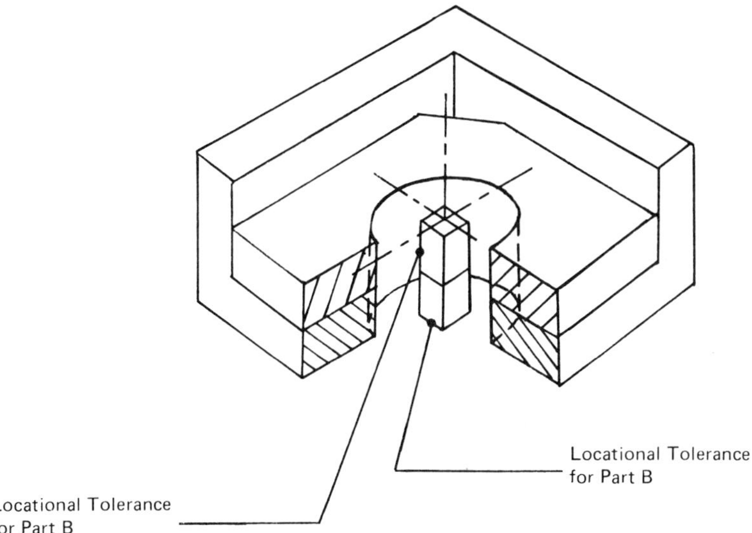

Figure 2.41 Locational tolerances of parts A and B.

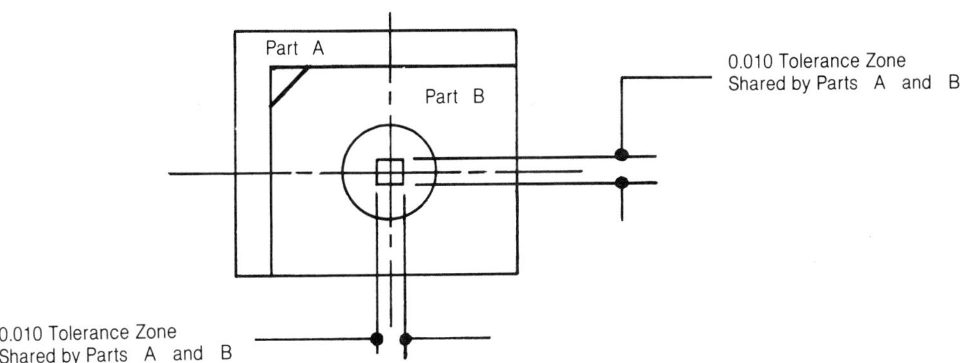

Figure 2.42 0.010 tolerance zone shared by parts A and B.

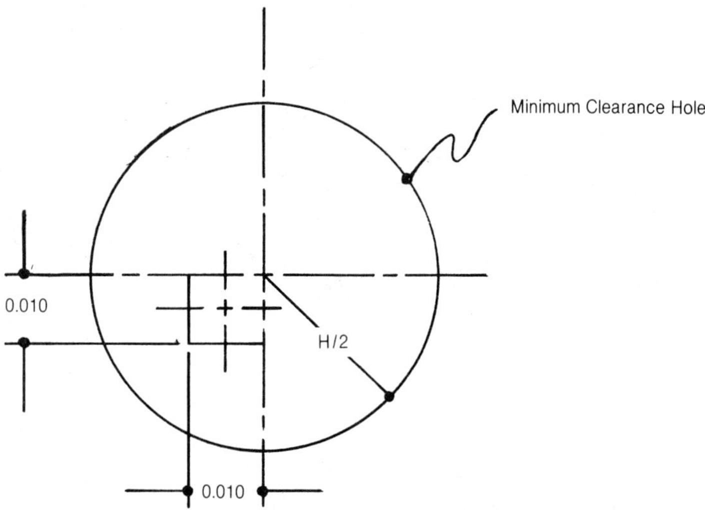

Figure 2.43 One extreme position of the minimum clearance hole.

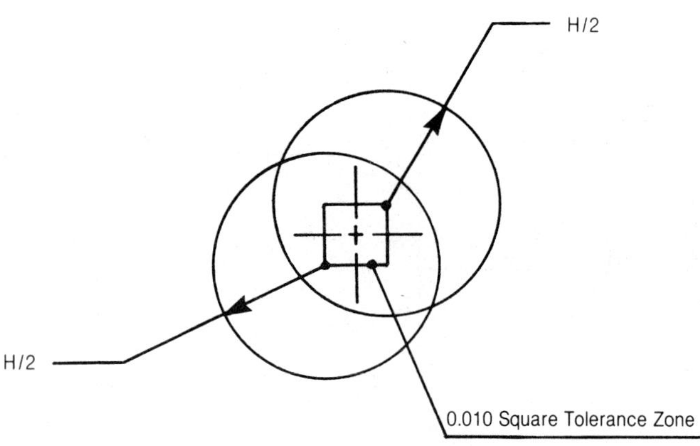

Figure 2.44 Superimposing both holes at their extreme positions on the 0.010 square tolerance zone.

The distance noted as B represents the maximum space available for placing through a maximum diameter bolt. Distance B would be equal to the maximum bolt diameter to guarantee 100% interchangeability. The bolt appears as the small circle in Figure 2.45. The bolt's center coincides with the tolerance zone's center in the worst condition of assembly.

From Figure 2.45, it can be seen that the radius of the minimum clearance hole minus one-half the diagonal of the tolerance zone is equal to the radius of the bolt or $B/2 = H/2 - T/2$, where B = the maximum diameter of the bolt, H = the diameter of the minimum hole, and T = the diagonal of the locational tolerance zone (Fig. 2.46). All measurements are in inches. Multiplying all the terms of the equation by 2 yields $H = B + T$. This equation is used for floating fastener applications.

Now one is able to calculate the minimum hole diameter of the hole if the hole's diameter and the diagonal of the tolerance zone is unknown. For example, if a square tolerance zone of 0.010 in. per side is to be used with a 0.500-in.diameter bolt, one can calculate the minimum hole. The diagonal of this square tolerance zone would be 0.014 inches. Using the floating fastener equation, $H = 0.500 + 0.014$, the minimum hole diameter H must be 0.514 inches.

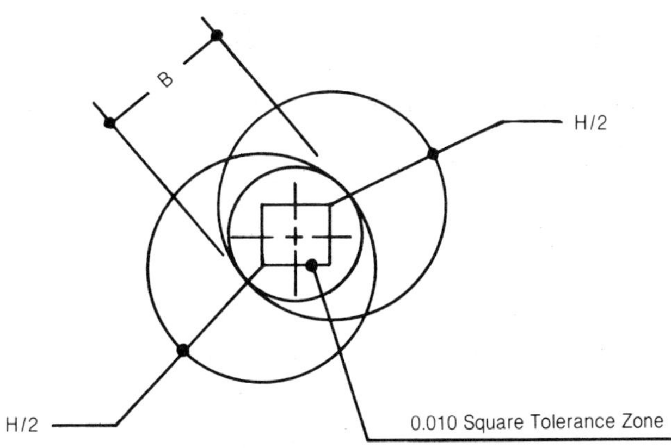

Figure 2.45 Maximum diameter bolt for 100% interchangeability.

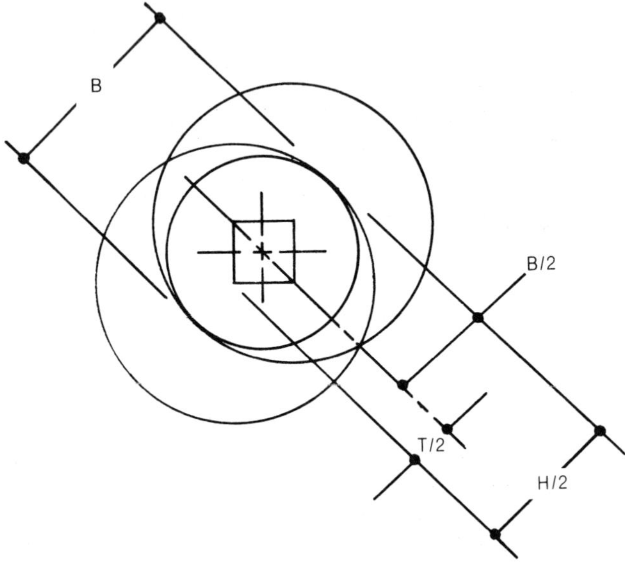

Figure 2.46 Diagram used for floating fastener equation.

2.5.2. Using Fixed Fasteners

The type of assembly where fasteners (bolt, screws, studs, etc.) are fixed in position on one of the parts is referred to as a fixed fastener condition. Assembled parts having the holes threaded (tapped) or pins pressed-fitted in one part and the clearance provided by the holes in the other part defines this condition. Figure 2.47 shows such an assembly.

The equation for determining the minimum size of this clearance hole is given as $H = F + T_1 + T_2$ or, if the same tolerance is specified for both parts, the equation becomes $H = F + 2T$, where H = minimum diameter of hole, F = maximum diameter of fastener, T_1 = hole's diagonal locational tolerance on part 2. These terms are illustrated in Figure 2.48.

For example, a 0.250 diameter fastener (F) is to be passed through a clearance hole (H) of part 1 into a threaded hole in part 2. The locational tolerance on part 1 is ±0.002 while the locational tolerance on part 2 is ±0.004 of an inch. The tolerances are 0.0028 and 0.0057, respectively, Substituting these values into the equation, $H = F + T_1 + T_2$ or $H = 0.250 + 0.0028 + 0.0057$, which yields 0.2585 inches for the minimum diameter of the hole (H) in part 1.

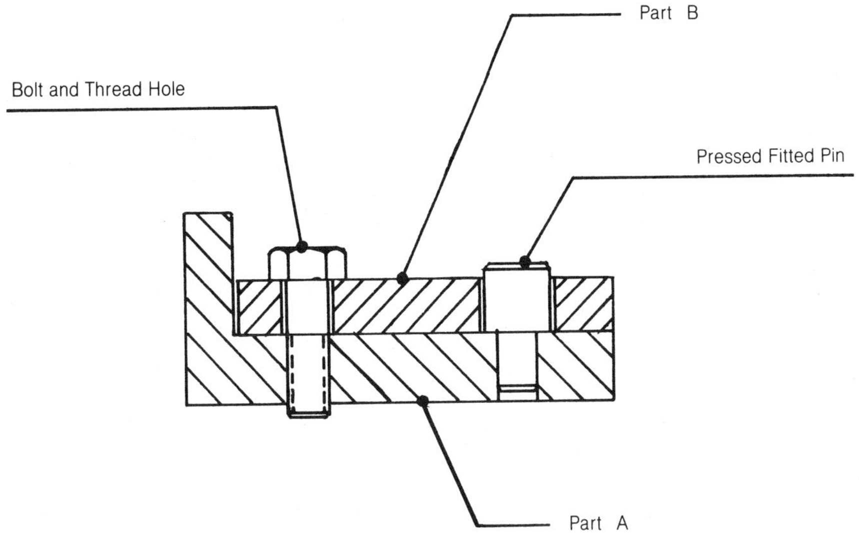

Figure 2.47 Parts A and B fastened together using fixed fasteners.

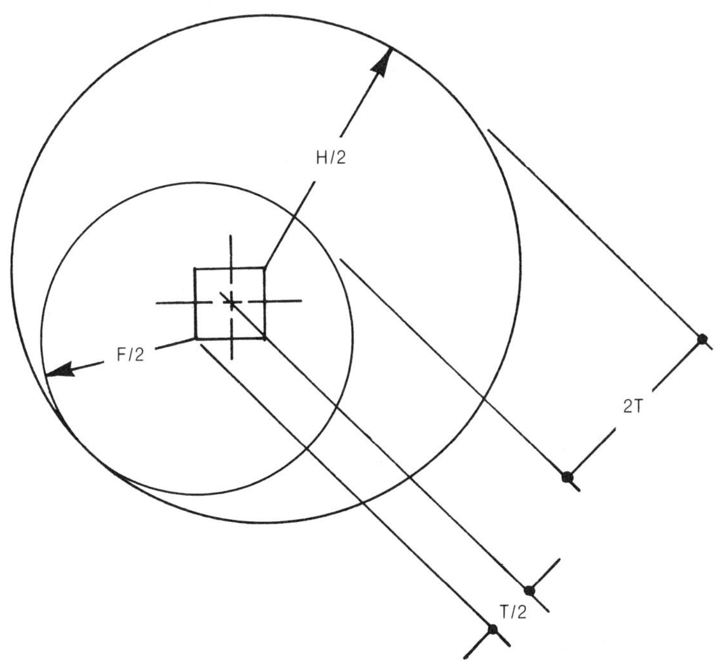

Figure 2.48 Diagram used for fixed fastener equation.

2.6. CONCENTRICITY

Tolerance applied to concentricity is a condition in which there is a coincidence of centers. In most cases, concentric cylinders, cones, etc. generated about common axes in manufacturing, will be concentric to a degree of precision more than required by the design. Therefore, no callout or note is required on the drawing concerning the allowable variation. However, mating pairs of two or more precise, close-fitting, machined cylindrical surfaces must have the axes of adjoined cylinders closely coinciding in order to permit assembly of the part. Thus, it is sometimes necessary to give the permissible deviation from concentricity.

Since the centerline of adjoining cylinders coincides on a drawing, the tolerance cannot be given as a dimension. One method of indicating it is to mark the diameters with reference letters and give the tolerance in note form as in Figure 2.49, or the reference letters can be omitted if the note is applied directly to the surfaces as in Figure 2.50.

In Figure 2.51, the two centerlines of the concentric shafts must both lie within a cylindrical tolerance zone of 0.008 in. in diameter.

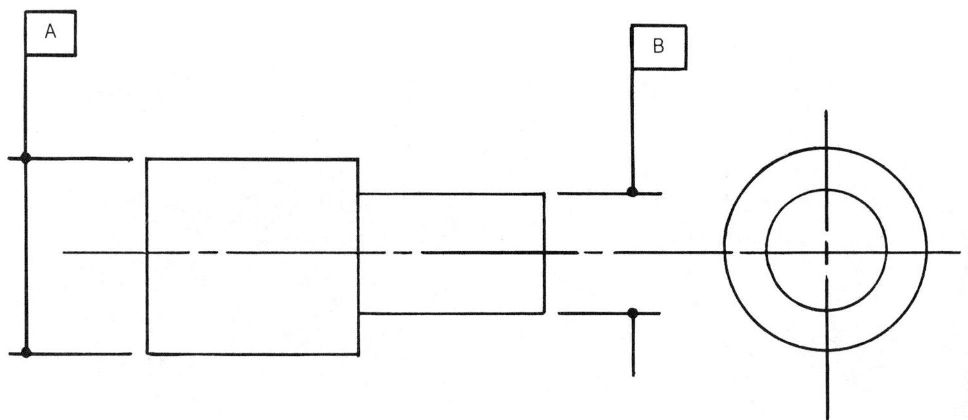

Note: Diameter A must be concentric to Diameter B within 0.008 T I R

Figure 2.49 Concentricity callout using a note.

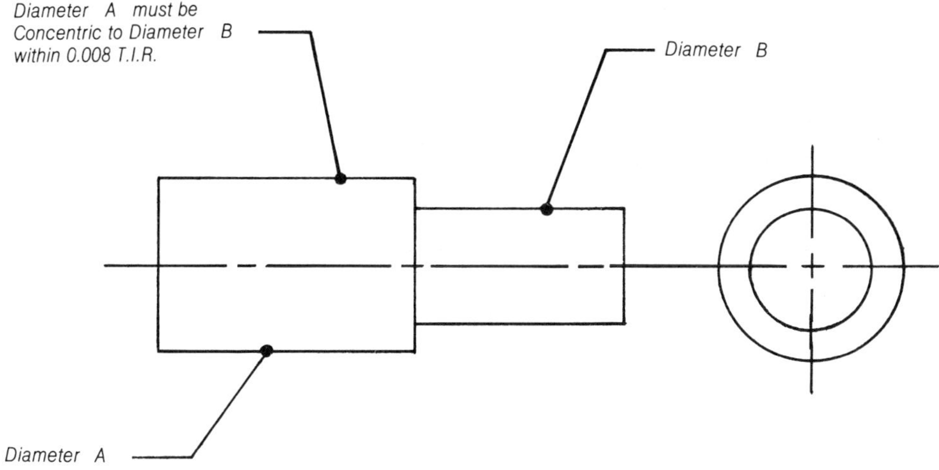

Figure 2.50 Concentricity callout using a note applied directly to the surfaces.

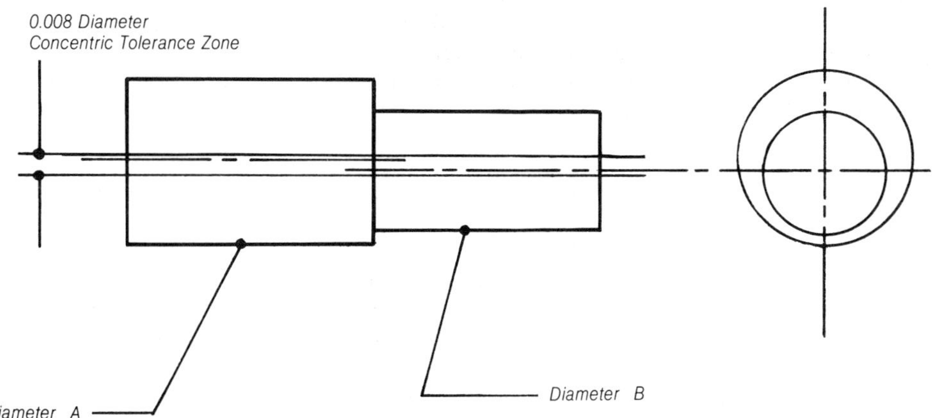

Figure 2.51 Concentricity interpretation.

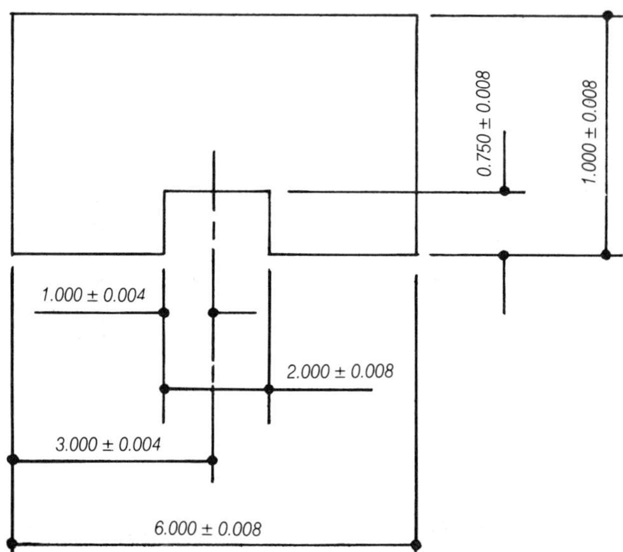

Figure 2.52 A slot shown symmetrically about its centerline.

2.7. SYMMETRY

Symmetry is a condition in which the feature or features are identical about a centerline or a centerplane. In Figure 2.52, a slot is shown as symmetrical about a centerline, with a centerline-locating dimension controlling the symmetry of the slot. This symmetrical tolerance is one-half the total size tolerance of the slot. In this case, the symmetrical tolerance is ±0.004 which is one-half the total size tolerance of ±0.008.

2.8. THE INTERRELATIONSHIP BETWEEN FORM AND LOCATION TOLERANCING

The interrelationship between form and location tolerancing on a part is difficult to understand and just as difficult to obtain in the shop. To illustrate this interrelationship, the part in Figure 2.53 will be used as an example.

The two 0.312 diameter holes have a square tolerance zone of 0.006 in. per side and are influenced by the perpendicularity tolerance of 0.001. Figure 2.54 shows how form tolerances affect the hole locations. The 0.010 square tolerance zone has a 0.001-in. horizontal movement due to the form tolerance of perpendicularity. Similarly, Figure 2.55 show two pins being located from a datum pin.

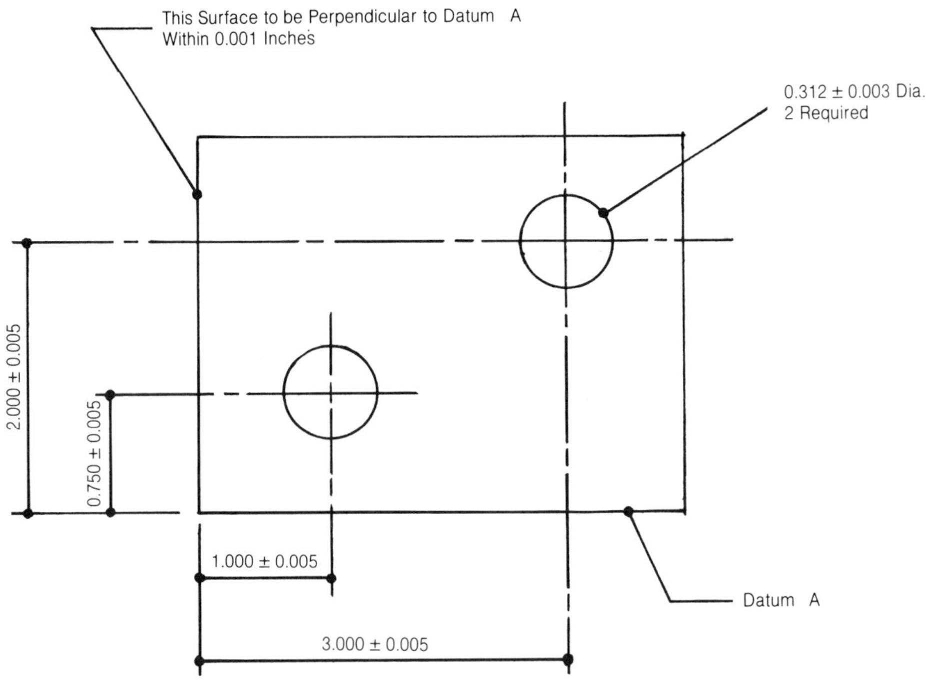

Figure 2.53 An example illustrating the interrelationship between form and position tolerancing.

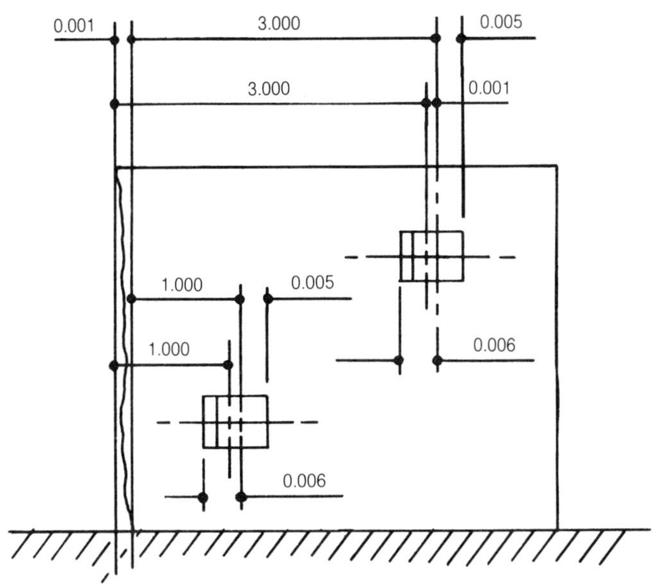

Figure 2.54 Hole location being affected by form tolerances.

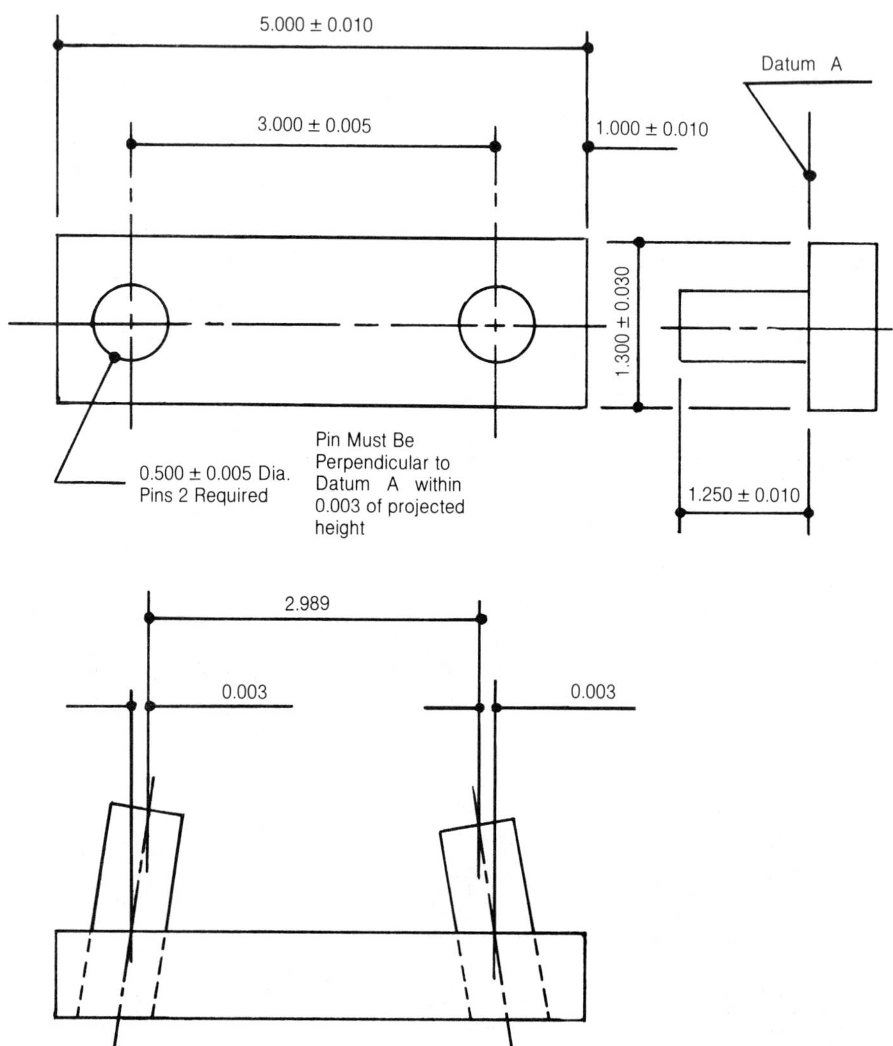

Figure 2.55 Two pins located from a datum pin.

2.9. COMPUTER GRAPHICS APPLICATIONS

Computer graphics has been successfully applied to rectangular coordinate dimensioning and tolerancing when calculating clearances and/or interferences between mating parts (tolerance stackups). The theoretical center of a feature can be moved about within the rectangular tolerance zone very easily with the use of computer graphics.

The various points, lines, and circles drawn with computer graphics are defined as explicit mathematical elements located within a three-dimensional coordinate system. These elements can be manipulated (moved, rotated, copied, erased, etc.) in space. This capability can be a great help to designers because now they have a tool that can place the parts' theoretical positions anywhere within their respective rectangular tolerance zones. The results are displayed on the computer graphics cathode-ray tube (CRT) screen or on a plotter.

To illustrate, how rectangular coordinate tolerance stackups are calculated through the use of computer graphics, the cam, rocker, and push rod system shown in Figure 2.56 will be used. The assembly drawing of this cam, rocker, and push rod assembly is shown in Figure 2.57 with its detail drawings appearing in Figures 2.58—2.63.

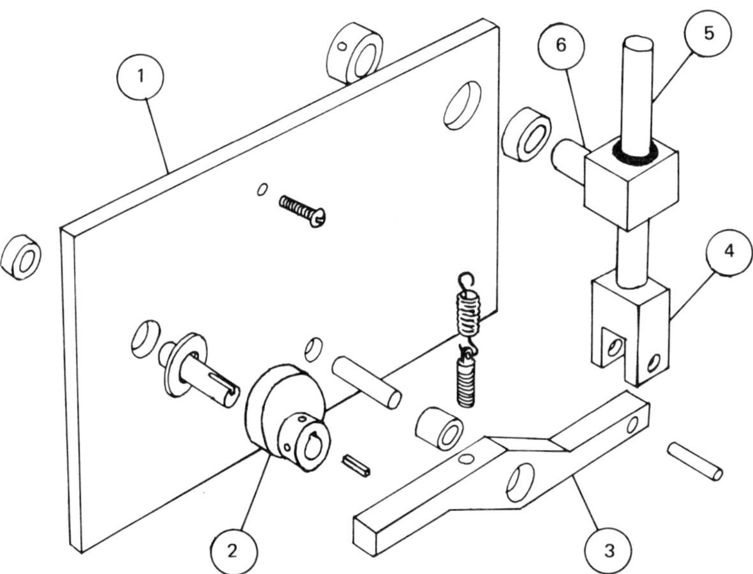

Figure 2.56 Cam, rocker, and push rod system.

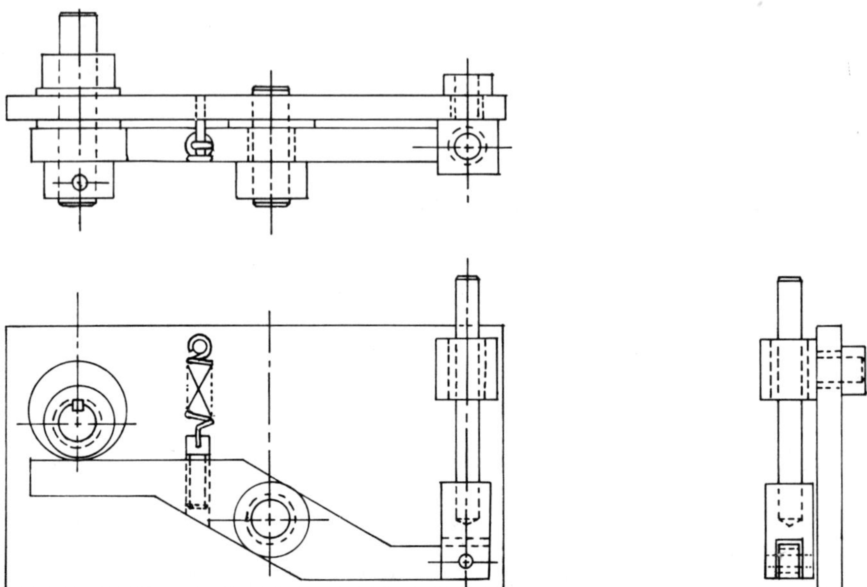

Figure 2.57 Cam, rocker, and push rod assembly.

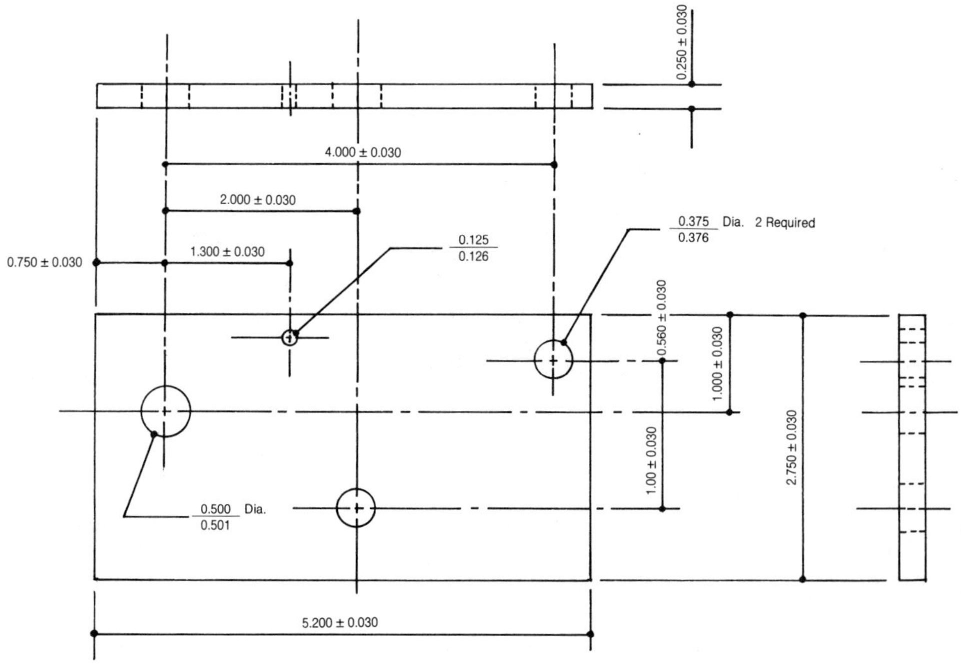

Figure 2.58 Plate detail.

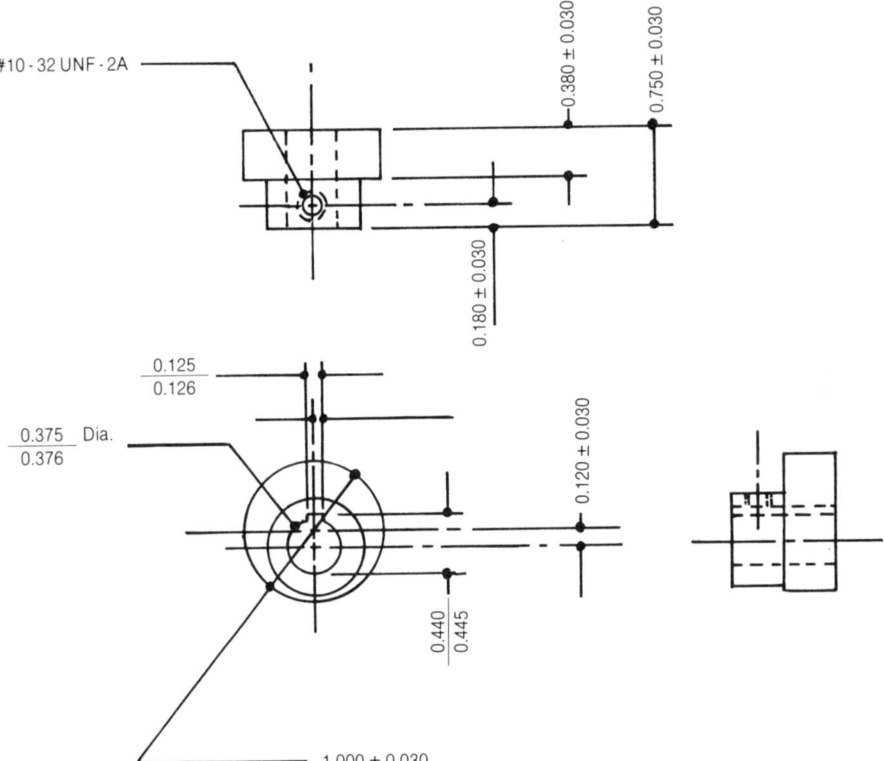

Figure 2.59 Cam detail.

The designer would draw these parts (Figs. 2.58—2.63) on a computer graphic system by using a digitizing pen, a tablet with a menu, and a CRT screen. As each part is drawn, it is stored in the computer's files under its respective file name.

Each part would be drawn on the computer graphics' CRT screen using the nominal dimensions that appear on their respective detail drawings (Figs. 2.58—2.63). The tolerances associated with these nominal dimensions cannot be inputted into the computer because geometric elements are defined and located explicitly (one numerical value) in space. For example, a circle with a radius of 2.163 in. and its center located at X = 8.752 in., Y = −3.565 in., and Z = 0.000 in., defines one, and only one, possible circle in space using the cartesian coordinate system.

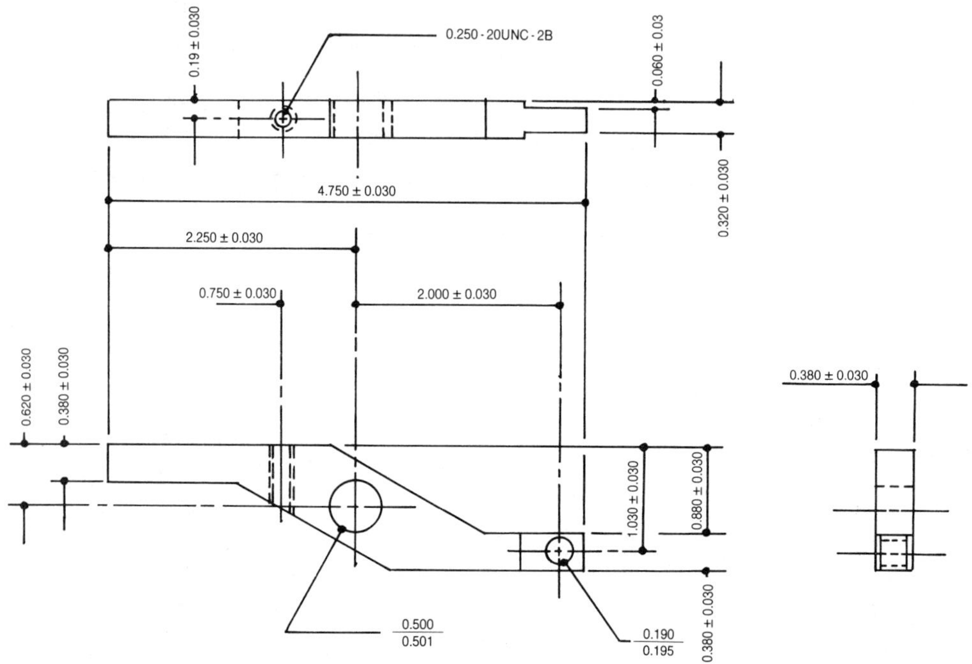

Figure 2.60 Rocker detail.

To illustrate how a part would be drawn on a computer graphics system, the detailed drawing of the cam in Figure 2.59 will be used as an example. A special drafting and design menu (Fig. 2.64) will be used for selecting the commands for drawing the part on the CRT screen. This particular menu contains various commands for placing points, lines, and circles on the CRT screen.

Since tolerances (manufacturing variances) are associated with each dimension appearing on the cam detail drawing, the designer must be able to account for them in any tolerance stack-up calculations. The most convenient way of handling tolerances on a computer graphics system is by, first, drawing the cam using its nominal dimensions. This drawing would be filed under the file name: nominal cam. Then the cam would be redrawn using the maximum material conditions (MMC) and redrawn again, using the least material conditions (LMC) and both these drawings would be filed under the file names: MMC cam and LMC cam, respectively. A nominal, a MMC, and a LMC drawing of each of the other parts: plate, rocker, clevis, push rod and slider block is drawn on the computer graphics system.

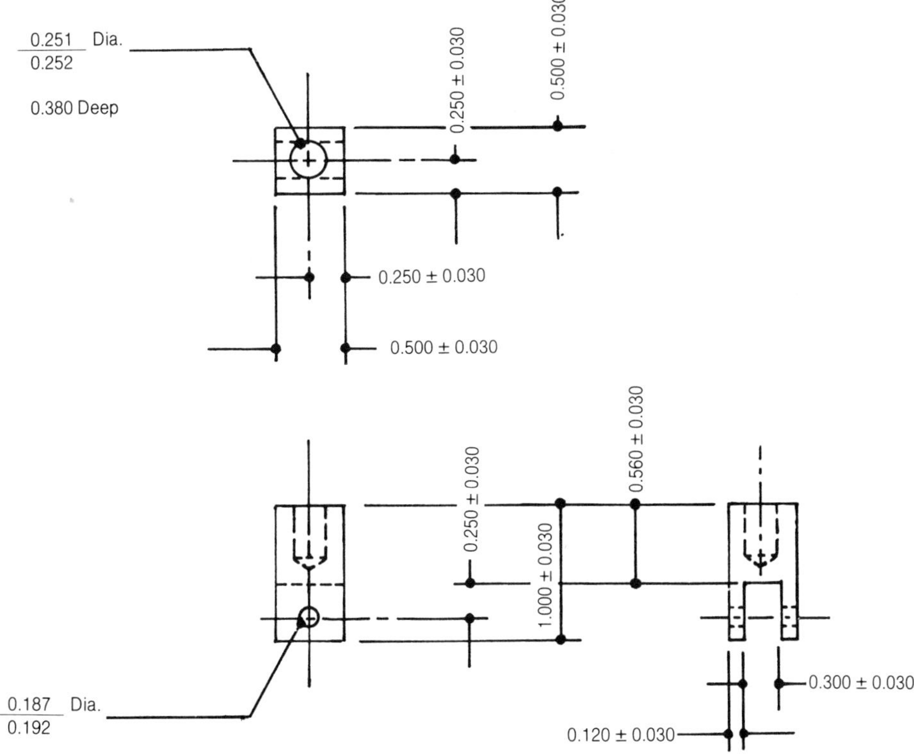

Figure 2.61 Clevis detail.

The profile of the cam is made up of circles with various radii with three intersecting lines representing the keyseat. The commands needed to draw the cam on the computer graphic's CRT screen are obtained by pushing the correct buttons on the menu and returning this completed command back to the computer.

The 0.375-inch diameter hole in the cam is located at the origin (0,0,0) of the cartesian coordinate system which is in the center of the CRT screen. The center of this hole will be used as the datum from which the other parts are located. The command used for locating this 0.375 hole is:

```
①     ②         ③         ④⑥⑧    ⑩       ⑫⑭⑯
PLACE CIRCLE DIAMETER 0.375 / INPUT X0Y0Z0 CR
                      ⑤⑦  ⑨         ⑪ ⑬⑮  ⑰
```

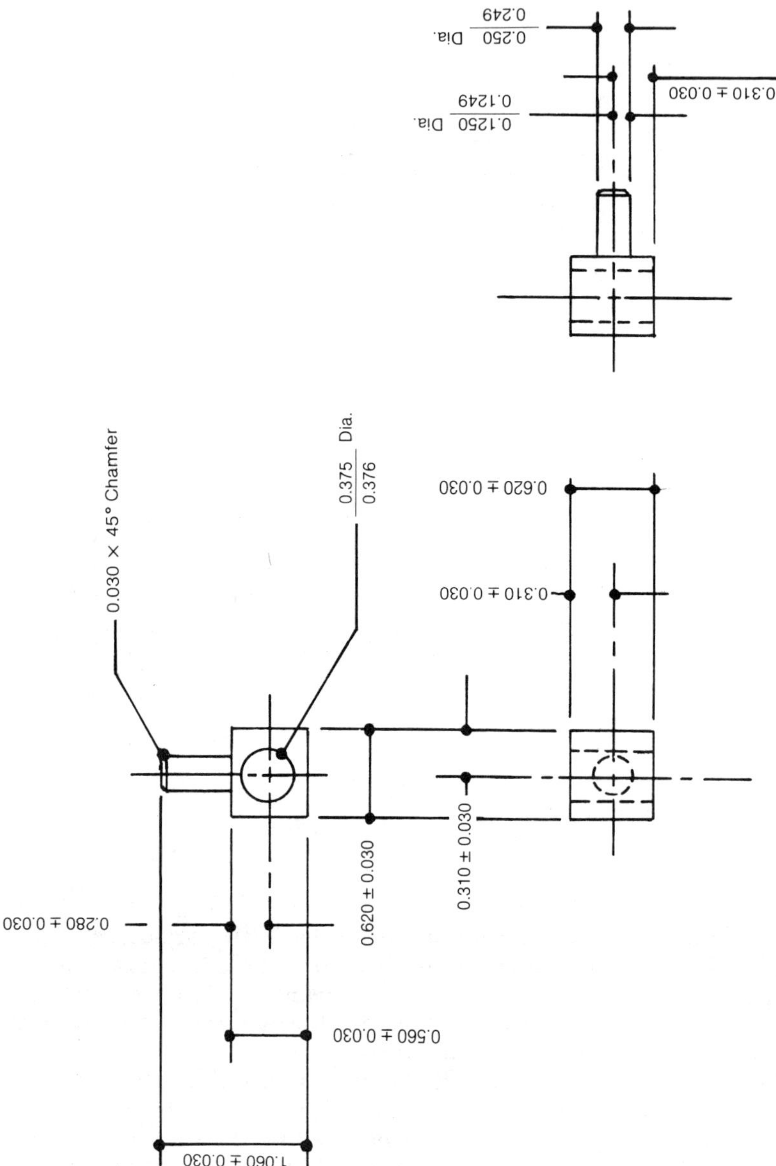

Figure 2.62 Slider block detail.

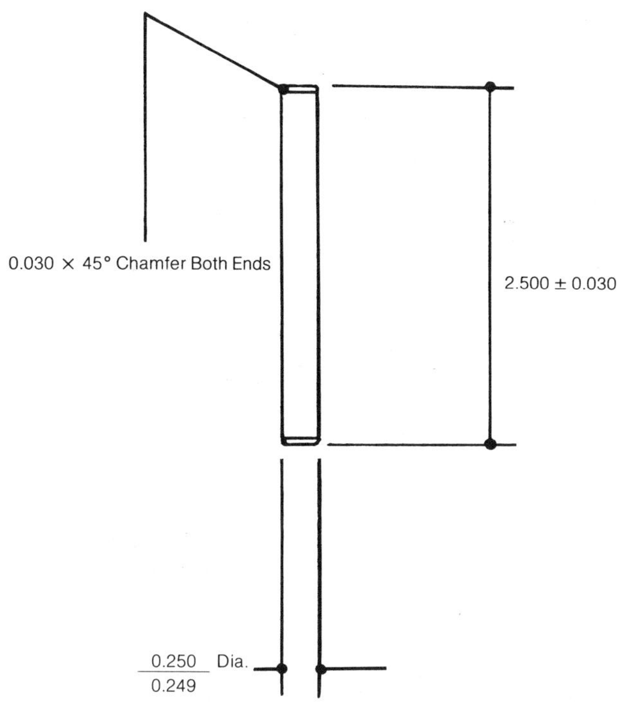

Figure 2.63 Push rod detail.

PLACE	POINT	DIA.	/	UNIT	INPUT	CR
ERASE	LINE	RADIUS	.	1	6	END
MOVE	CIRCLE	TAN. TO	;	2	7	ON
COPY	ARC	PERP.	X	3	8	ORIGIN
ROTATE	DISTANCE	PRL.	Y	4	9	HORZ.
MEASURE	LONG	ANGLE	Z	5	0	VERT.

Figure 2.64 A special drafting and design menu.

1 PLACE	POINT	3 DIA.	9 /	UNIT	10 INPUT	17 CR
ERASE	LINE	RADIUS	5 .	1	6	END
MOVE	2 CIRCLE	TAN. TO	;	2	7 7	ON
COPY	ARC	PERP.	11 X	6 3	8	ORIGIN
ROTATE	DISTANCE	PRL.	13 Y	4	9	HORZ.
MEASURE	LONG	ANGLE	15 Z	8 5	4 14 0 12 16	VERT.

Figure 2.65 Identifying the buttons pushed on the menu.

The buttons that were pushed on the menu to input the previous command into the computer are shown in Figure 2.65 with numbers placed on the menu's buttons to identify the buttons that were pushed and in what order they were pushed.

The command used for placing the 0.78-in. diameter circle at the same location is done in a similar manner:

① ② ③ ④ ⑥ ⑧ ⑩ ⑫ ⑭ ⑯
PLACE CIRCLE DIAMETER 0.78 / INPUT X0Y0Z0 CR
⑤ ⑦ ⑨ ⑪ ⑬ ⑮

The buttons that were pushed on the menu to input the previous command into the computer are shown in Figure 2.66 with numbers placed on the menu's buttons to identify the buttons that were pushed and in what order they were pushed.

The command used for placing the 1.000-in. diameter circle at X = 0.000, Y = 0.125, and Z = 0.000 would be:

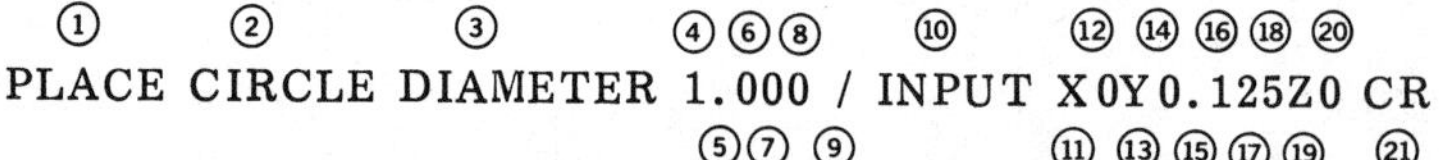

The buttons that were pushed on the menu to input the previous command into the computer are shown in Figure 2.67 with numbers placed on the menu's buttons to identify the buttons that were pushed in what order they were pushed.

1 PLACE	POINT	3 DIA.	8 /	UNIT	9 INPUT	16 CR
ERASE	LINE	RADIUS	.	5 1	6	END
MOVE	2 CIRCLE	TAN. TO	;	2	6 7	ON
COPY	ARC	PERP.	10 X	3	7 8	ORIGIN
ROTATE	DISTANCE	PRL.	12 Y	4	9	HORZ.
MEASURE	LONG	ANGLE	14 Z	5	4 13 0 11 15	VERT.

Figure 2.66 Identifying the buttons pushed on the menu.

1 PLACE	POINT	3 DIA.	9 /	UNIT	10 INPUT	21 CR
ERASE	LINE	RADIUS	5 15 .	4 16 1	6	END
MOVE	2 CIRCLE	TAN. TO	;	17 2	7	ON
COPY	ARC	PERP.	11 X	3	8	ORIGIN
ROTATE	DISTANCE	PRL.	13 Y	4	9	HORZ.
MEASURE	LONG	ANGLE	19 Z	18 5	6 7 8 0 12 14 20	VERT.

Figure 2.67 Identifying the buttons pushed on the menu.

An assembly drawing of these six parts can be constructed on a computer graphics system by calling up and placing each part at its proper location on the CGS display. These parts can be drawn by using their nominal or maximum or minimum material dimensions.

Perhaps, the designer would like to know the variation of movement the push rod will have due to the location tolerances and the maximum and minimum material conditions of the parts making up the assembly. This type of calculation is sometimes referred to as a tolerance stackup in industry. In this case, this calculation would be very difficult or almost impossible to do without the use of computer graphics.

The first task to be performed on the computer graphics system is to place the part which is filed under the file name, MMC plate, onto its display screen. The three holes in this plate are used to locate the other five parts of the assembly; therefore, this plate would be considered the starting point when calculating the tolerance stackup for this assembly.

These three holes are located in the plate by using a rectangular tolerance of ±0.03 of an inch associated with each locating dimension. This tolerance produces a square tolerance zone with sides equal to 0.06 inches and a diagonal of 0.085 inches. These three holes can be moved about on the display screen by using the move command on the menu. The command, MOVE /UNIT D_1 INPUT X0Y0Z0, IX-0.03IY-0.03IZ0 CR, will place the center of the hole (used in locating the cam) at one of its extreme positions. This same procedure is performed on the rocker. The slider block, clevis, and push rod are not placed on the display because the tolerance stackup can be done without them.

A point is placed at the center of each of these holes by using the command PLACE POINT / INPUT ORIGIN D_1 CR, and digitizing each hole. Now the cam and the rocker can be placed about their respective points of locations by using the commands PLACE MMC CAM / INPUT POINT D_1 CR and PLACE MMC ROCKER / INPUT POINT D_2 CR. As previously mentioned, there is no need to place the slider block, clevis, and push rod on the display, since a straight line can represent these parts on the display.

A point is placed at the center of the 0.255/0.260-diameter hole in the rocker and another point is placed at the end of the straight line. Using the rotate command, the cam is rotated 180 degrees in the clockwise direction which would yield the greatest movement of the push rod, in this case, the straight line. The command ROTATE ANGLE −180/INPUT D_1 D_2 D_3 D_4 D_5 D_6 CR, would be used to rotate the cam. The rocker's geometry is simplified by constructing a stick model of it. This stick model is constructed by using the commands PLACE CIRCLE DIAMETER 1.250/INPUT POINT D_1 CR, PLACE LINE TANGENT

TO / UNIT D_2 D_3 CR, and PLACE LINE PARALLEL LONG 2 / UNIT D_4 INPUT END D_5 D_6 CR. A straight line with a length of 3 inches is placed from the end of the rocker's output through the point representing the slider block's pivot center. A point is placed at the end of this 3-in. line.

The distance between these two points can be measured by using the command MEASURED DISTANCE / UNIT POINT D_1 D_2 CR. This process is repeated when the position of the cam and rocker is moved to the opposite end of the tolerance zone's diagonal. The distance the push rod moves, in each case, can be compared to each other to determine what effects the various variables (±0.03 tolerance zone, MMC, or LMC) have on the output of the push rod.

2.10. WORK PROBLEMS

1. Make detailed working drawings of the parts of the safety valve shown in Figure 2.68 in assembly. Supply additional detail dimensions where necessary and dimension according to the rectangular coordinate system using datums, appropriate place decimals, form tolerancing, etc. Change all fractions to decimals with one place decimals equal to ±0.1, two place decimals equal to ±0.01, and three place decimals equal to ±0.001 inches. Use tables from Chapter 1 for the fits.

2. Make detailed working drawings of the parts of the channel pipe hanger shown in Figure 2.69 in assembly. Supply additional detail dimensions where necessary and dimension according to the rectangular coordinate system using datums, appropriate place decimals, form tolerancing, etc. Change all fractions to decimals with one place decimals equal to ±0.1, two place decimals equal to ±0.01, and three place decimals equal to ±0.001 inches. Use tables in Chapter 1 for the fits.

3. Make detailed working drawings of the chain pipe vise shown in Figure 2.70 in assembly. Supply additional detail dimensions where necessary and dimension according to the rectangular coordinate system using datums, appropriate place decimals, form tolerancing, etc. Change all fractions to decimals with one place decimals equal to ±0.1, two place decimals equal to ±0.01, and three place decimals equal to ±0.001 inches. Use tables in Chapter 1 for the fits.

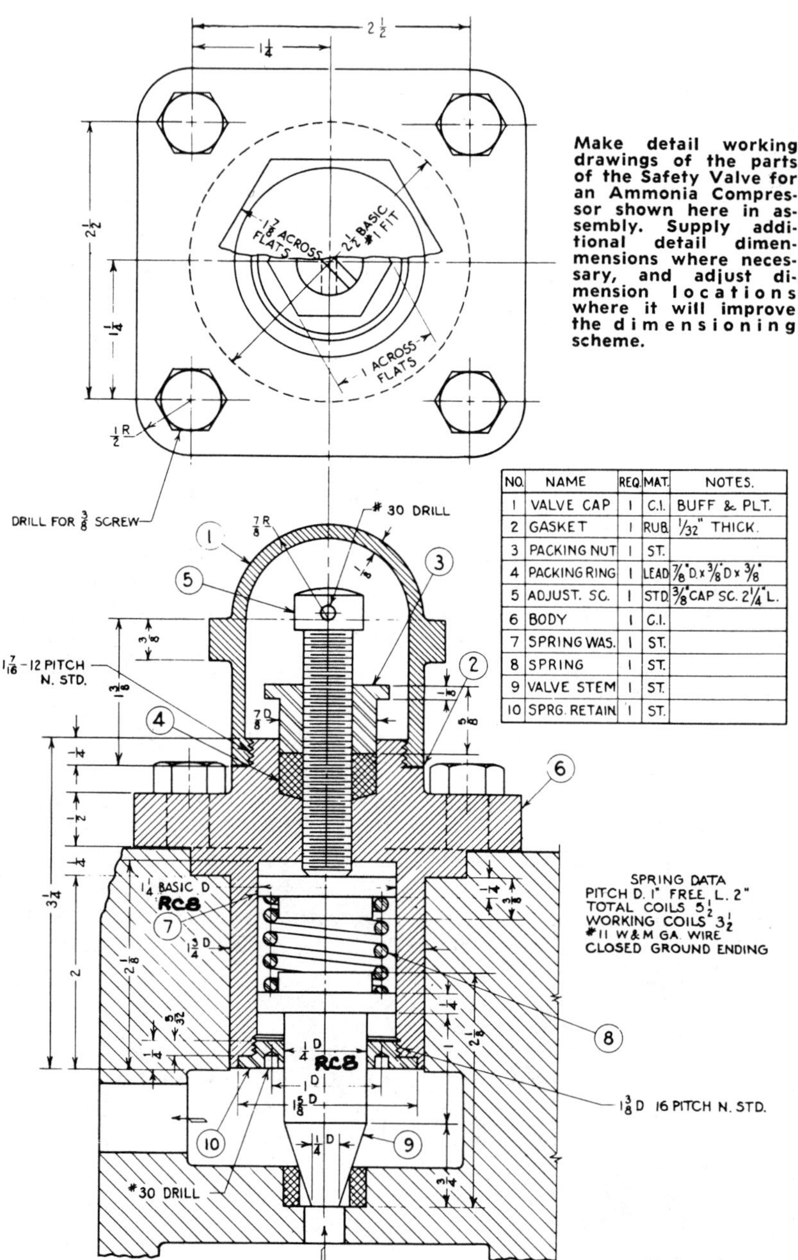

NO.	NAME	REQ.	MAT.	NOTES.
1	VALVE CAP	1	C.I.	BUFF & PLT.
2	GASKET	1	RUB.	1/32" THICK.
3	PACKING NUT	1	ST.	
4	PACKING RING	1	LEAD	7/8" D. x 3/8" D x 3/8"
5	ADJUST. SC.	1	STD.	3/8" CAP SC. 2 1/4" L.
6	BODY	1	C.I.	
7	SPRING WAS.	1	ST.	
8	SPRING	1	ST.	
9	VALVE STEM	1	ST.	
10	SPRG. RETAIN.	1	ST.	

Figure 2.68 Safety valve. (From H. D. Orth, R. R. Worsencroft, and H. B. Doke, *Theory and Practice of Engineering Drawing*, William C. Brown, Dubuque, Iowa, 1959.)

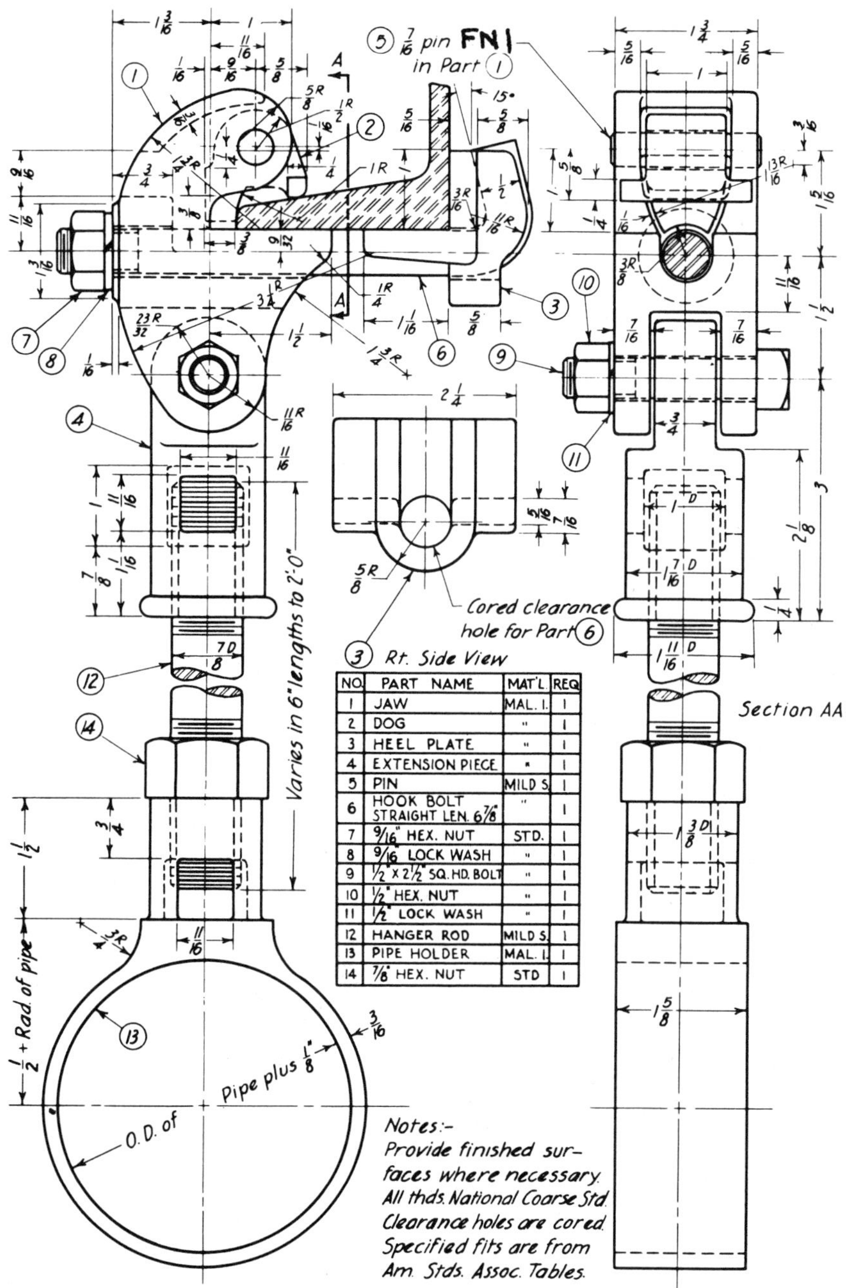

NO.	PART NAME	MAT'L.	REQ.
1	JAW	MAL. I.	1
2	DOG	"	1
3	HEEL PLATE	"	1
4	EXTENSION PIECE	"	1
5	PIN	MILD S.	1
6	HOOK BOLT STRAIGHT LEN. 6 7/8"	"	1
7	9/16" HEX. NUT	STD.	1
8	9/16" LOCK WASH	"	1
9	1/2" X 2 1/2" SQ. HD. BOLT	"	1
10	1/2" HEX. NUT	"	1
11	1/2" LOCK WASH	"	1
12	HANGER ROD	MILD S.	1
13	PIPE HOLDER	MAL. I.	1
14	7/8" HEX. NUT	STD	1

Figure 2.69 Channel pipe hanger. (From H. D. Orth, R. R. Worsencroft, and H. B. Doke, *Theory and Practice of Engineering Drawing*, William C. Brown, Dubuque, Iowa, 1959.)

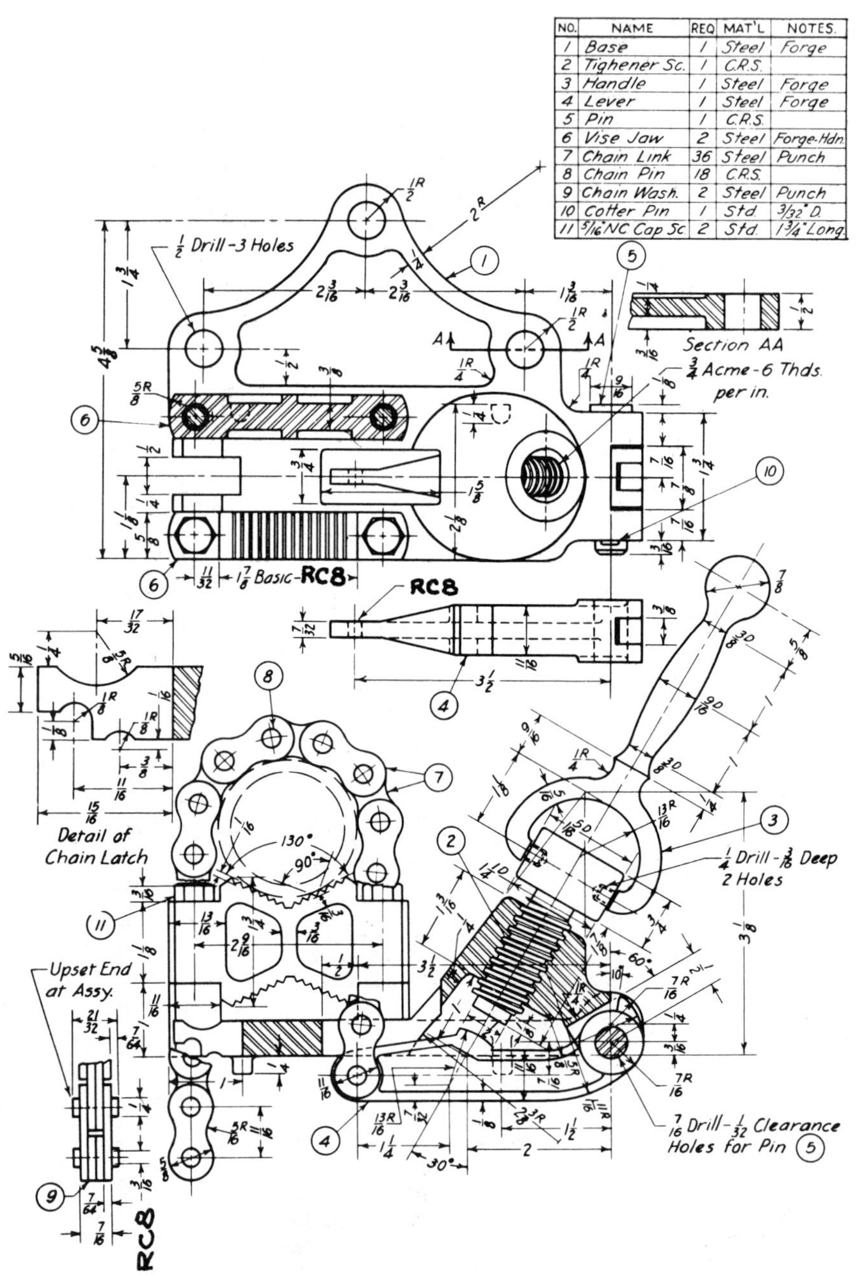

NO.	NAME	REQ	MAT'L	NOTES
1	Base	1	Steel	Forge
2	Tighener Sc.	1	C.R.S.	
3	Handle	1	Steel	Forge
4	Lever	1	Steel	Forge
5	Pin	1	C.R.S.	
6	Vise Jaw	2	Steel	Forge-Hdn.
7	Chain Link	36	Steel	Punch
8	Chain Pin	18	C.R.S.	
9	Chain Wash.	2	Steel	Punch
10	Cotter Pin	1	Std.	3/32" D.
11	5/16" NC Cap Sc.	2	Std.	1 3/4" Long

Figure 2.70 Chain pipe vise. (From H. D. Orth, R. R. Worsencroft, and H. B. Doke, *Theory and Practice of Engineering Drawing*, William C. Brown, Dubuque, Iowa, 1959.)

3

Form and Position Dimensioning and Tolerancing

3.1. GENERAL RULES, TERMS, AND PHILOSOPHY

Form and position dimensioning and tolerancing is sometimes referred to as geometric dimensioning and tolerancing. In either case, it is a means of dimensioning and tolerancing a drawing of a part with respect to the actual function of the part's features and how it can be most economically manufactured.

It is a method of dimensioning and tolerancing that provides, one and only one, interpretation. It assures interchangeability of mating parts at assembly, provides for maximum producibility of the part through maximum production tolerances, and ensures design integrity of the part throughout manufacturing.

Dimensioning and tolerancing methods such as the rectangular coordinate method covered in Chapter 2, lack the conciseness, uniformity, and clarity that is being demanded from our high technological designs of today. A method which provides the means of stating necessary dimensional or tolerance requirements on the drawing not otherwise covered by standard interpretation is needed. Form and position dimensioning and tolerancing satisfy these requirements. This method should be used if: (1) part features are critical to function or interchangeability, (2) functional gaging techniques are required, (3) datums are used to ensure consistency between manufacturing and gaging operations, and (4) when standard interpretation of tolerancing is vague.

The characteristics and symbols used in form and position dimensioning and tolerancing are:

⏥	=	Flatness
—	=	Straightness
○	=	Roundness (circularity)
⌭	=	Cylindricity
⌒	=	Profile of a line
⌓	=	Profile of a surface
⊥	=	Perpendicularity (squareness)
∠	=	Angularity
//	=	Parallelism
↗	=	Runout
⌖	=	Position
◎	=	Concentricity
⌯	=	Symmetry

The symbols used in form and position tolerancing came from the American Standard Y14.5 document titled "Dimensioning and Tolerancing." In the last several years this document has become the most important drafting, design, and engineering standard for domestic and military use.

The revised ANSI Y14.5M—1982 standard including metrics was released early in 1982. Further improvements and expansion into the symbology have been made to enhance universal understanding of the information contained on technical drawings worldwide and due to the ever-increasing use of computer graphics in this field.

The concepts from the ANSI Y14.5 standards are covered in the following sections.

3.2. DEFINING THE FORM OF INDIVIDUAL FEATURES WITH RESPECT TO NO DATUMS

Flatness, straightness, roundness, and cylindricity are form characteristics which are independent of any other characteristic; therefore, they are not used with datums. These form characteristics are used with size tolerances which do not provide adequate control with respect to the function and/or interchangeability of the part.

The form characteristics symbols for flatness, straightness, roundness, and cylindricity are listed below.

⏥ = flatness

— = straightness

○ = roundness

⌭ = cylindricity

This symbol is placed in a rectangular box (Fig. 3.1) with its tolerance. This box is attached to a leader with an arrowhead pointing to the particular feature that it refers to on the drawing.

3.2.1. Flatness

Definition. Flatness is the condition of a surface where all of the elements that make up the surface lie in one plane.

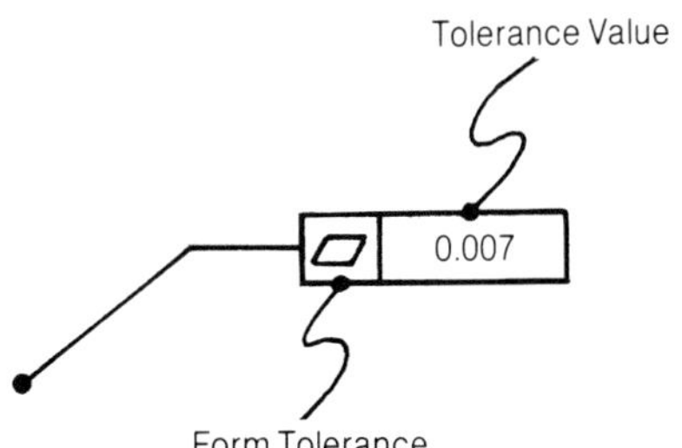

Figure 3.1 Form tolerance symbol placed in its box.

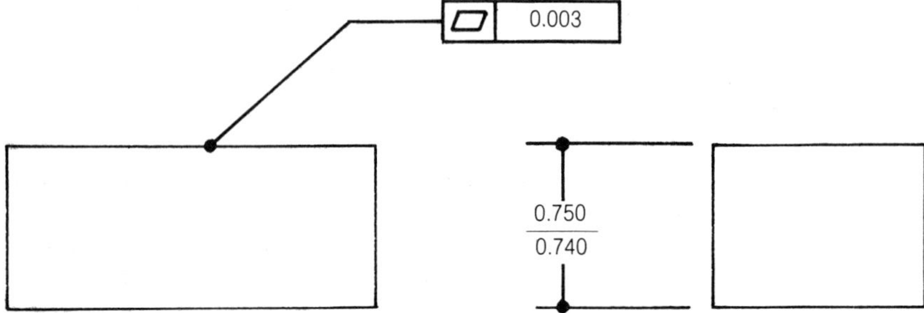

Figure 3.2 Flatness callout.

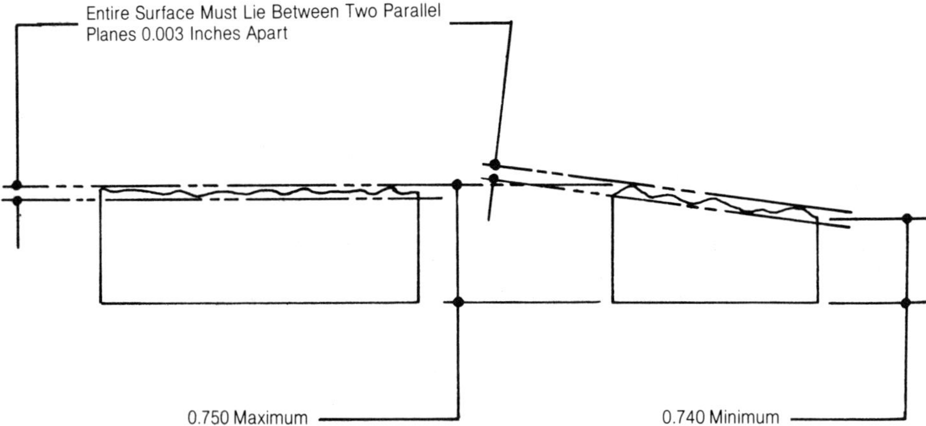

Figure 3.3 Flatness tolerance zone.

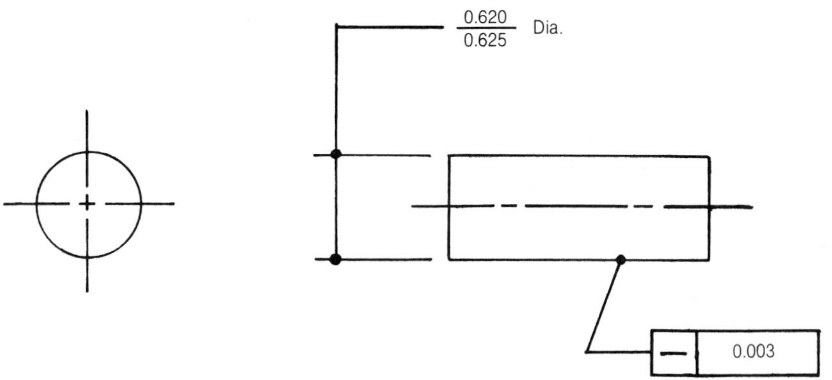

Figure 3.4 Straightness callout.

Tolerance. A flatness tolerance applied to a surface means that the entire surface must lie between two parallel planes separated by a distance equal to the tolerance specified.

Explanation. Figure 3.2 shows the flatness tolerance callout on a drawing. Any longitudinal element of the surface must lie between these two parallel and straight planes which define the flatness tolerance zone. This flatness tolerance zone is shown in Figure 3.3.

3.2.2. Straightness

Definition. Straightness is the condition in which the longitudinal elements of a surface are compared to a straight line.

Tolerance. A straight tolerance specifies a tolerance zone of uniform width along a straight line, within which all points of the considered line must lie.

Explanation. Figure 3.4 shows the straightness tolerance callout on a drawing. Any longitudinal element of the surface must lie between two parallel straight lines which define the straightness tolerance zone and where these two lines, including the axis of the features, share a common plane. This straightness tolerance is shown in Figure 3.5.

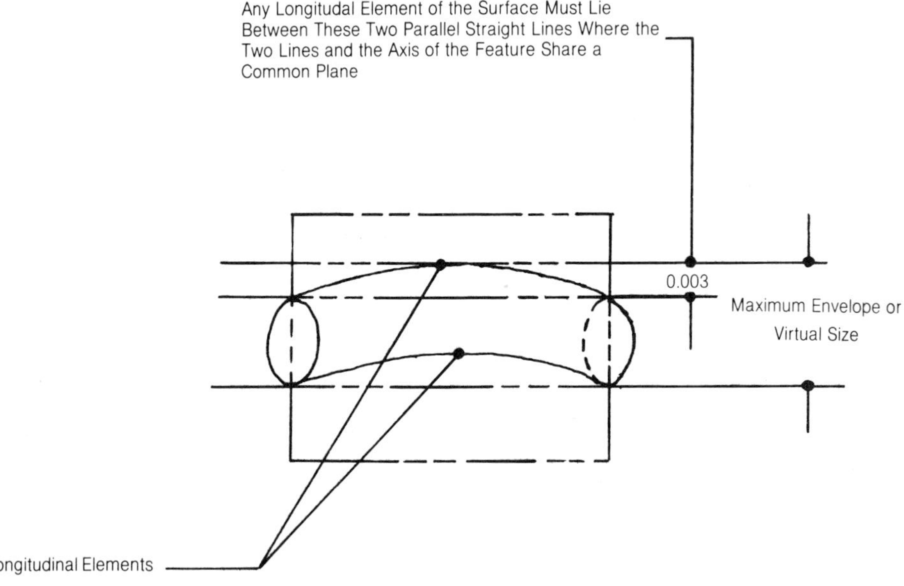

Figure 3.5 Straightness tolerance zone.

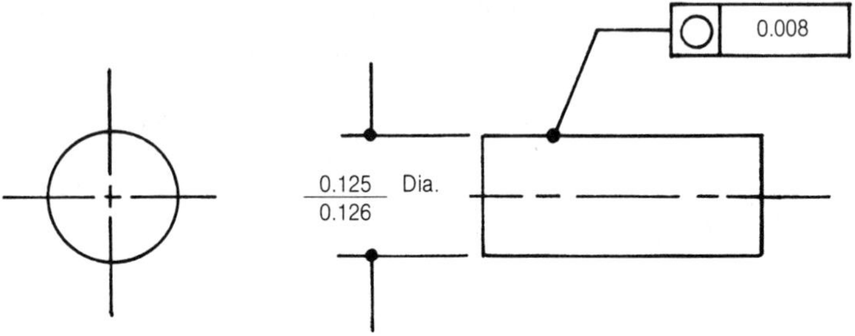

Figure 3.6 Roundness callout.

3.2.3. Roundness

Definition. Roundness is a condition of a surface of revolution such as a cylinder, cone, or sphere, where all points of the surface intersected by any plane, (1) perpendicular to a common axis (cylinder, cone) or (2) passing through a common center (sphere), are equidistant from the axis.

Tolerance. The roundness tolerance applied to a surface means that every point in any cross section perpendicular to a reference axis must lie between two concentric circles separated by a distance equal to the tolerance specified.

Explanation. Figure 3.6 shows the roundness tolerance callout on a drawing. The concentric circles specify a circular cross section that represents the tolerance zone within which the cross section of the surface of revolution must lie. Tolerances of roundness are always specified along a radial line; therefore, the thickness of the tolerance zone would be divided equally on each side of the mean radial dimension. This zone is shown in Figure 3.7 and applies only to the cross-sectional profile of the surface of revolution at that particular cross section.

3.2.4. Cylindricity

Definition. Cylindricity is a condition of a surface of revolution in which a solid figure, described by the edge of a rectangle, is rotated around its parallel edge and this edge becomes the axis of the formed cylinder.

Tolerance. A cylindricity tolerance specifies a tolerance zone confined between two coaxial cylinders within which the surface must lie.

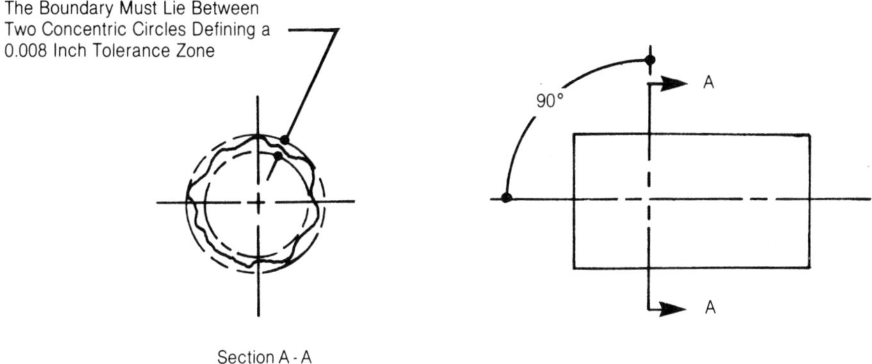

Figure 3.7 Roundness tolerance zone.

Explanation. Figure 3.8 shows the cylindricity tolerance callout on a drawing. The two concentric cylinders form a tubular type of tolerance zone within which the entire surface of revolution must lie. Tolerances of cylindricity are always along a radial line; therefore, the thickness of the tolerance zone would be divided equally on each side of the mean radial dimension. This zone is shown in Figure 3.9 and applies to the entire surface of revolution.

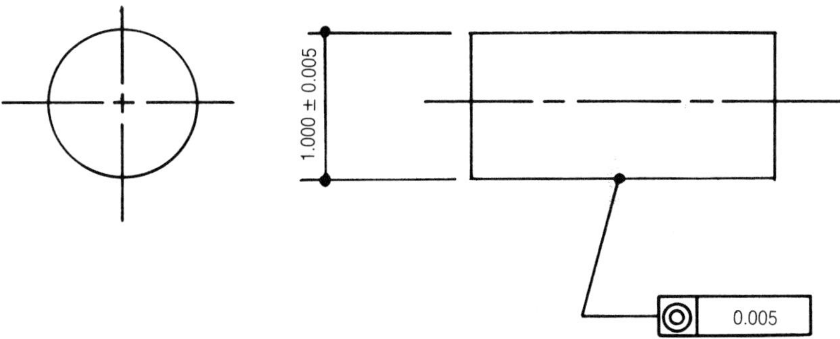

Figure 3.8 Cylindricity callout.

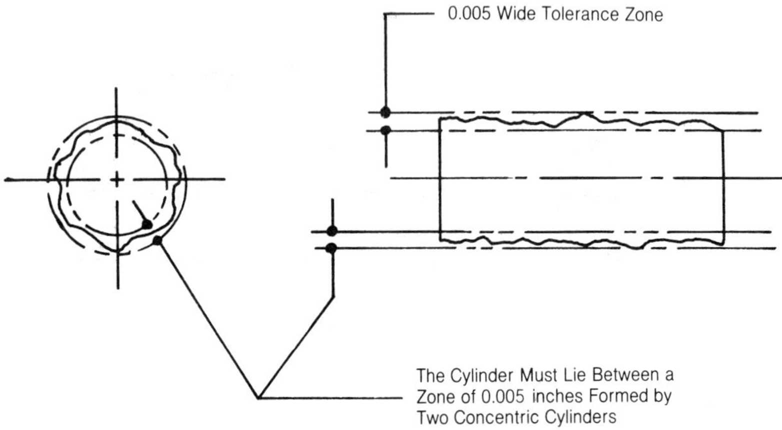

Figure 3.9 Cylindricity tolerance zone.

3.3. DEFINING THE ORIENTATION OF FEATURES WITH RESPECT TO DATUMS

The orientation of one feature relative to another feature (datum) is referred to as perpendicularity, angularity, parallelism, profile, runout, concentricity, and symmetry tolerances. These tolerances are specified when there is a required relationship between surfaces or size features, or a need for close tolerance control of a feature during its location placement.

The orientation symbols for perpendicularity, angularity, parallelism, profile, runout, concentricity, and symmetry are listed below.

⊥ = Perpendicularity

∠ = Angularity

// = Parallelism

⌒ = Profile of a line

⌓ = Profile of a surface

↗ = Runout

⌰ = Total runout

◎ = Concentricity

⌯ = Symmetry

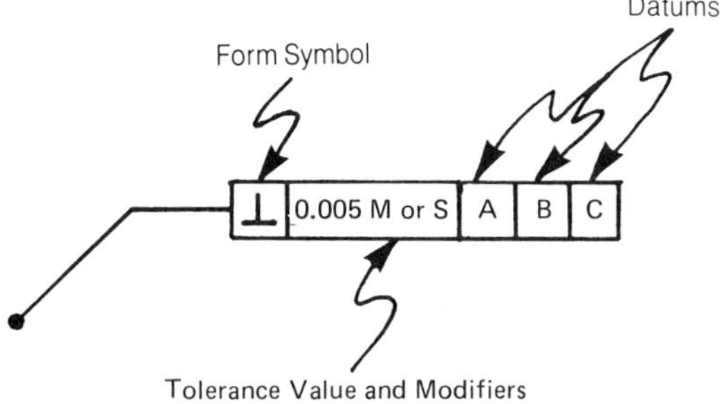

Figure 3.10 Form symbol with its tolerance value and respective datums.

The appropriate symbol is placed in a rectangular box (Fig. 3.10) with its tolerance referenced to a datum or datums. The letters M or S are modifiers and they refer to the maximum material condition (M) or regardless of feature size (S) of the particular feature.

3.3.1. Perpendicularity

Perpendicularity is the condition of a surface, axis, or line which is at right angles (90 degrees) to a datum plane or axis.

Perpendicularity of a Surface to a Datum Plane

Tolerance. A perpendicularity tolerance applied to a surface (with respect to a datum plane) means that the entire surface must lie between two parallel planes perpendicular to the datum plane. The two parallel planes are separated by a distance equal to the tolerance specified.

Explanation. Figure 3.11 shows the perpendicularity tolerance callout on a drawing. The surface of a feature must lie between the two parallel planes which are perpendicular to the tolerance zone as shown in Figure 3.12.

Perpendicularity of a Surface to Two Datum Planes

Tolerance. The perpendicularity tolerance zone for a noncylindrical feature of size is the distance between two parallel planes with the specified tolerance apart and perpendicular to the datum plane.

Explanation. Figure 3.13 shows the perpendicularity tolerance callout on a drawing. A tolerance zone confined by two parallel planes perpendicular to a datum plane within which the centerplane of a feature must lie. This perpendicularity tolerance is shown in Figure 3.14.

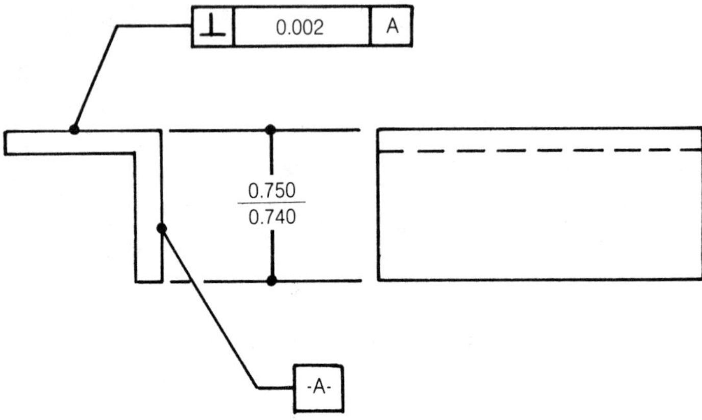

Figure 3.11 Perpendicularity of a surface to a datum plane callout.

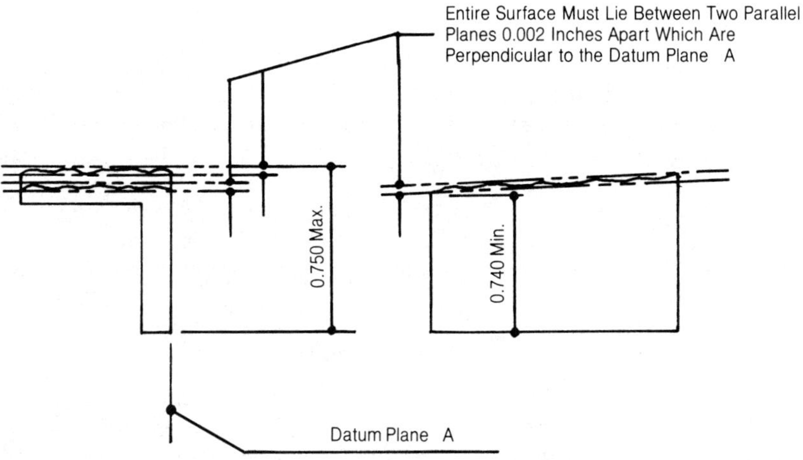

Figure 3.12 Perpendicularity of a surface to a datum plane tolerance zone.

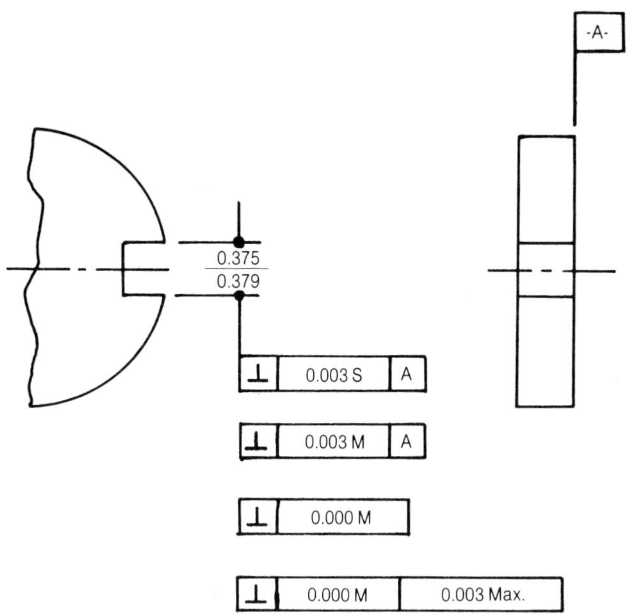

Figure 3.13 Perpendicularity of a surface to two datum planes callout.

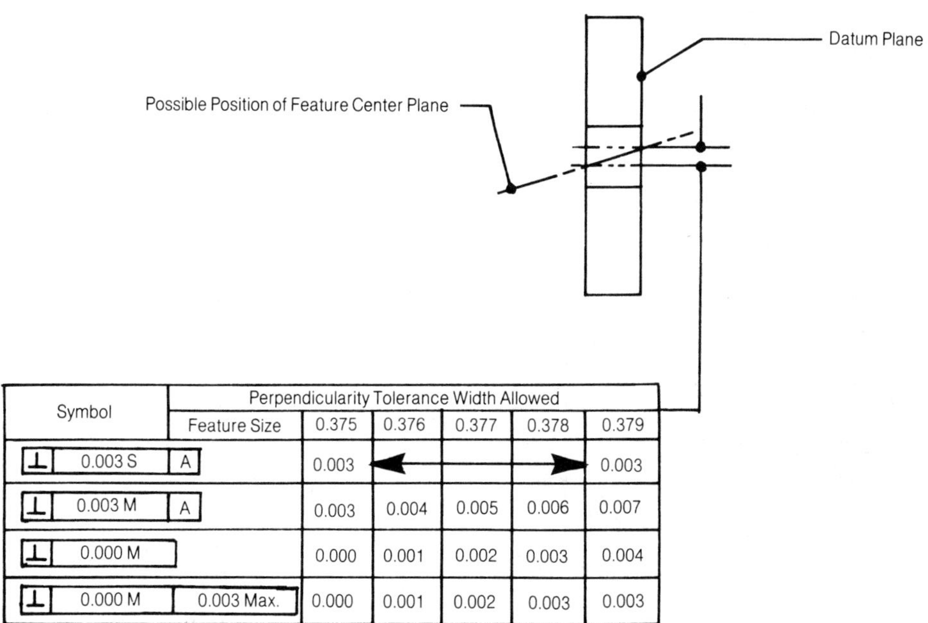

Symbol	Perpendicularity Tolerance Width Allowed					
	Feature Size	0.375	0.376	0.377	0.378	0.379
⊥ 0.003 S A		0.003	←		→	0.003
⊥ 0.003 M A		0.003	0.004	0.005	0.006	0.007
⊥ 0.000 M		0.000	0.001	0.002	0.003	0.004
⊥ 0.000 M 0.003 Max.		0.000	0.001	0.002	0.003	0.003

Figure 3.14 Perpendicularity of a surface to two datum planes tolerance zone.

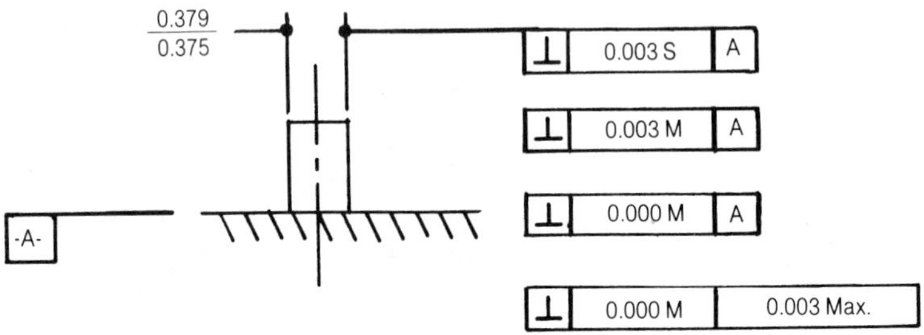

Figure 3.15 Perpendicularity of an axis to a datum plane callout.

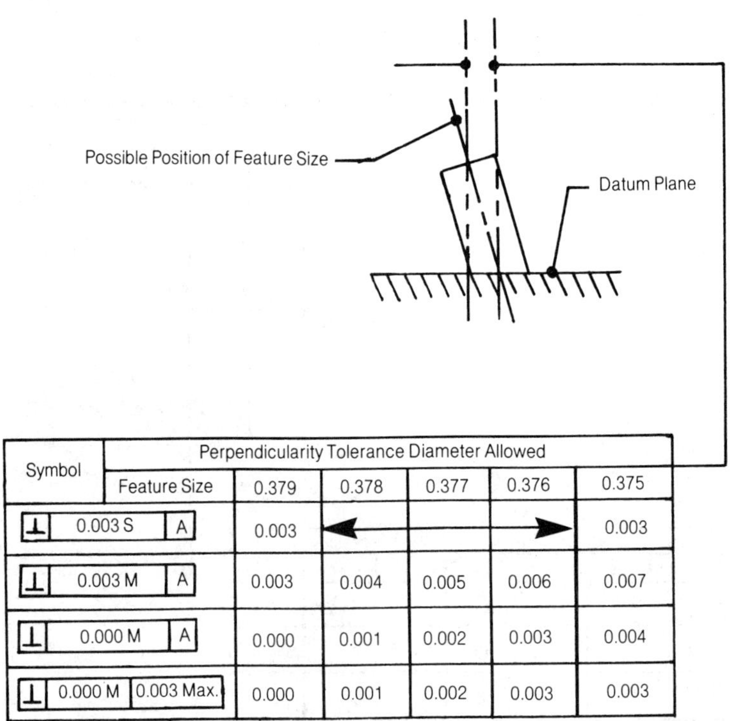

Symbol	Perpendicularity Tolerance Diameter Allowed					
	Feature Size	0.379	0.378	0.377	0.376	0.375
⊥ 0.003 S A		0.003	←		→	0.003
⊥ 0.003 M A		0.003	0.004	0.005	0.006	0.007
⊥ 0.000 M A		0.000	0.001	0.002	0.003	0.004
⊥ 0.000 M 0.003 Max.		0.000	0.001	0.002	0.003	0.003

Figure 3.16 Perpendicularity of an axis to a datum plane tolerance zone.

Perpendicularity of an Axis to a Datum Plane

Tolerance. A perpendicularity tolerance applied to an axis (with respect to a datum plane) means that the axis must fall within a cylindrical tolerance zone whose diameter is equal to the specified perpendicularity tolerance.

Explanation. Figures 3.15, 3.17, 3.19, and 3.21 show the perpendicularity tolerance callout on several drawings. A cylindrical tolerance zone is perpendicular to a datum plane within which the axis of the feature must lie. This perpendicularity tolerance is shown in Figures 3.16, 3.18, 3.20, and 3.22.

Perpendicularity of a Surface to a Datum Axis

Tolerance. A perpendicularity tolerance applied to a surface (with respect to a datum axis) means that the entire surface must lie between two parallel planes perpendicular to the datum axis. The parallel planes are separated by a distance equal to the tolerance specified.

Explanation. Figure 3.23 shows the perpendicularity tolerance callout on a drawing. A tolerance zone is confined by two parallel planes perpendicular to a datum axis within which the axis of a feature must lie. This perpendicularity tolerance is shown in Figure 3.24.

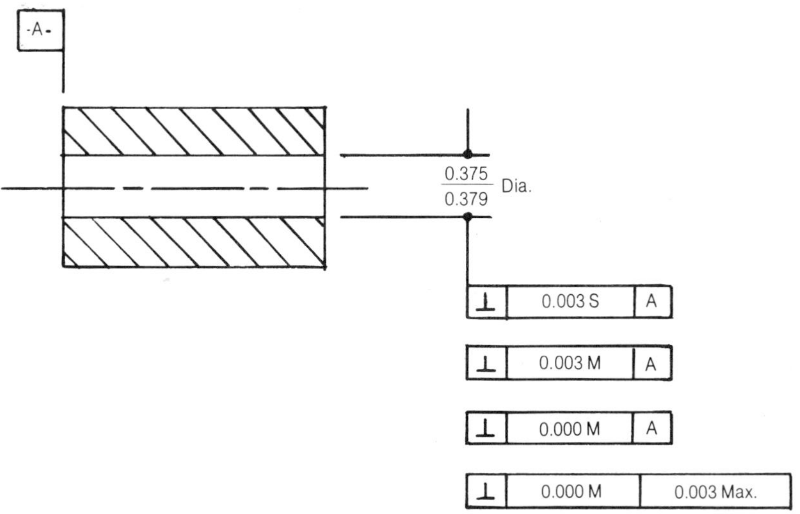

Figure 3.17 Perpendicularity of an axis to a datum plane callout.

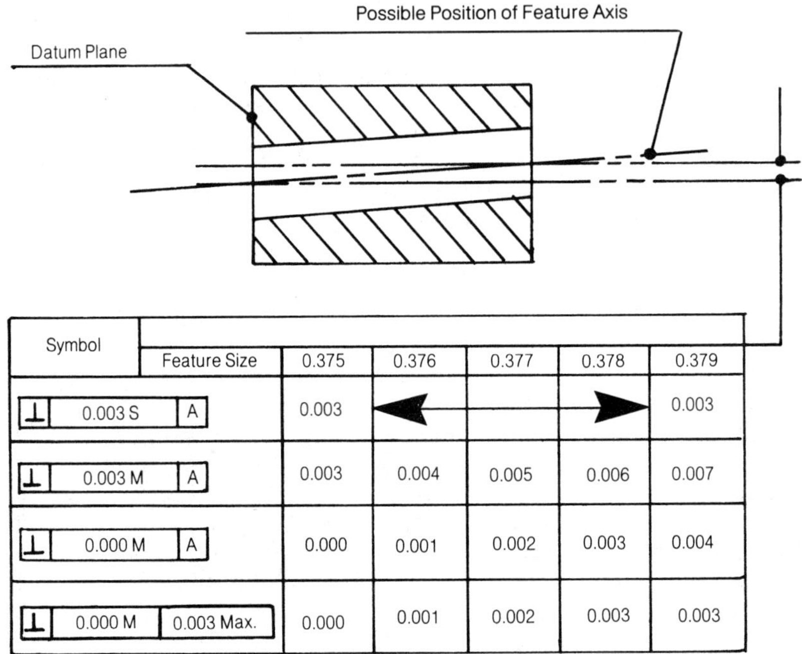

Symbol	Feature Size 0.375	0.376	0.377	0.378	0.379
⊥ 0.003 S A	0.003	←		→	0.003
⊥ 0.003 M A	0.003	0.004	0.005	0.006	0.007
⊥ 0.000 M A	0.000	0.001	0.002	0.003	0.004
⊥ 0.000 M 0.003 Max.	0.000	0.001	0.002	0.003	0.003

Figure 3.18 Perpendicularity of an axis to a datum plane tolerance zone.

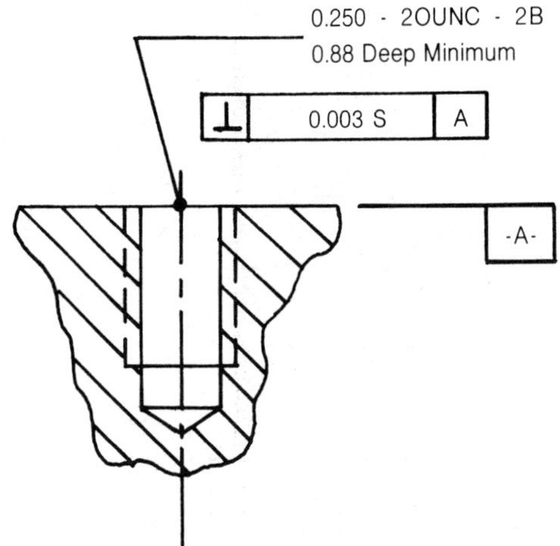

Figure 3.19 Perpendicularity of an axis to a datum plane callout.

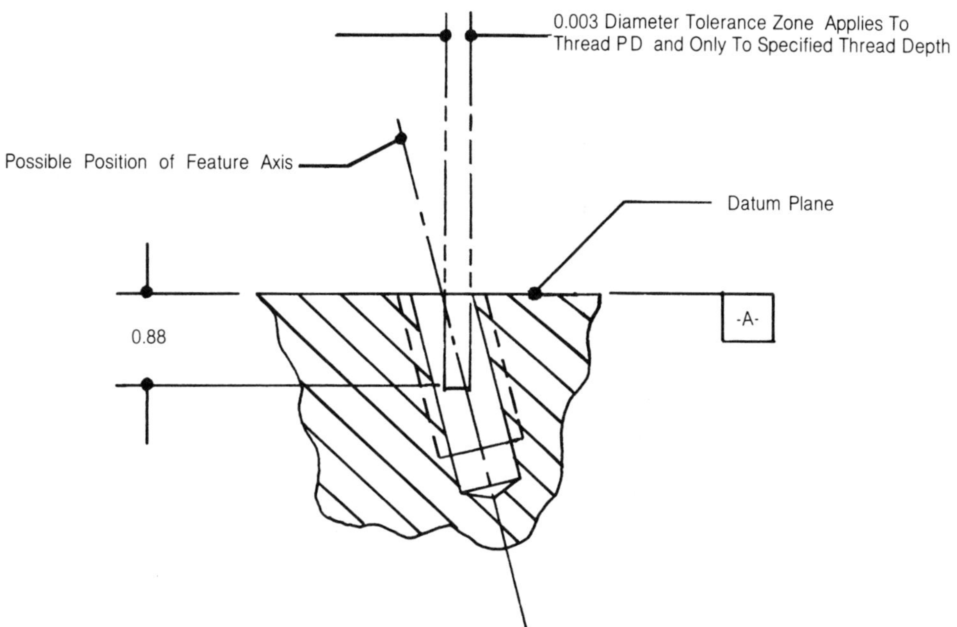

Figure 3.20 Perpendicularity of an axis to a datum plane tolerance zone.

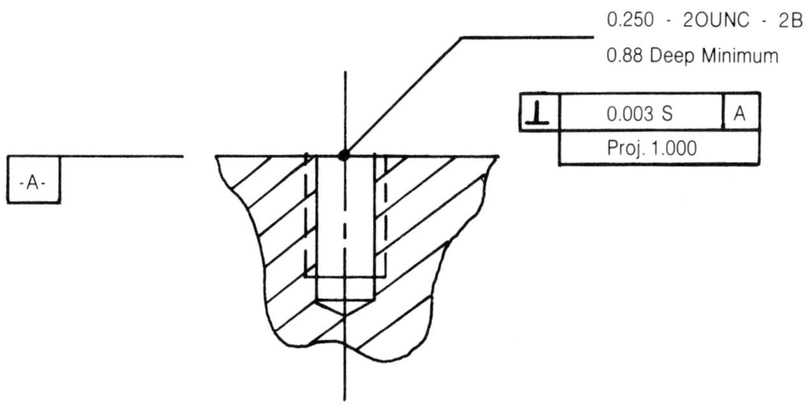

Figure 3.21 Perpendicularity of an axis to a datum plane callout.

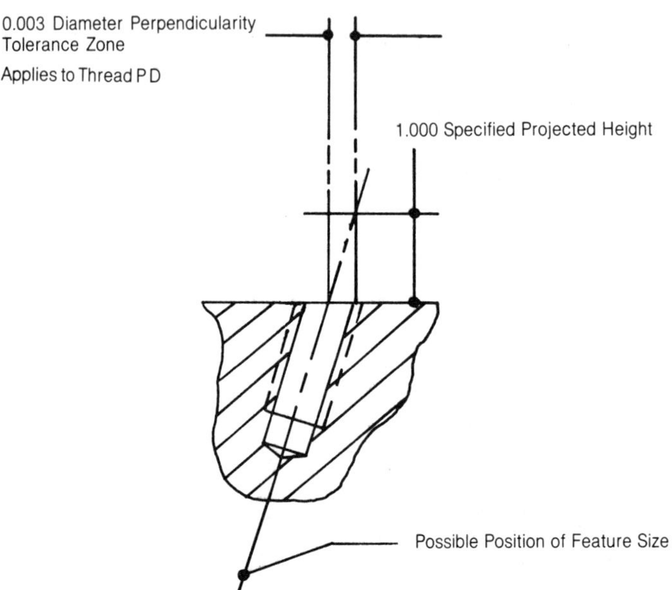

Figure 3.22 Perpendicularity of an axis to a datum plane tolerance zone.

3.3.2. Angularity

Definition. Angularity is the condition of a surface or line which is at the specified angle (other than 90 degrees) from a datum plane or axis.

Tolerance. The angularity tolerance applied to a surface (with respect to datum planes) means that the entire surface must lie between two parallel planes which are at a true angular relation to the specified datum planes and are separated by a distance equal to the tolerance specified.

Explanation. Figure 3.25 shows the angularity tolerance callout on a drawing. An angularity tolerance for a surface specifies a tolerance zone confined by two parallel planes, inclined at the specified angle to a datum plane within which the toleranced surface must lie. Note that the angularity tolerance when applied to a plane surface controls flatness if a flatness tolerance is not specified. This angularity tolerance is shown in Figure 3.26.

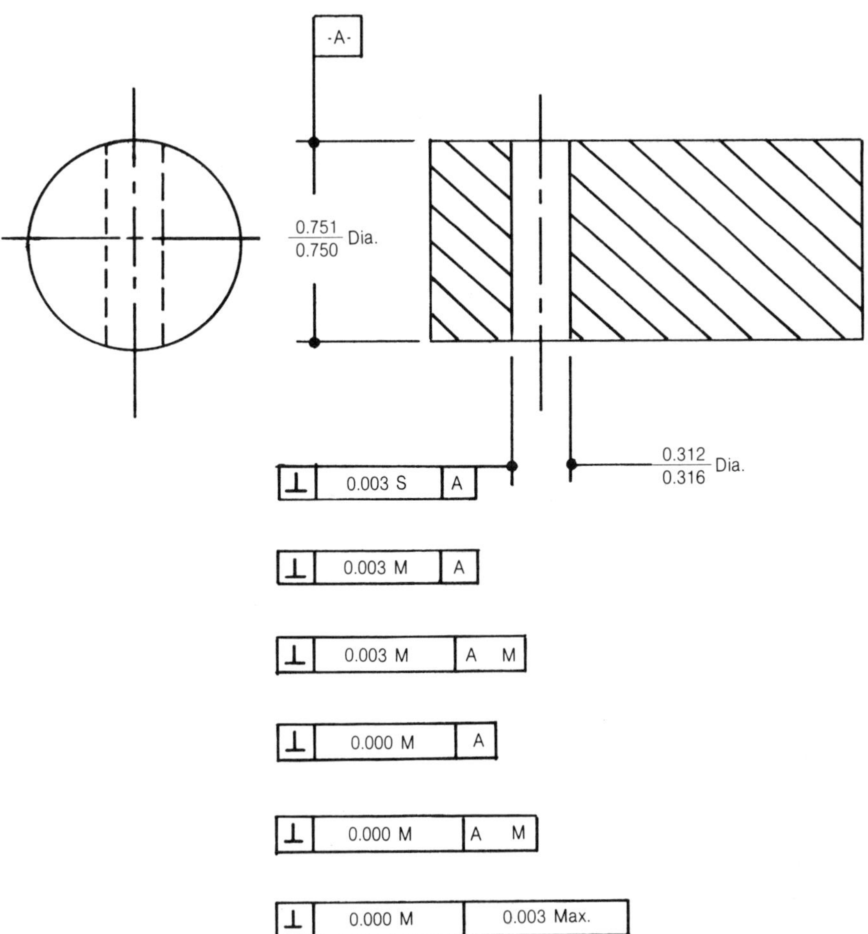

Figure 3.23 Perpendicularity of a surface to a datum axis callout.

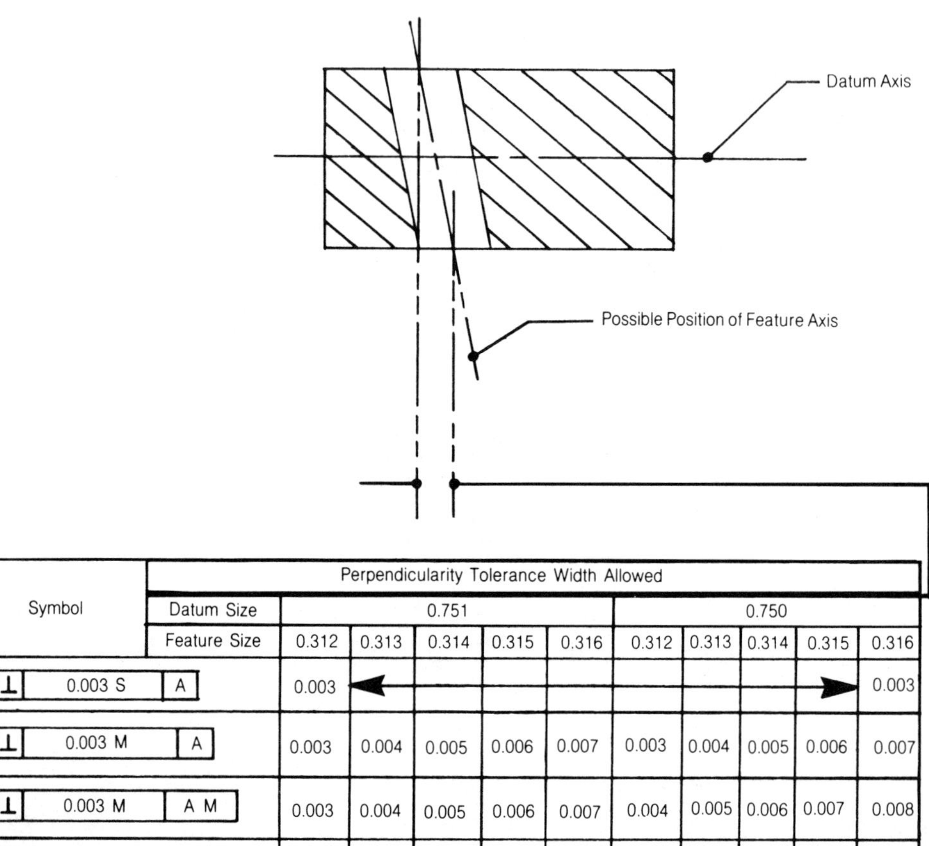

Symbol	Perpendicularity Tolerance Width Allowed										
	Datum Size	0.751					0.750				
	Feature Size	0.312	0.313	0.314	0.315	0.316	0.312	0.313	0.314	0.315	0.316
⊥ 0.003 S A		0.003	←							→	0.003
⊥ 0.003 M A		0.003	0.004	0.005	0.006	0.007	0.003	0.004	0.005	0.006	0.007
⊥ 0.003 M A M		0.003	0.004	0.005	0.006	0.007	0.004	0.005	0.006	0.007	0.008
⊥ 0.000 M A		0.000	0.001	0.002	0.003	0.004	0.000	0.001	0.002	0.003	0.004
⊥ 0.000 M A M		0.000	0.001	0.002	0.003	0.004	0.001	0.002	0.003	0.004	0.005
⊥ 0.000 M 0.003 Max.		0.000	0.001	0.002	0.003	0.003	0.000	0.001	0.002	0.003	0.003

Figure 3.24 Perpendicularity of a surface to a datum axis tolerance zone.

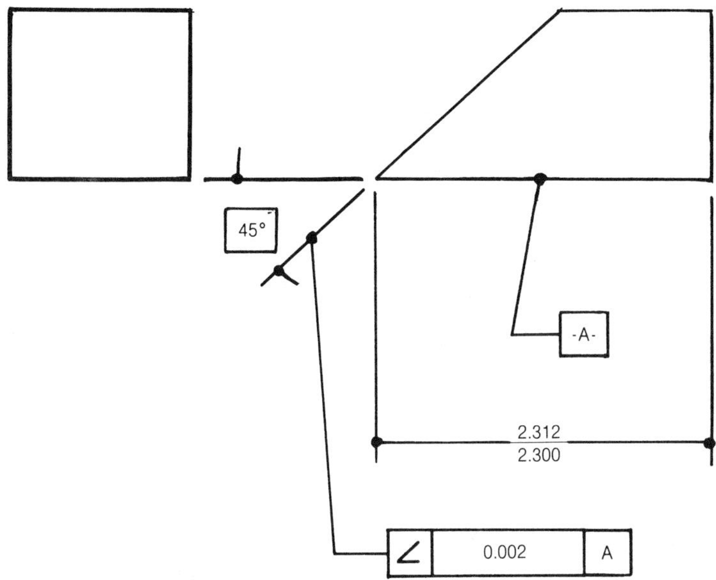

Figure 3.25 Angularity callout.

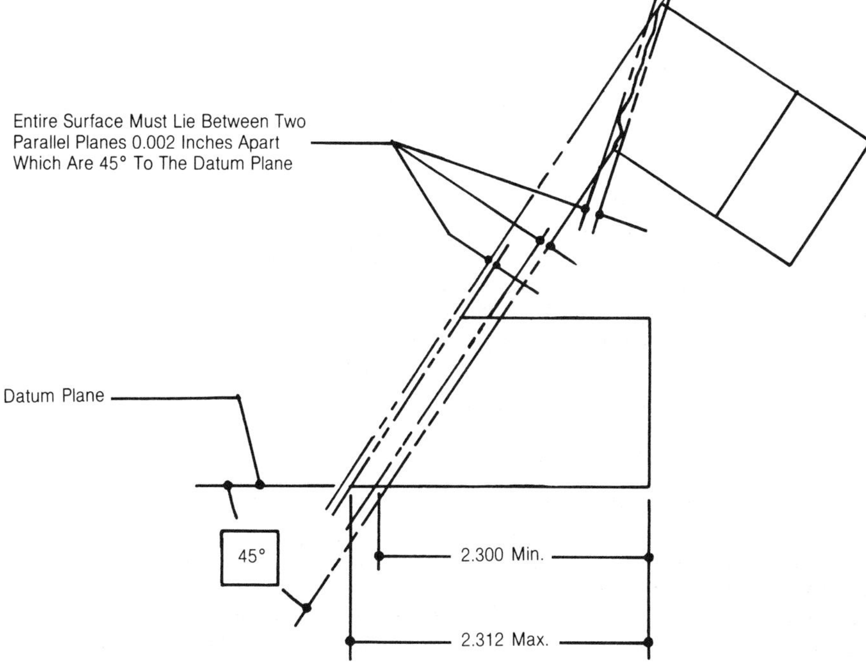

Figure 3.26 Angularity tolerance zone.

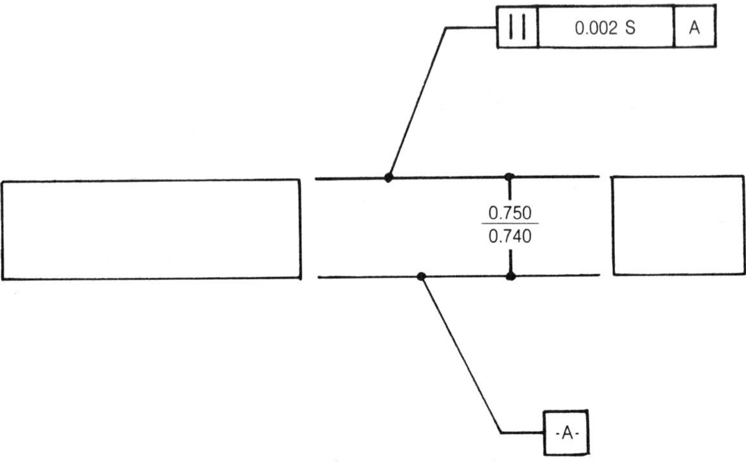

Figure 3.27 Parallelism of a surface to a datum plane callout.

3.3.3. Parallelism

Parallelism is the condition of a surface, axis, or line which is equidistant at all points from a datum plane or axis.

Parrallelism of a Surface to a Datum Plane

Tolerance. A parallelism tolerance applied to a surface (with respect to a datum plane) means that all points on that surface must lie between two parallel planes. These two planes are separated by a distance equal to the specified tolerance and are parallel to the datum plane.

Explanation. Figure 3.27 shows the parallelism tolerance callout on a drawing. The tolerance zone is confined by two planes parallel to a datum plane within which the considered feature (surface or axis) must lie. This parallel tolerance is shown in Figure 3.28.

Parallelism of an Axis to a Datum Axis

Tolerance. A parallelism tolerance applied to an axis (with respect to a datum axis) means that the feature axis must lie within a cylindrical tolerance zone whose diameter is equal to the specified parallelism tolerance. This tolerance zone is parallel to the datum axis.

Explanation. Figure 3.29 shows the parallelism tolerance callout on a drawing. A cylindrical tolerance zone parallels a datum feature axis, within which the axis of a feature must lie. This parallelism tolerance zone is shown in Figure 3.30.

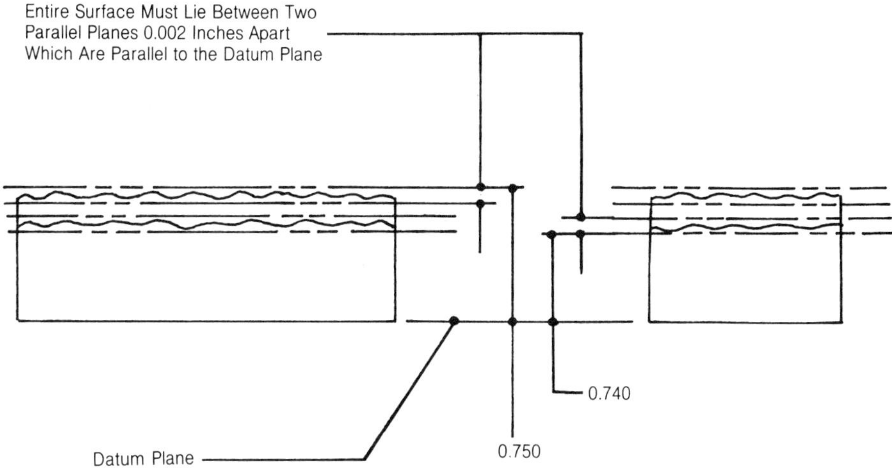

Figure 3.28 Parallelism of a surface to a datum plane tolerance zone.

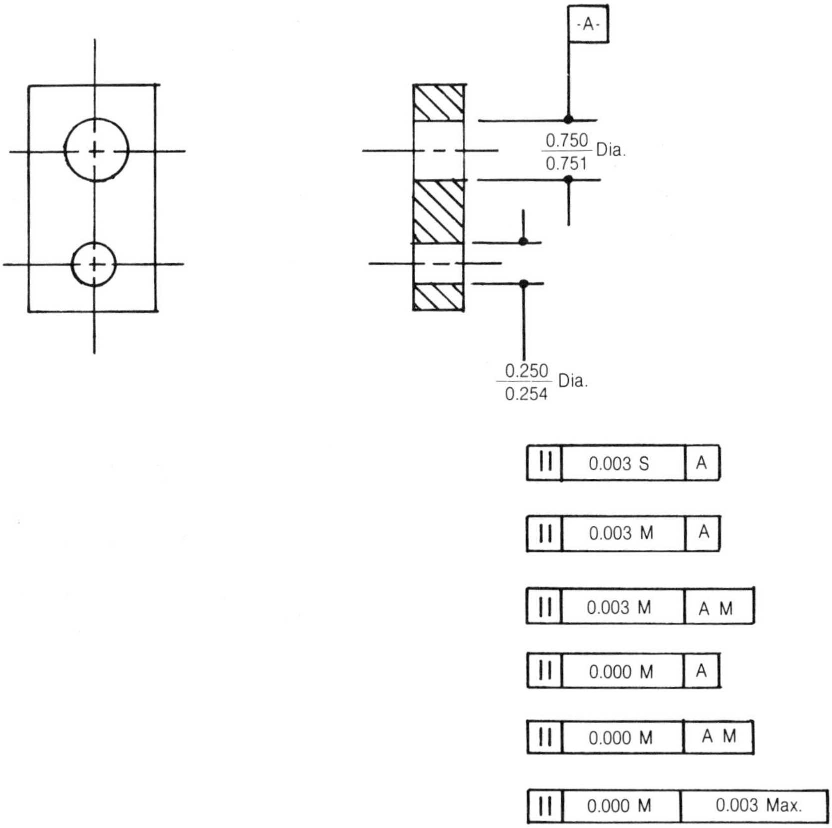

-A-

0.750 / 0.751 Dia.

0.250 / 0.254 Dia.

∥	0.003 S	A
∥	0.003 M	A
∥	0.003 M	A M
∥	0.000 M	A
∥	0.000 M	A M
∥	0.000 M	0.003 Max.

Figure 3.29 Parallelism of an axis to a datum axis callout.

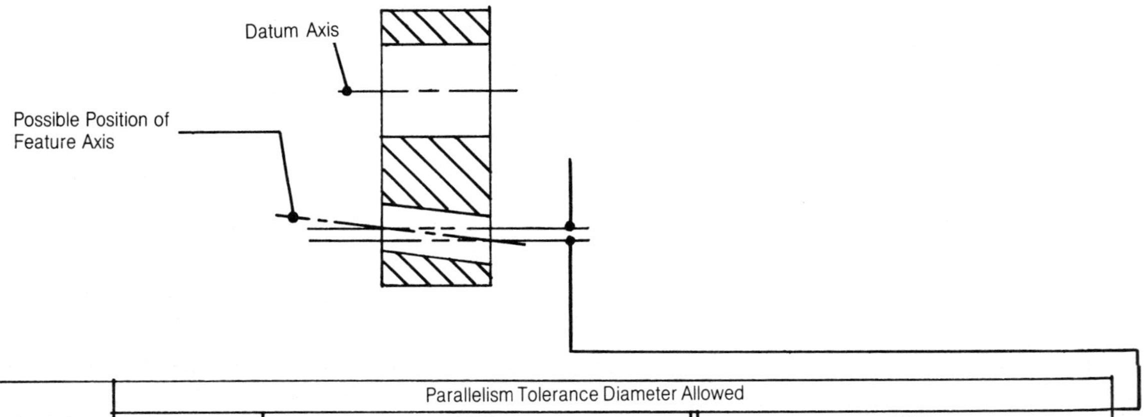

Datum Axis

Possible Position of Feature Axis

Symbol	Parallelism Tolerance Diameter Allowed									
	0.750					0.751				
Feature Size	0.250	0.251	0.252	0.253	0.254	0.250	0.251	0.252	0.253	0.254
0.003 S | A	0.003									0.003
0.003 M | A	0.003	0.004	0.005	0.006	0.007	0.003	0.004	0.005	0.006	0.007
0.003 M | A M	0.003	0.004	0.005	0.006	0.007	0.004	0.005	0.006	0.007	0.008
0.000 M | A	0.000	0.001	0.002	0.003	0.004	0.000	0.001	0.002	0.003	0.004
0.000 M | A M	0.000	0.001	0.002	0.003	0.004	0.001	0.002	0.003	0.004	0.005
0.000 M | 0.003 Max.	0.000	0.001	0.002	0.003	0.003	0.000	0.001	0.002	0.003	0.003

Figure 3.30 Parallelism of an axis to a datum axis tolerance zone.

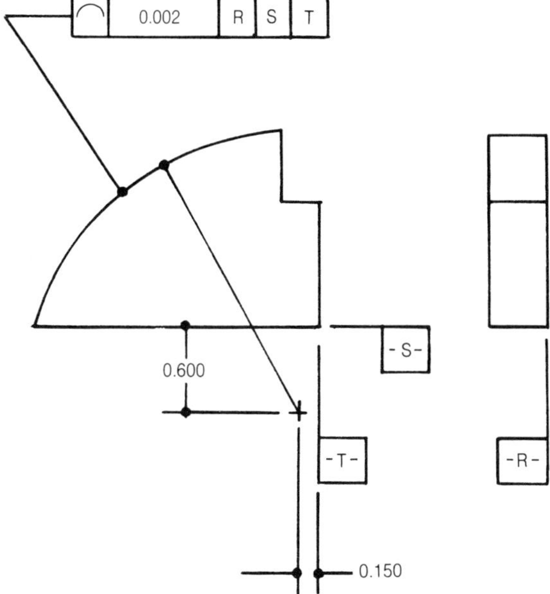

Figure 3.31 Profile callout.

3.3.4 Profile

Definition. Profile tolerancing is a method used where a uniform amount of variation may be permitted along a line or surface. The line or surface may consist of straight lines or curved lines, the latter being either arcs or irregular curves.

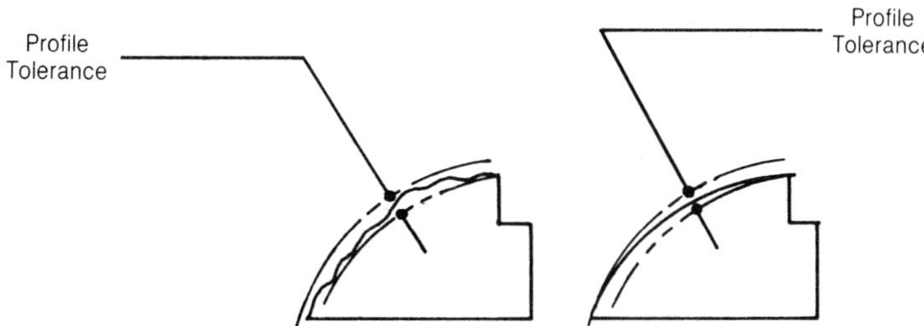

Figure 3.32 Profile tolerance zone.

Tolerance. A profile tolerance (either bilateral or unilateral) specifies a tolerance zone, always measured normal to the profile at all points of the profile, within which the specified line or surface must lie.

Explanation. Figure 3.31 shows the profile tolerance callout on a drawing. A profile tolerance is a zone of specified width along a true contour within which the entire designated surface must lie. This profile tolerance zone is shown in Figure 3.32.

3.3.5. Runout

Definition. Runout is the condition of perfect form and axial alignment of two or more surfaces of revolution such as cylinders, cones, or contours and may include plane surfaces perpendicular to and generated about a common axis.

Tolerance. A runout tolerance controls the relationship of two or more features within the allowable errors of concentricity, perpendicularity, and alignment of the features. It also controls variations in roundness, straightness, flatness, angularity, and parallelism of individual surfaces. In essence, runout establishes composite-form control of those features of a part having a common axis.

Explanation. Figure 3.33 shows the runout tolerance callout on a drawing. Runout tolerance applied to a part surface is a means of controlling the composite form of that surface. For example, a runout tolerance applied to a cylindrical surface (with respect to a datum axis) controls straightness, cylindricity, and angularity of its surface elements. This runout tolerance zone is shown in Figure 3.34.

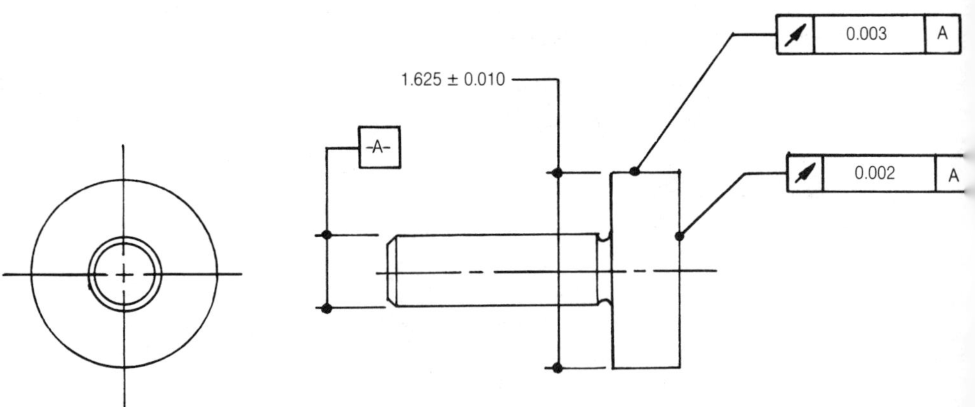

Figure 3.33 Runout callout.

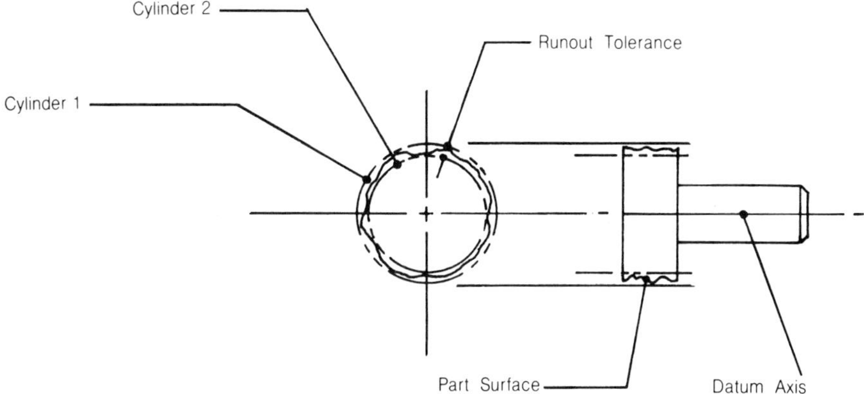

Figure 3.34 Runout tolerance zone.

3.3.6. Concentricity

Definition. Concentricity is the condition of surfaces of revolution, such as cylinders, cones, or spheres, wherein they have a common axis.

Tolerance. A concentricity tolerance is the diameter of the cylindrical tolerance zone within which the axis of the feature(s) so toler-

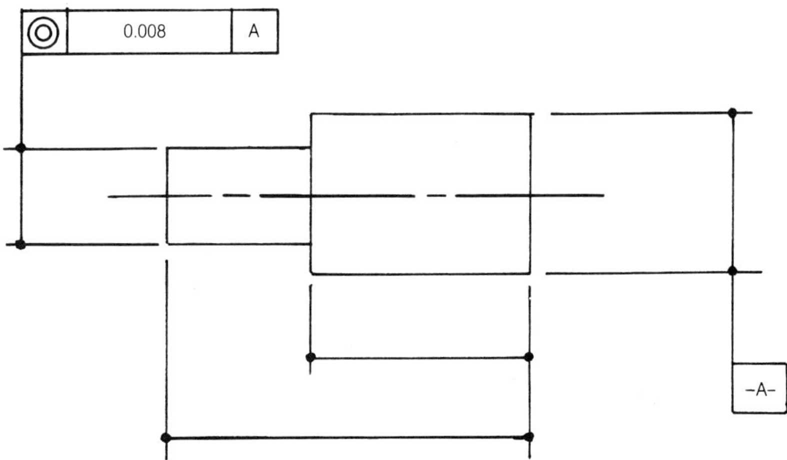

Figure 3.35 Concentricity callout.

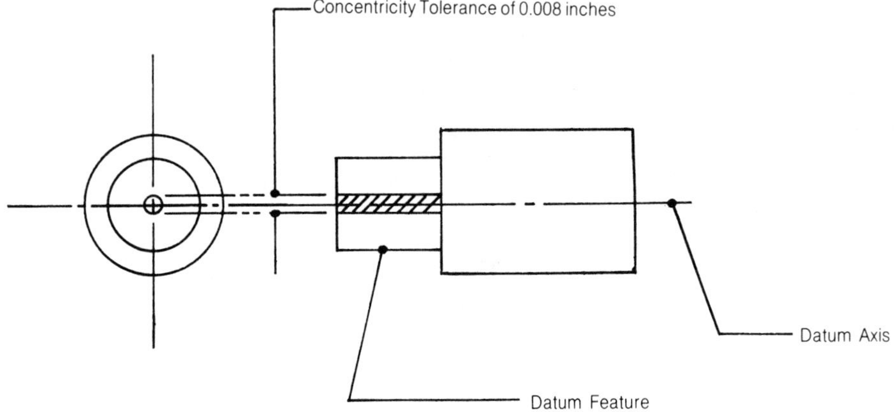

Figure 3.36 Concentricity tolerance zone.

anced must lie. The positional relationship of features on a common axis is not controlled unless concentricity, runout, or true position is specified on the drawing.

Explanation. Figure 3.35 shows the concentricity tolerance callout on a drawing. A concentricity tolerance applied to a feature means that the axis of that feature is concentric (coaxial) with a datum axis. The cylindrical zone is equal in diameter to the specified concentricity tolerance. This concentricity tolerance zone is shown in Figure 3.36.

3.3.7. Symmetry

Definition. Symmetry is a condition wherein a part or a feature has the same contour and size on opposite sides of a central plane or a condition in which a feature is symmetrically disposed about the central plane of a datum feature.

Tolerance. A positional tolerance applied to a noncylindrical feature means that the centerplane of that feature must lie between two parallel planes. These two planes are separated by a distance equal to the specified tolerance and are equally disposed about a datum plane.

Explanation. Figure 3.37 shows the symmetry tolerance callout on a drawing. A symmetry tolerance is placed on a drawing where it is required that a feature be located symmetrically with respect to a datum feature. This symmetry tolerance zone is shown in Figure 3.38.

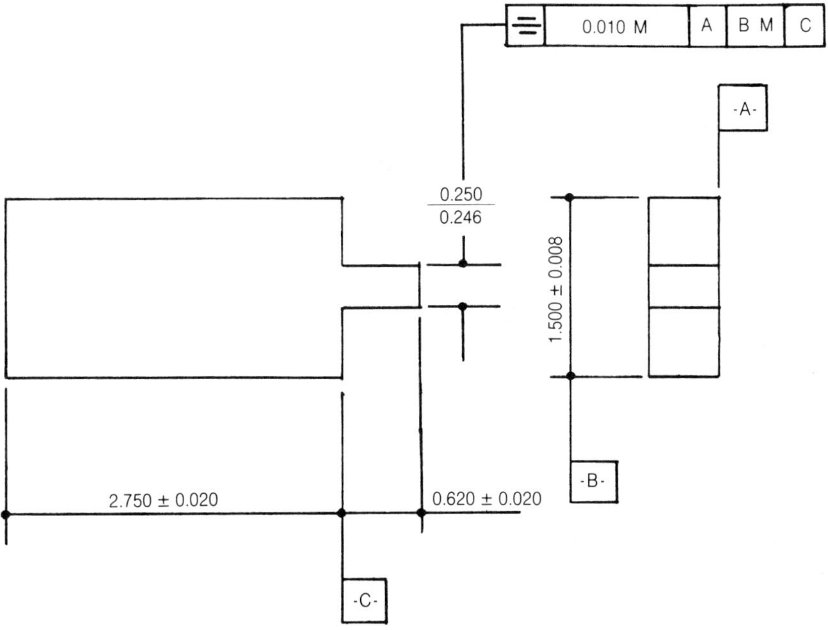

Figure 3.37 Symmetry callout.

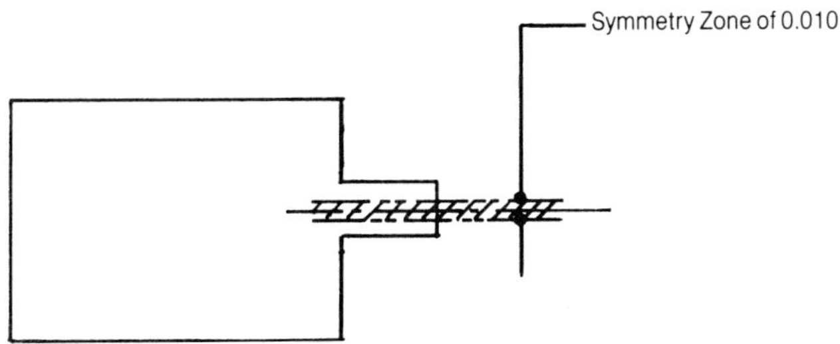

Figure 3.38 Symmetry tolerance zone.

3.4. PROCEDURE FOR INSPECTING FORM TOLERANCES

Form tolerances should be used only when the part's shape or form would affect the function and/or interchangeability of the part. Form tolerances should be approximately 50% of the feature's size tolerance band.

Placing form tolerances on drawings is useless if there is no method of ensuring that these tolerances are being met during manufacturing. Therefore, procedures for checking these form tolerances are necessary. The inspection procedure for each of the form tolerances in Section 3.2 and 3.3 are covered below.

First, the size dimension of the feature with its tolerance must be checked to ensure that it falls within the upper and lower limits designated. This is done by using a caliper or an open-end-type of gage (Fig. 3.39). After satisfying this inspection, one would inspect the feature to determine if the feature's form is within the specified tolerance. This is usually done by using a dial indicator which is attached to a post. The post rests on a surface plate with the part that is being inspected. A dial indicator with its post and surface plate is shown in Figure 3.40.

This is the procedure used for each form tolerance in the following discussions.

3.4.1. Flatness

1. Check the distance between the surfaces with a caliper or a GO-NO GO open-end gage (Fig. 3.39). The measured values must be between 0.747 and 0.753 in. for the part to be acceptable.

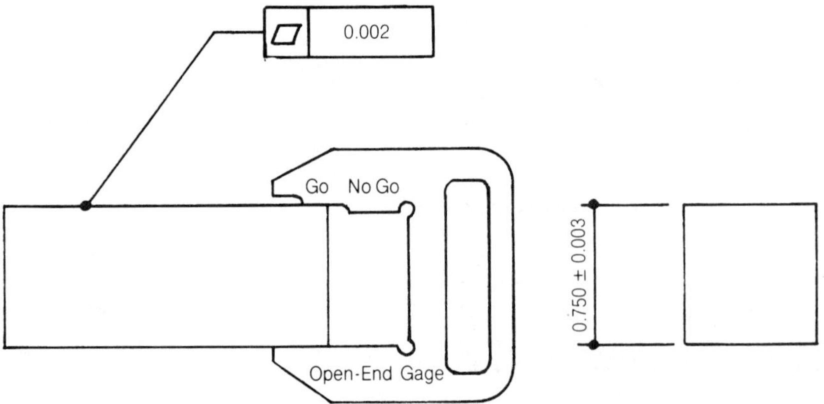

Figure 3.39 Open-end GO-NO GO gage.

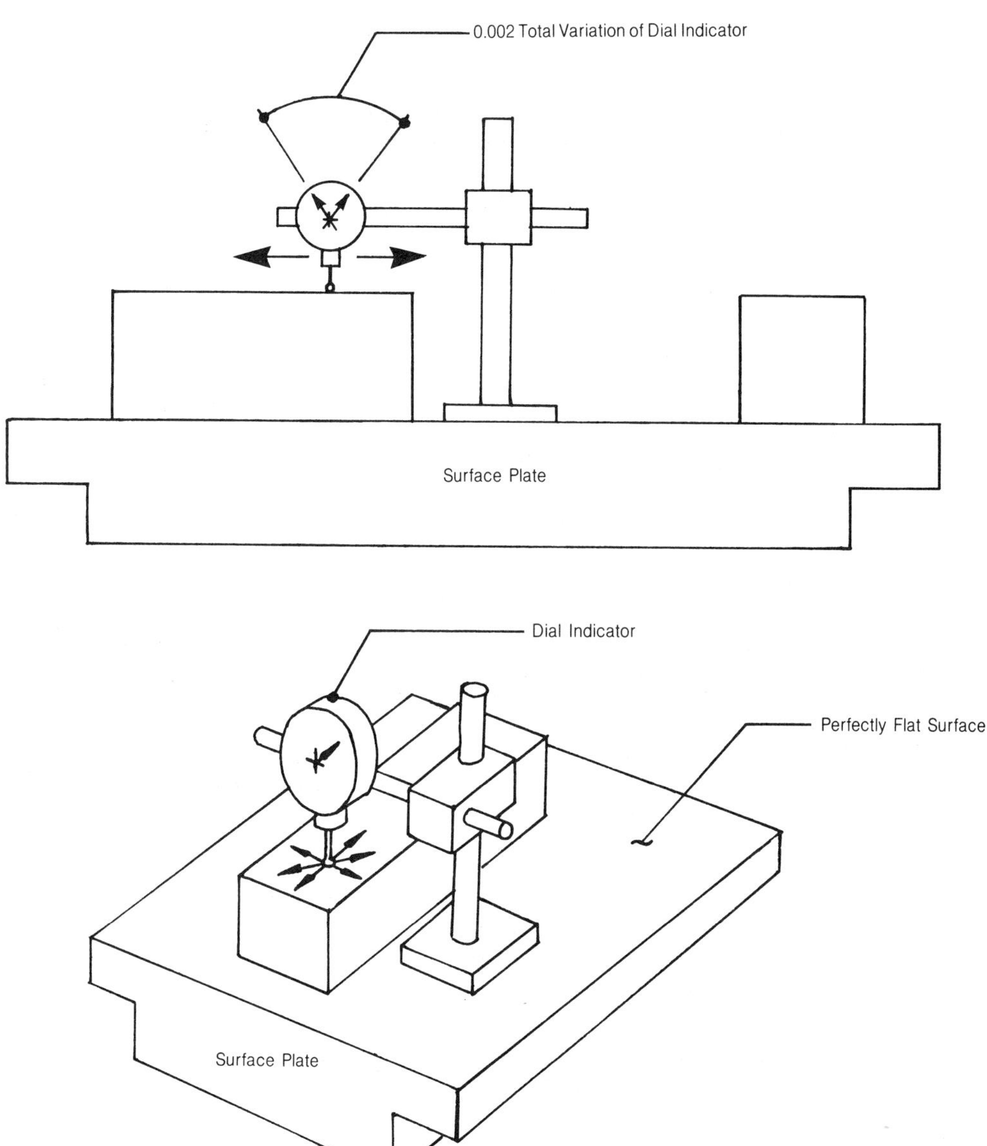

Figure 3.40 A dial indicator with its post and a surface plate.

2. Check the variation of the entire surface with a dial indicator (Fig. 3.40). The part is placed on a perfectly flat surface on which the base of the indicator's support rests. The dial indicator's arm is moved in all directions on the surface. The dial's variation must be no more than 0.002 of an inch of movement.

3.4.2. Straightness

1. Check diameter of shaft along its axis with calipers or open-end gage (Fig. 3.41). The measured values must be between 0.499 and 0.511 inches.
2. The shaft is placed on a perfectly flat surface on which the base of the indicator's support rests. The indicator's sensitive arm moves along on one line element of the shaft lying in a plane (Fig. 3.42). The total dial variation cannot exceed 0.004 of an inch.

3.4.3. Roundness

1. Check the shaft's diameter along its axis with a caliper or an open-end gage (Fig. 3.43). The measured values must be between 0.867 and 0.877 inches.
2. Place shaft in a V-type fixture as shown in Figure 3.44. Touch the dial indicator's arm on the shaft and rotate the shaft one full turn. Observe the dial indicator's total variation. This variation must not exceed 0.003 inches.

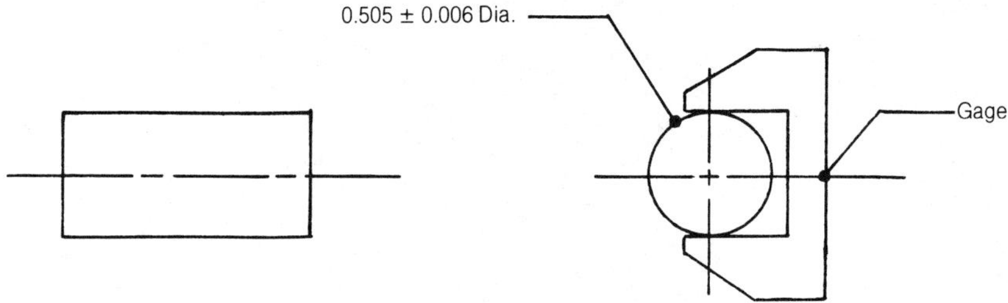

Figure 3.41 Open-end gage measuring the shaft's diameter.

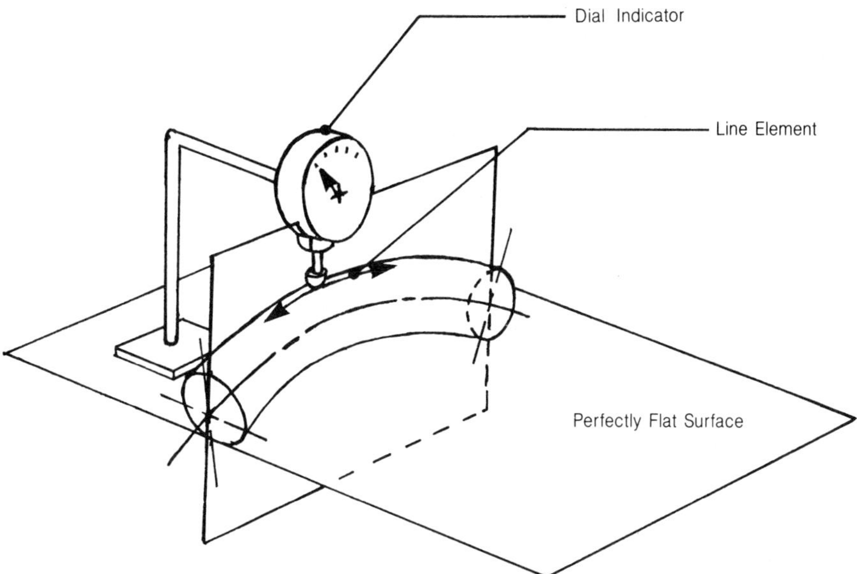

Figure 3.42 Inspecting the shaft for straightness.

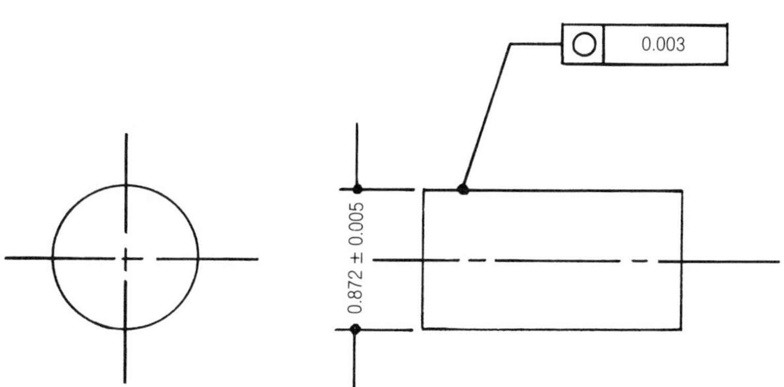

Figure 3.43 Open-end gage measuring the shaft's diameter.

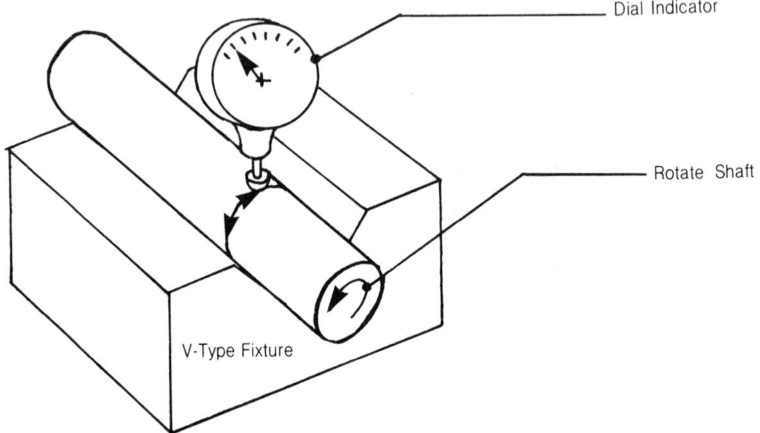

Figure 3.44 Inspecting the shaft for roundness.

3.4.4. Cylindricity

1. Check the shaft's diameter along its axis with a caliper or an open-end gage (Fig. 3.45). The measured values must be between 0.527 and 0.533 inches.
2. Place shaft between two V-type fixtures as shown in Figure 3.46. Touch the dial indicator's arm on the shaft and rotate the shaft

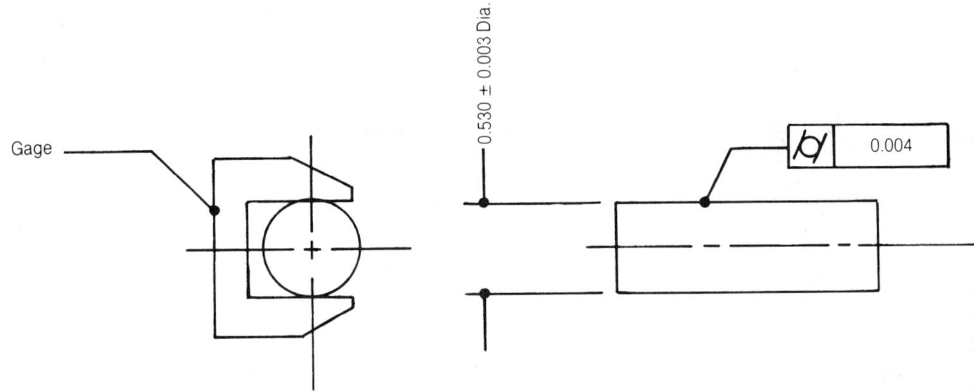

Figure 3.45 Open-end gage measuring the shaft's diameter.

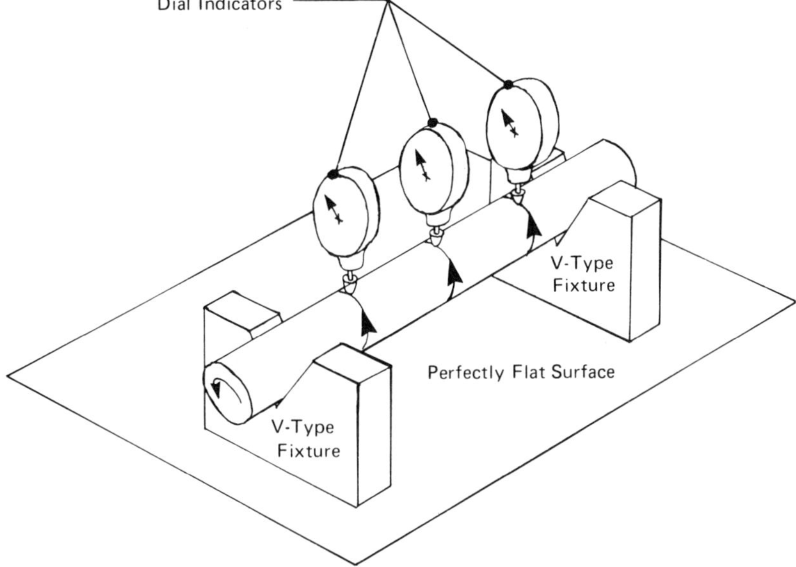

Figure 3.46 Inspecting the shaft for cylindricity.

one full turn. Observe the dial indicator's total variation. The variation must not exceed 0.004 inches.

3.4.5. Profile of a Line or a Surface

1. Check the individual feature sizes of the part shown in Figure 3.47 by using a caliper or a gage. These sizes must be between the upper and lower limits of each feature.
2. The outline of the profile tolerance is shown in Figure 3.48. An optical comparator (Fig. 3.49) projects a greatly enlarged image of the work onto a screen where it may then be conveniently examined. This involves the use of a profile drawing of the part on the screen. It is made to the same scale as the magnification. The shadow is then moved to coincide with the profile drawing. Any deviation is then quickly noted and can be measured.

3.4.6. Perpendicularity

1. Check the thickness of each of the legs of the part with a caliper or an open-end gage (Fig. 3.50). The measured values must

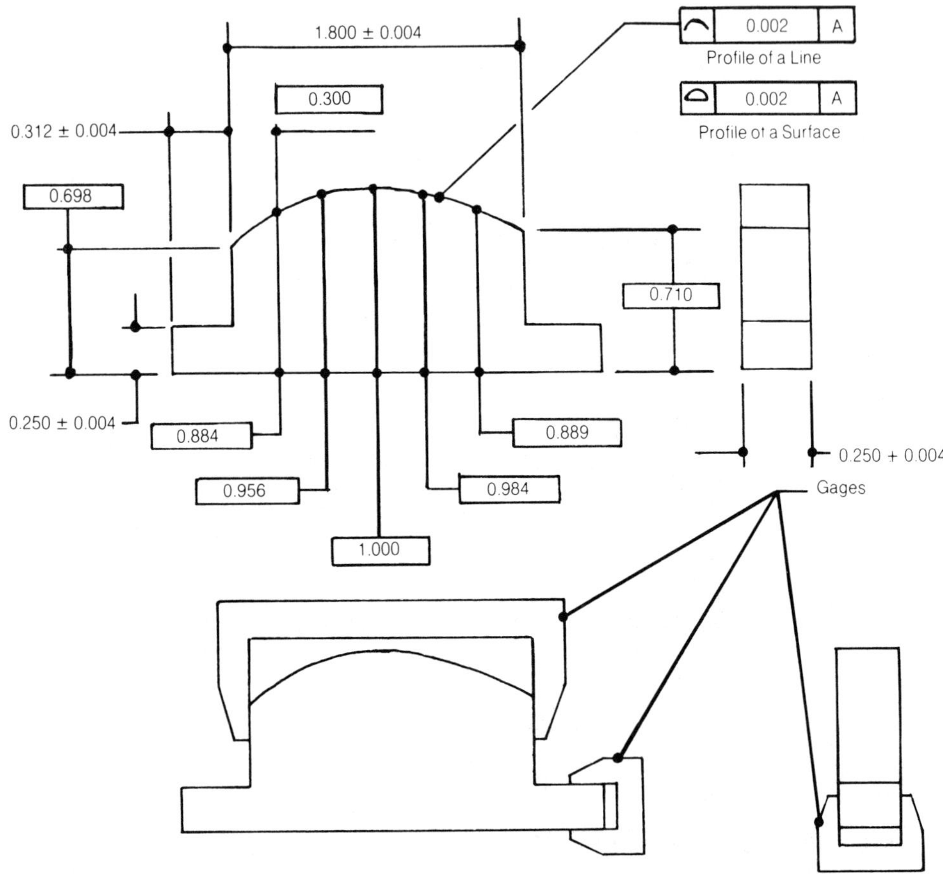

Figure 3.47 Open-end gage measuring the part's size.

be between 0.255 and 0.245 in. for one leg and between 0.380 and 0.370 in. for the other leg.

2. Check the variation of the entire surface with a dial indicator (Fig 3.51). The part is placed on a perfectly flat surface on which the base of the indicator's support rests. The dial indicator's arm is moved in all directions on the surface. The dial variation must be no more than 0.007 of an inch movement.

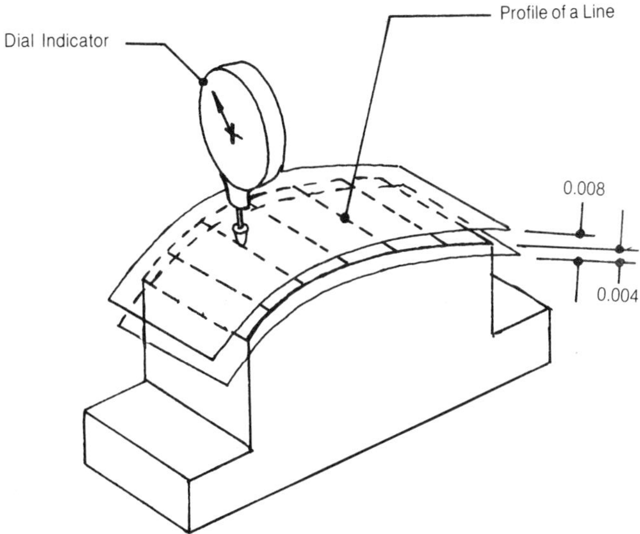

Figure 3.48 Outline of the profile tolerance.

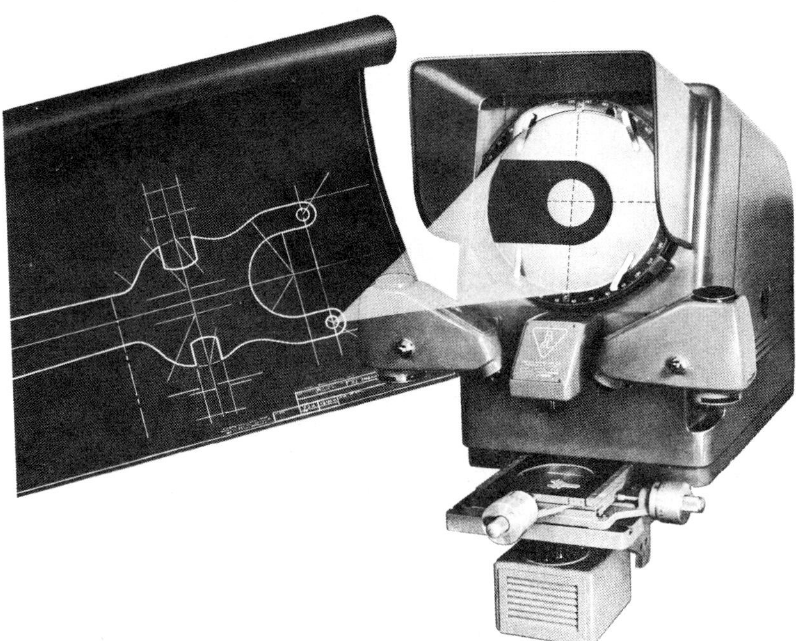

Figure 3.49 Optical comparator.

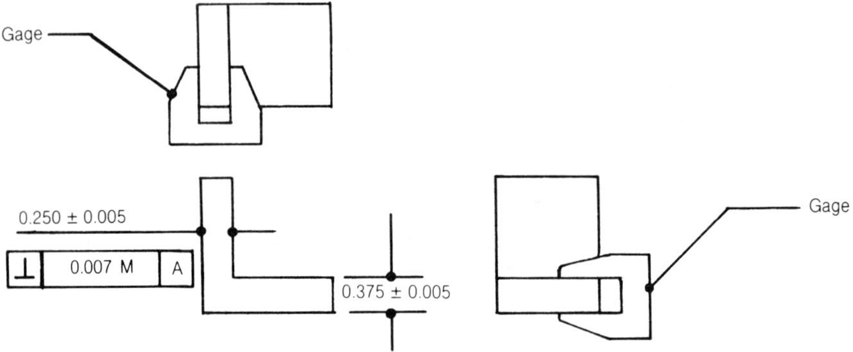

Figure 3.50 Open-end gage measuring the part's size.

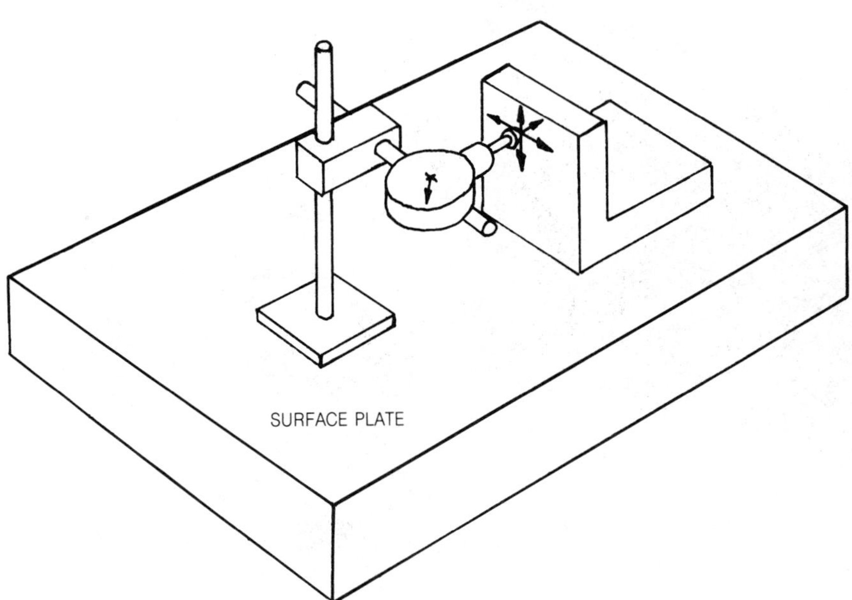

Figure 3.51 Inspecting the perpendicularity of the surface.

3.4.7. Angularity

1. Check all the linear dimensions of the part in Figure 3.52 with a caliper or an open-end gage. The measured values must be between the upper and lower limits placed on each feature.
2. Check the deviation from angularity of the surface specified in Figure 3.52. A sine table is used and set to the exact angle of 30 degrees by means of gage blocks as shown in Figure 3.53. The part is placed on the sine table so that datum surface A contacts the table and datum surface B contacts the parallel bar. This positions the surface to be inspected in a plane parallel to the surface plate. The total indicator reading of the entire surface represents the deviation from angularity.

3.4.8. Parallelism

1. Check the various features' sizes on the parts shown in Figures 3.54 and 3.55 by using calipers or gages. These sizes must lie between the upper and lower limits of the respective dimensions.

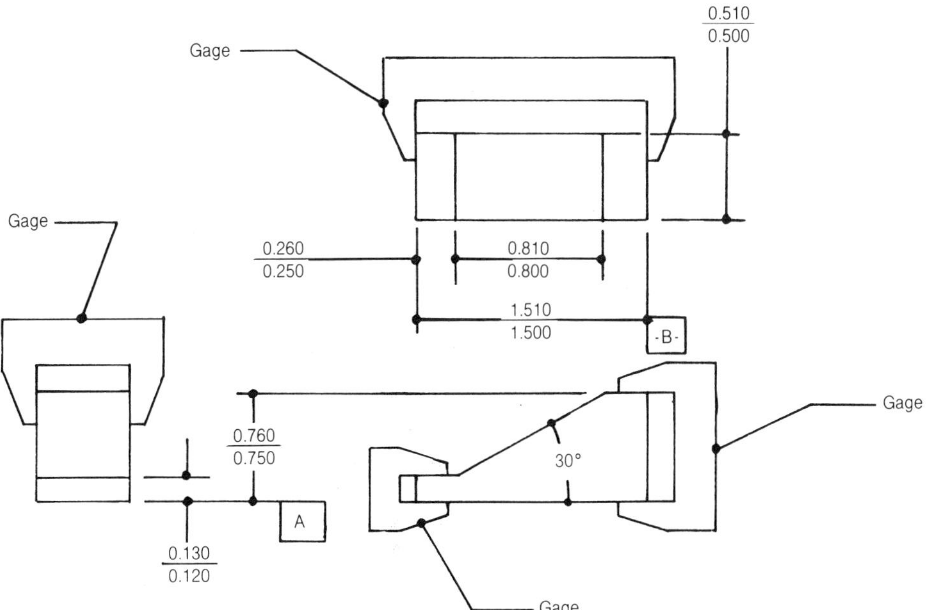

Figure 3.52 Open-end gage measuring the part's sizes.

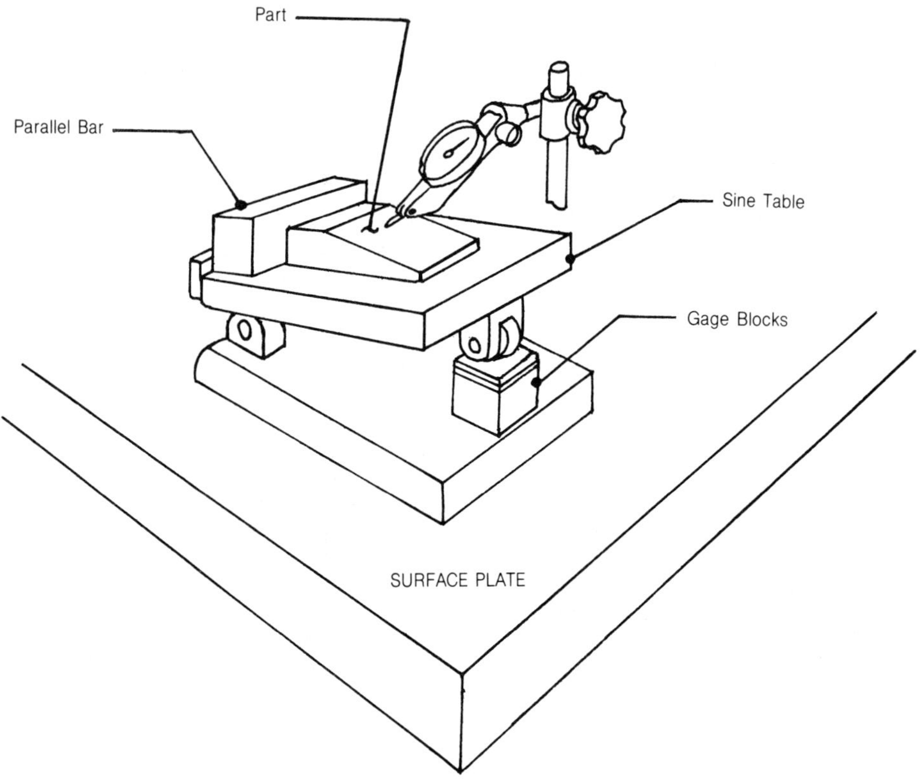

Figure 3.53 Inspecting the angularity of the surface.

2. The upper surface of the part in Figure 3.56 is parallel to the datum surface within 0.005 of an inch. This parallel tolerance can be checked by using a dial indicator and a surface plate (Fig. 3.56). The movement of the indicator's arm over this upper surface will determine if the parallel tolerance is satisfied. In Figure 3.55, the 0.500/0.495-in. diameter shaft is parallel to the datum shaft with the diameter of 0.750/0.746 inches within 0.005 of an inch. The procedure is shown in Figure 3.57 in which a dial indicator's arm is moved along on the 0.500/0.495 diameter shaft while the 0.750/0.746 diameter shaft is held in a V groove.

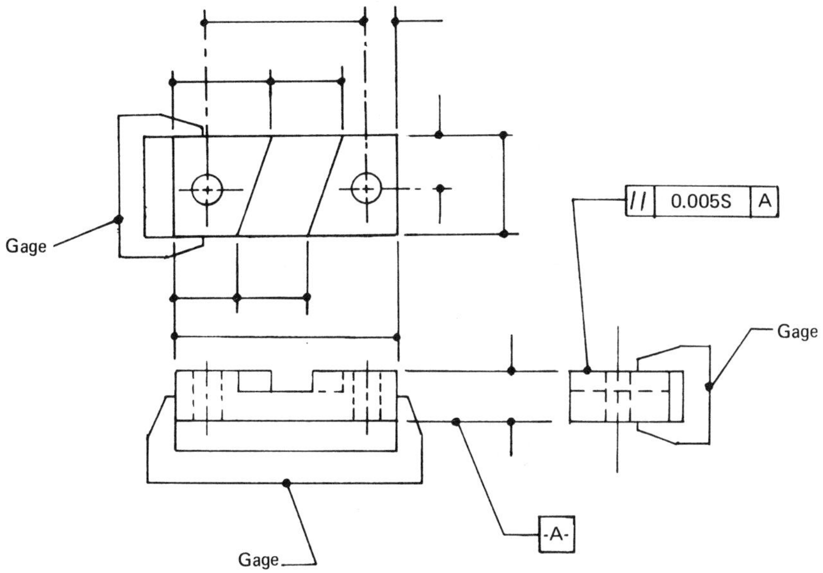

Figure 3.54 Open-end gages measuring various features of one of the mating parts.

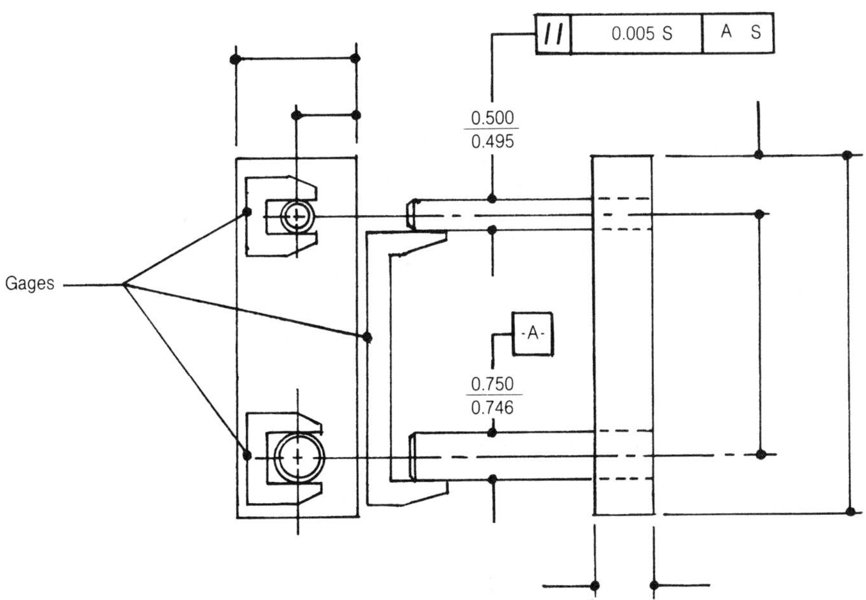

Figure 3.55 Open-end gages measuring the location and sizes of the shafts on the other mating part.

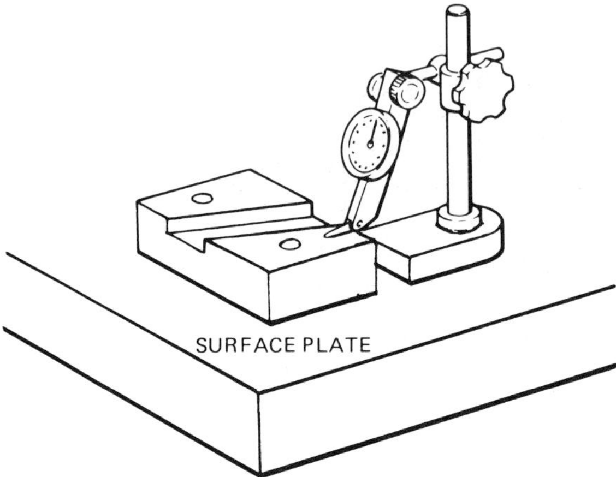

Figure 3.56 Checking the surface for parallelism.

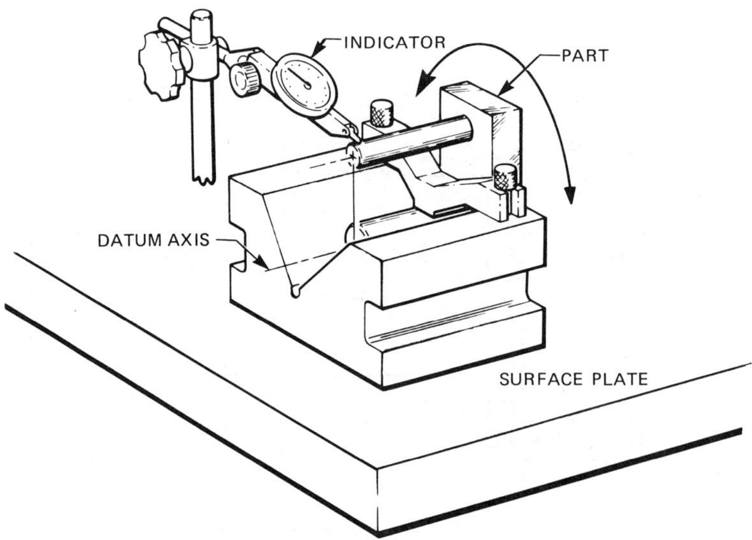

Figure 3.57 Checking the shafts for parallelism.

3.4.9. Runout

1. Check the diameter of the shaft shown in Figure 3.58 by using caliper or gage. This size must lie between the upper and lower limits of the shaft's diameter.
2. The dial indicator's arm cannot deviate more than 0.005 of an inch throughout the length of the shaft while the shaft is being rotated. Figure 3.59 shows the shaft being held in V grooves which are sitting on a surface plate. The base of the dial indicator is also resting on the surface plate. The shaft is rotated while the dial indicator's arm touches the shaft's surface and is moved parallel to the shaft's axis.

3.4.10. Concentricity

1. The part shown in Figure 3.60 has two diameters that must lie between their respective upper and lower limits. These diameters are checked by using calipers or gages.
2. The eccentricity or concentricity between the centers of the two diameters is checked by the same procedure for checking the runout of a shaft outlined in the previous section. Figure 3.61 shows the stepped shaft with one of its diameters lying in V grooves and the other diameter being checked by a dial indicator's arm. Both the V grooves and dial indicator rest on a surface plate.

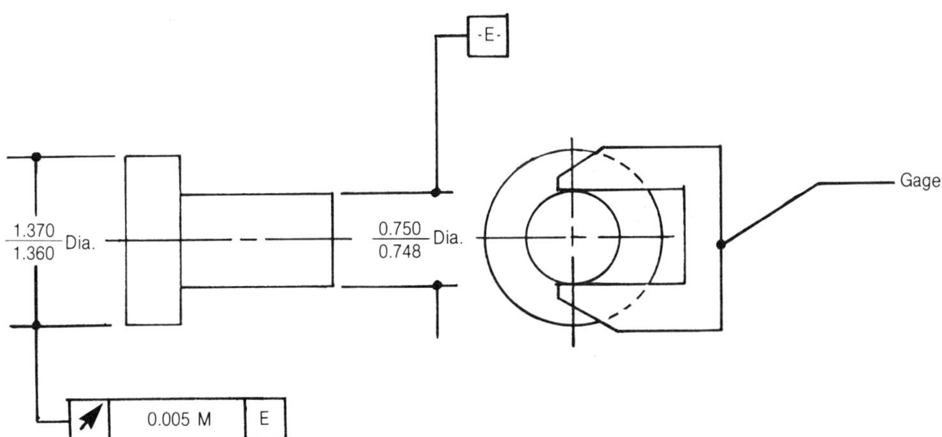

Figure 3.58 Open-end gages measuring the shaft's diameters.

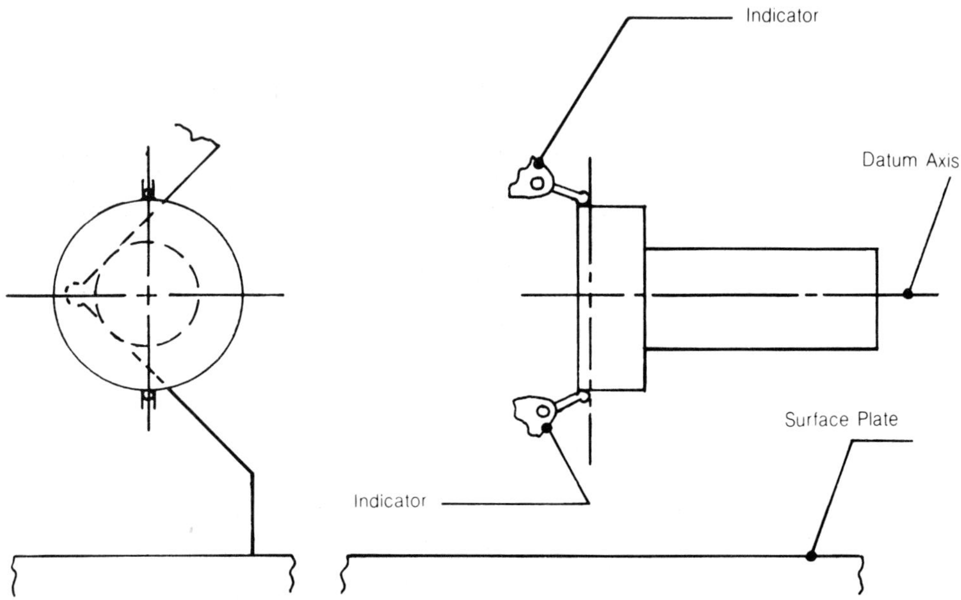

Figure 3.59 Inspecting the concentricity of the shaft's diameters.

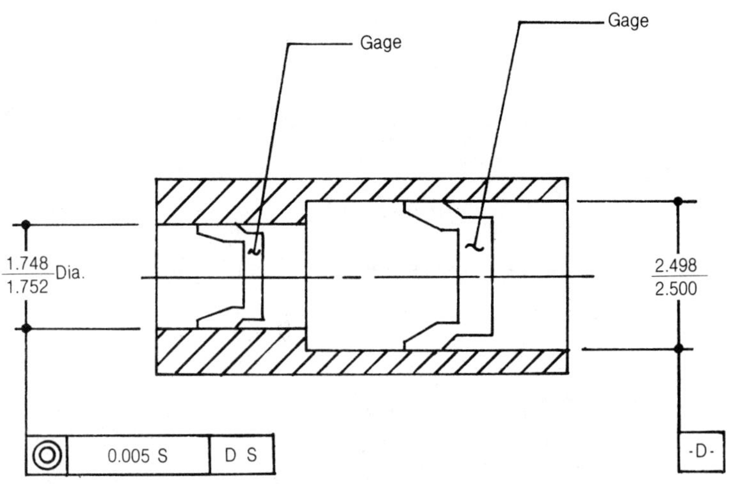

Figure 3.60 Checking the sizes of the holes using gages.

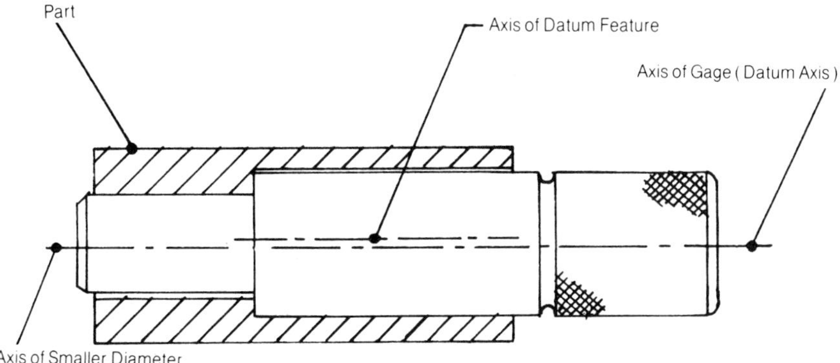

Figure 3.61 Checking the holes for concentricity.

3.4.11. Symmetry

1. The width of the keyway in the shaft shown in Figure 3.62 is measured either by using a caliper or gage. The width of the keyway must lie between the upper and lower limits of the dimension.

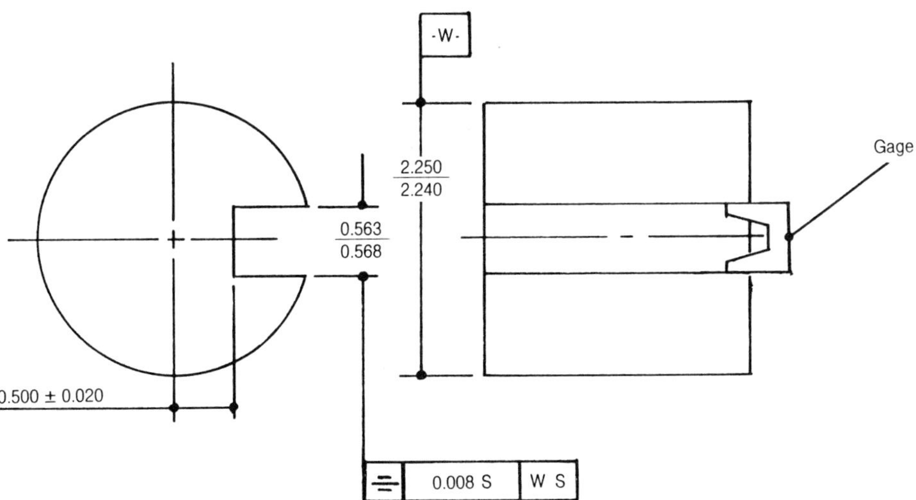

Figure 3.62 Checking the keyway's size using a gage.

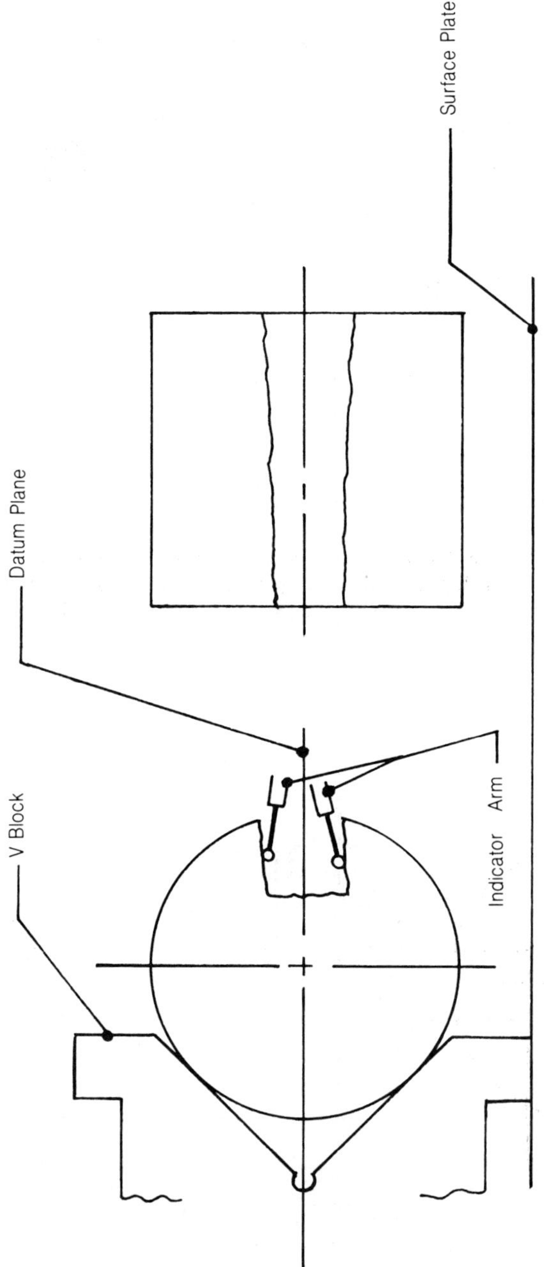

Figure 3.63 Inspecting the keyway's symmetry using a dial indicator.

2. The symmetry tolerance of 0.008 of an inch could be measured by placing the shaft in V grooves and passing the dial indicator's arm along the side of the keyway. Both the V grooves and dial indicator rest on a surface plate. The movement of the dial indicator's arm is observed. Then the shaft is rotated 180 degrees and the arm of the indicator is moved along the other side of the keyway (Fig. 3.63) and the arm's movement is observed. The difference between the maximum and minimum values of the readings must be less than or equal to 0.008 of an inch if the symmetry tolerance is to be satisfied.

3.5. FUNDAMENTALS OF POSITION TOLERANCES

In a rectangular tolerance zone, as shown in Figure 3.64, the allowable deviation from nominal (center of the tolerance zone) is greater along its diagonal axis than its sides. If the maximum allowable deviation from the nominal position is acceptable in the horizontal and vertical directions, it seems reasonable that the deviation along the diagonal direction should be acceptable too.

When the minimum clearance hole center is allowed to deviate from its nominal to all of the extreme possible positions, the maximum bolt diameter is created and a circular zone is the result (Fig. 3.65). A maximum diameter bolt will pass through all the minimum diameter

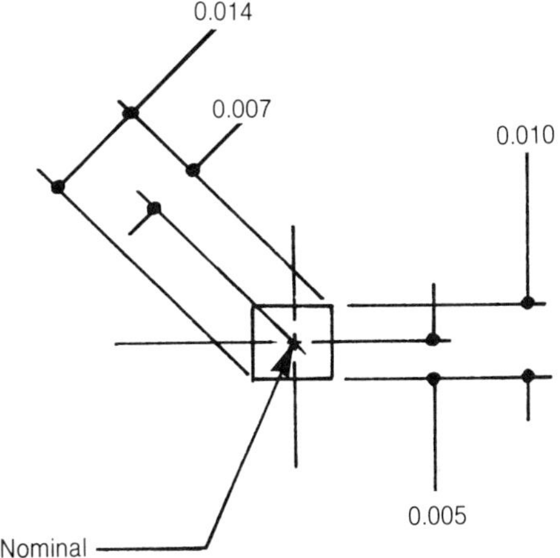

Figure 3.64 A rectangular tolerance zone.

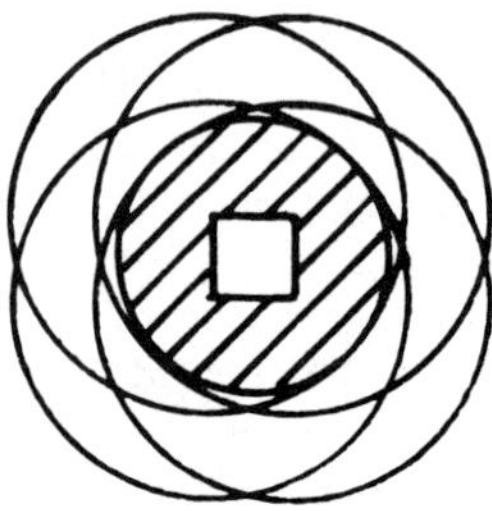

Figure 3.65 Creating the maximum bolt diameter.

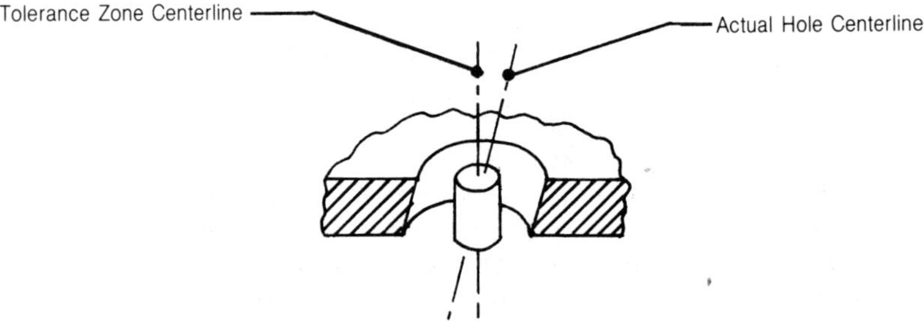

Figure 3.66 A position tolerance.

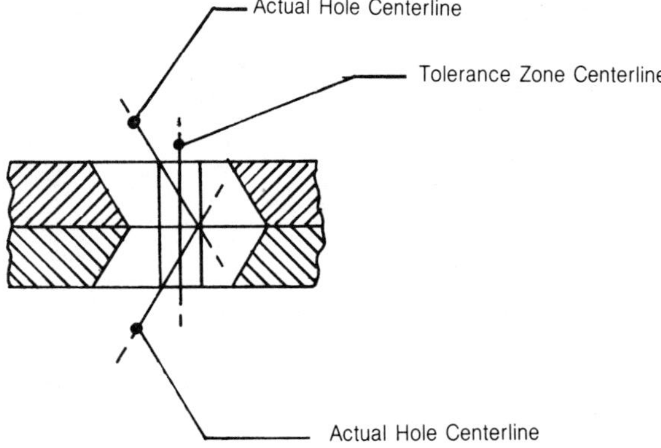

Figure 3.67 Controlling perpendicularity of holes using position tolerances.

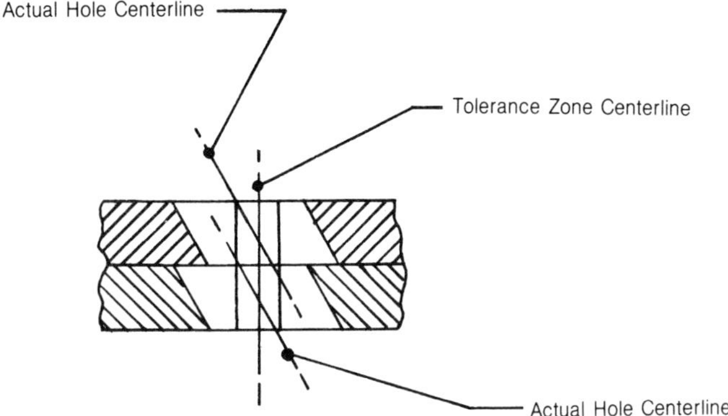

Figure 3.68 Controlling perpendicularity of holes using position tolerances.

clearance holes having centers within this circular zone. The diagonal of this rectangular zone would be the diameter of the circular zone and this circular zone permits the maximum deviation in any direction from the nominal without interfering with interchangeability (assembly).

This circular zone is referred to as a *position tolerance* zone which permits greater deviation of the hole location from its nominal dimension (Fig. 3.66). Also, this zone extends down into the part, a distance equal to the depth of the hole and the centerline of the hole must be within the cylindrical tolerance zone for the entire depth of the hole.

Positional tolerances control the perpendicularity of holes as well as their positions. Figures 3.67 and 3.68 show two possible arrangements.

3.5.1. Holes and Floating Fasteners

The equation $H = B + T$, derived in Chapter 2, Sect. 2.5.1 for floating fasteners stays the same for position tolerancing. This equation is valid only when the position tolerance is the same in both parts and identical minimum clearance holes are to be used in both parts. Many parts must be designed to mate with existing parts, thus preventing the use of identical clearance hole diameters and/or identical position tolerances. Another general equation is required for this condition. It can be shown that $H_1 + H_2 = 2B + T_1 + T_2$

where H_1 = minimum clearance hole diameter in part 1

H_2 = minimum clearance hole diameter in part 2

B = maximum bolt diameter

T_1 = position tolerance of part 1

T_2 = position tolerance of part 2

For example, if parts A and B shown in Figures 3.69 and 3.70 have a position tolerance of 0.015 in., and both parts must be assembled with a 0.500-in. diameter bolt and nut, then the minimum diameter of the holes in parts A and B should be H = F + T or H = 0.500 + 0.015 which yields H as 0.515 inches. This minimum hole diameter is added to parts A and B with a suitable maximum hole diameter.

Suppose there is a part (part 1) which has a pattern of clearance holes that are 0.786 in. minimum diameter, located within 0.021 in. positional tolerance diameter. Part 2 has a pattern of clearance holes that are 0.518 in. minimum diameter, located within 0.013-in. positional tolerance diameter.

If part 2 is to be assembled to part 1 with 0.750-in. maximum diameter bolts, what would the minimum redrilled diameter of the holes of part 2 have to be assembled with part 1? Using the equation, $H_1 + H_2 = 2B + T_1 + T_2$, and substituting the appropriate values into this equation results in $0.786 + H_2 = 2\,(0.750) + 0.021 + 0.013$ or H_2 equals 0.748 inches.

To inspect and verify the proper location of holes in a part, one can either use an open set-up procedure or gages. Open setup is a procedure in which the inspector physically measures the part using a surface plate, angle plates, height measuring devices, V blocks, precision measuring blocks, or other device.

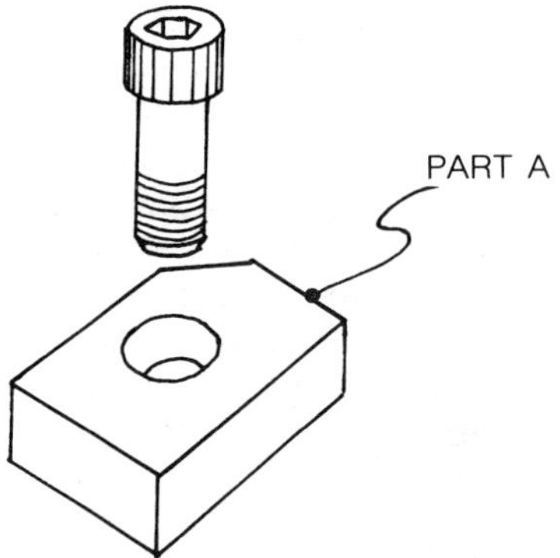

Figure 3.69 Part A with a bolt.

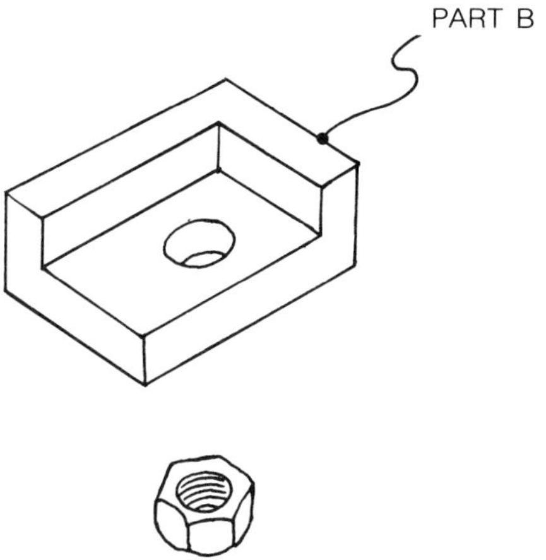

Figure 3.70 Part B with a nut.

3.5.2. Inspection of Holes Using Floating Fasteners

The inspection of holes using floating fasteners is done by using a gage with pins. The size of the pins on the gage are determined by subtracting the holes' position tolerance from the maximum material condition (MMC) of the holes or P = H − T. For example, in Figure 3.71 are two parts that are to be assembled with 0.500 diameter fasteners. The diameter of the holes in the two parts are $0.512\,^{+0.008}_{-0.000}$ and a 0.012 position tolerance is associated with them. The MMC of the gage pins' diameters would be 0.512 − 0.012 = 0.500, and this value would be placed on the pin gage shown in Figure 3.72. The least material condition (LMC) related to these gage pins is determined by using gagemakers' tolerances. Further discussion concerning these gagemaker's tolerances will be found in Chapter 5.

3.5.3. Holes and Fixed Fasteners

The equation $H = F + T_1 + T_2$, derived in Chapter 2, Sect. 2.5.2 for fixed fasteners stays the same for position tolerancing. If the same position tolerance is specified for both parts, the equation would become H = F + 2T, where H = minimum size of the clearance hole, F = MMC of the fixed fastener, $T = T_1 = T_2$ = position tolerance of fixed fasteners and holes in each part.

For example, if parts A and B shown in Figure 3.73 have a position tolerance diameter of 0.015 and both parts must be assembled with

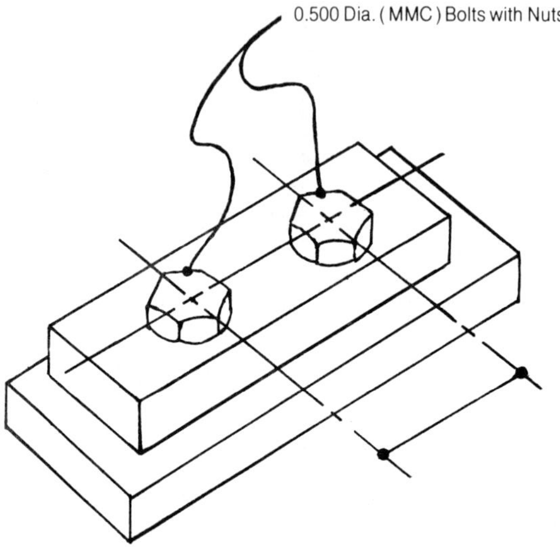

Figure 3.71 Parts 1 and 2 assembled with two 0.500 diameter bolts and nuts.

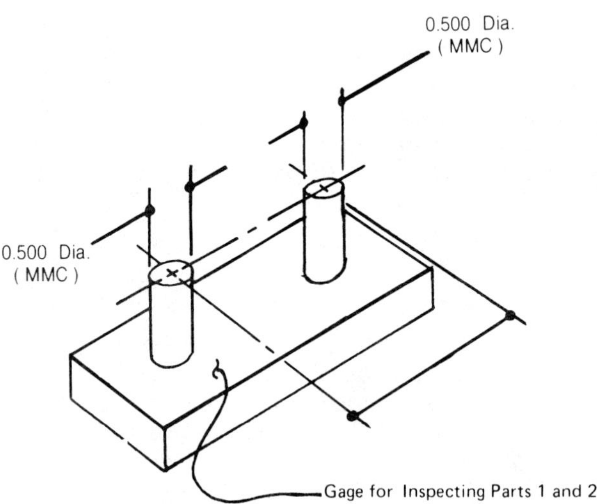

Figure 3.72 Gage used to inspect parts 1 and 2.

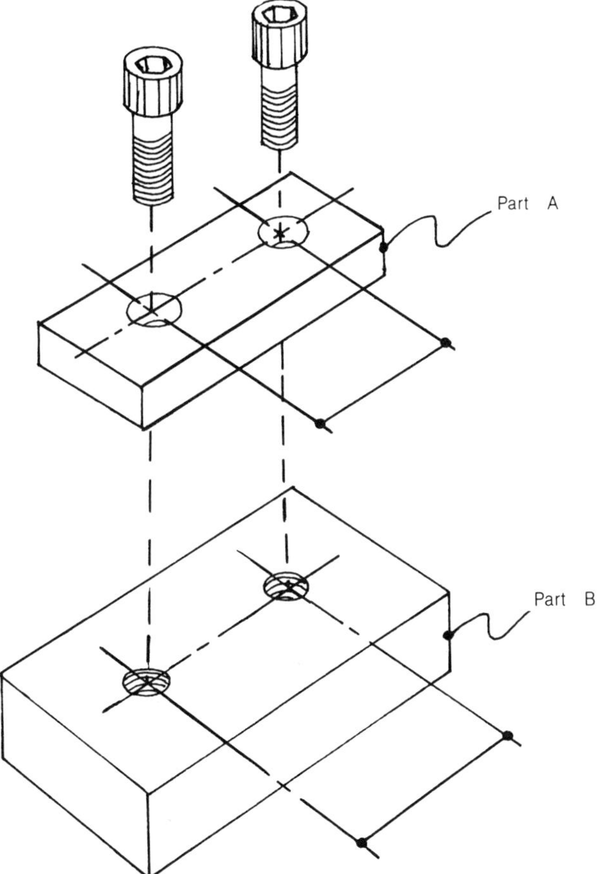

Figure 3.73 Parts A and B held together with fixed fasteners.

fixed 0.500 diameter bolts, then the minimum diameter of the holes in part A should be H = F + 2T or 0.500 + 2(0.015) which yields H as 0.530 inches. This minimum hole diameter is added to part A with a suitable maximum hole diameter limit.

Suppose there is a part (part 1) which has a pattern of clearance holes that are located within a position tolerance diameter of 0.021. Part 2 has a pattern of 0.750 maximum diameter threaded holes, located within a position tolerance diameter of 0.013. The minimum size of the clearance holes in part 1 is 0.750 + 0.021 + 0.013 = 0.784.

These formulas apply either to position tolerancing or to coordinate tolerancing where T is either the specified diameter of the position tolerance zone or the length of the diagonal or sides of the rectangular tolerance zone.

The perpendicularity of fixed fasteners when the holes for such fasteners are not made perpendicular to the surface of the holes are not compensated by these formulas. Extra hole clearances may be required to provide for such conditions. The projected tolerance zone described in Section 3.5.9 can be used to control the perpendicularity of holes.

3.5.4. Inspection of Holes Using Fixed Fasteners

The inspection of holes using fixed fasteners is done by using two gages. A pin gage would be used to inspect the holes in part A while a gage with holes in would be used to inspect the studs in part B.

The MMC gage pins would be determined by using the formula, P = H − T, where P = MMC of pin, H = MMC of hole, and T = position tolerance of hole. In this case, H = 0.530 and T = 0.015 or 0.530 − 0.015 yielding P as 0.515. The gage hole size would be determined by using the formula H = F + T, where H = MMC of hole, F = MMC of fixed fastener, and T = position tolerance of fixed fastener. In this case, 0.500 + 0.015 yielding H as 0.515. Note that the pin gage (P) value equals the hole gage (H) value.

3.5.5. Implied and Specified Datums Relative to Position Tolerancing

Implied datums are surfaces, centerplanes, etc. which are not explicitly called out on a drawing as datums or starting points for dimensioning and tolerancing. In this case, the method of dimensioning determines which surfaces are to be considered datums. Four aligned holes are dimensioned in Figure 3.74 by the traditional base line method using implied datums. One must remember that implied datums of any kind, tend to promote multiple interpretations. Placing specific datums, represented by A, B, and C on the appropriate surfaces (Fig. 3.75) with their respective dimensions and position tolerance of 0.015 eliminates this ambiguity. With explicit datums, form tolerancing is easier to use and understand. For example, the form tolerance of perpendicularity can be added to the datum surfaces without influencing the holes' dimensions and tolerances.

In Figure 3.74, the part with the four holes is shown with its implied datums. The high points or extremities of the part's datum surface rests against the theoretical flat surface. Since these same surfaces would be specified as datums A, B, and C, why not label them on the drawing if these surfaces are important for the location of interrelated features.

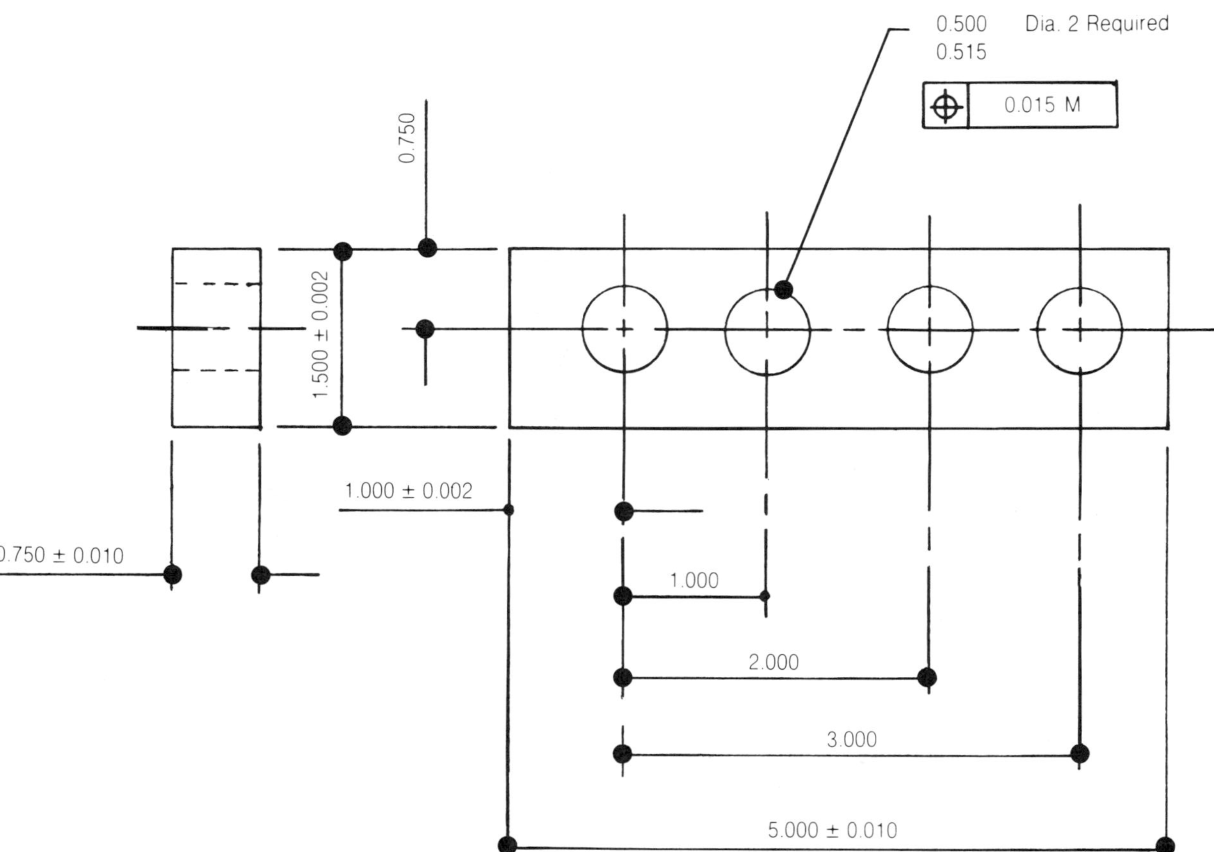

Figure 3.74 A part with four aligned holes.

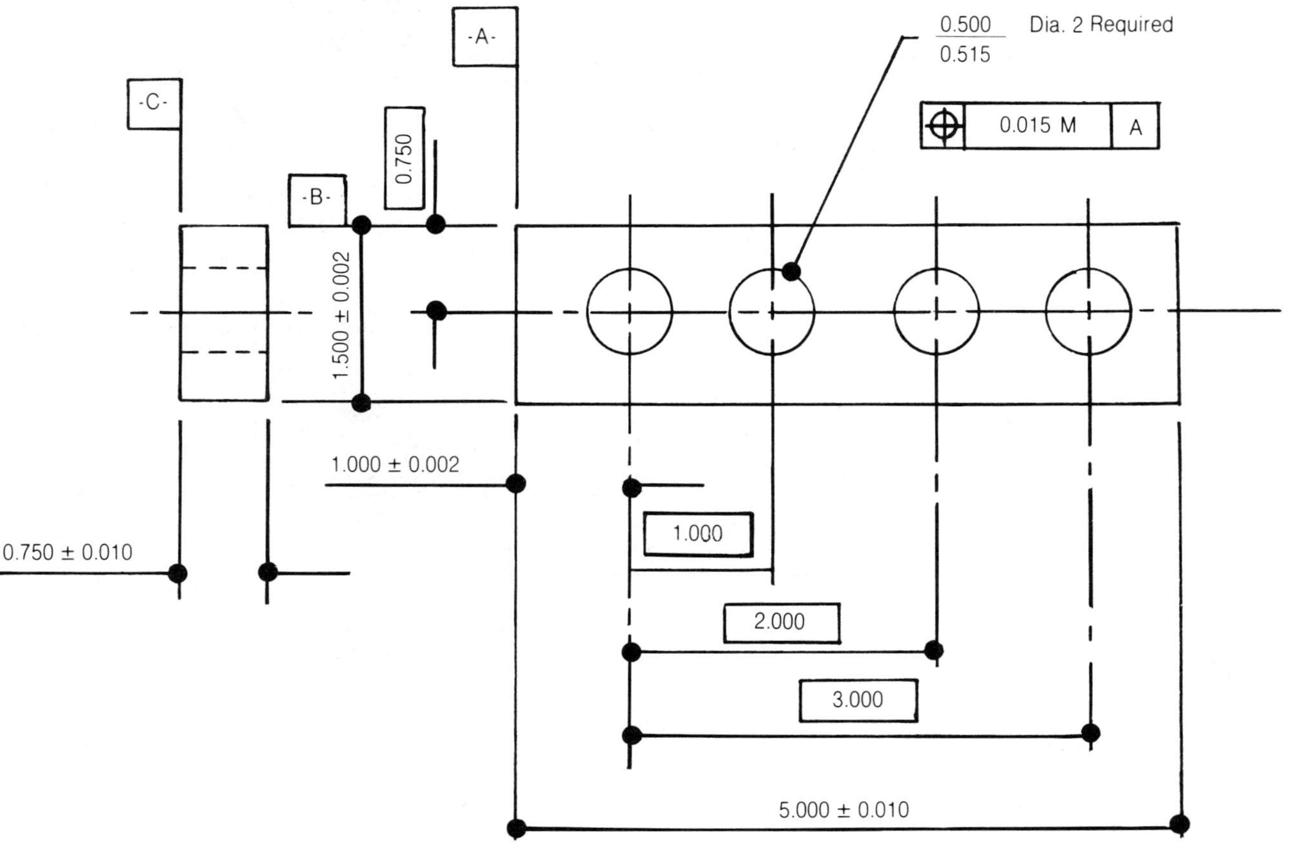

Figure 3.75 Placing specific datums on the appropriate surfaces.

By using datums, all the geometric controls are specifically stated, removing all doubt as to design intent, the manufacturing setups and processes, and inspection requirements and techniques. However, caution should be taken that these datums are placed only where needed on a drawing and are based on the specific requirements of the application and function of the part.

3.5.6. Combining Coordinate and Position Tolerancing

Figure 3.76 shows a part with four holes located around a bolt circle. If the location of the pattern of the four holes is less important than the accuracy required within the pattern of holes, it is permissible to use rectangular coordinate tolerancing with position tolerancing.

The rectangular tolerance of ±0.01 is used with the locating dimensions of 0.78 and 1.65. The primary datum (-A-) is specified, but the secondary and tertiary datums are implied; since 0.02 is a relatively large tolerance zone compared to the other specified tolerances. The theoretical centerline of the hole pattern lies in the center of the 0.20 square tolerance zone. Figure 3.77 shows several extreme positions of

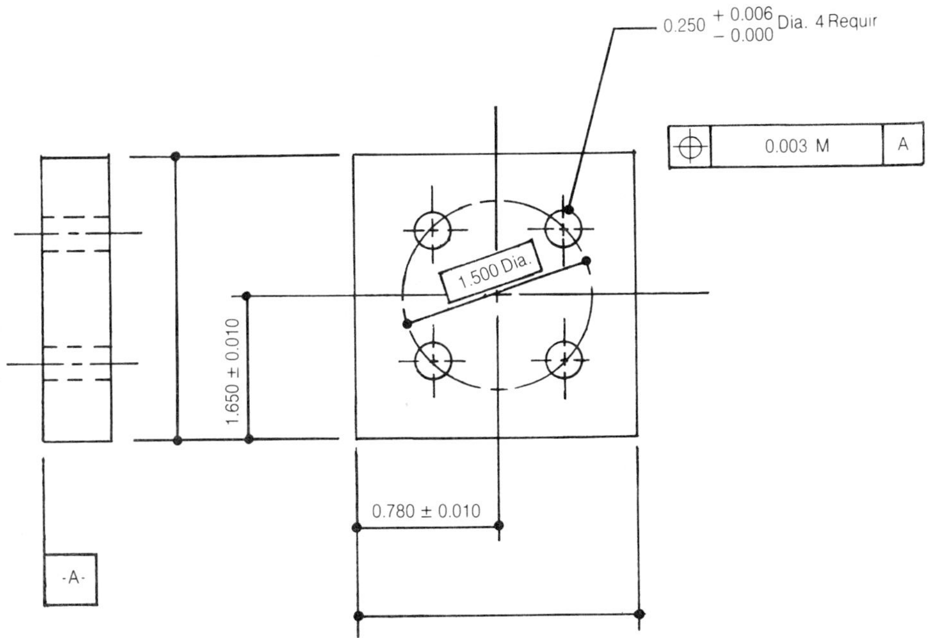

Figure 3.76 A part with four holes located around a bolt circle.

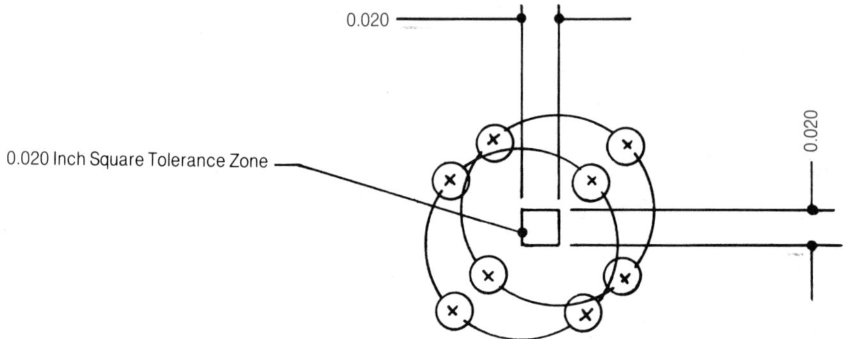

Figure 3.77 Several extreme positions of the four-hole bolt circle pattern.

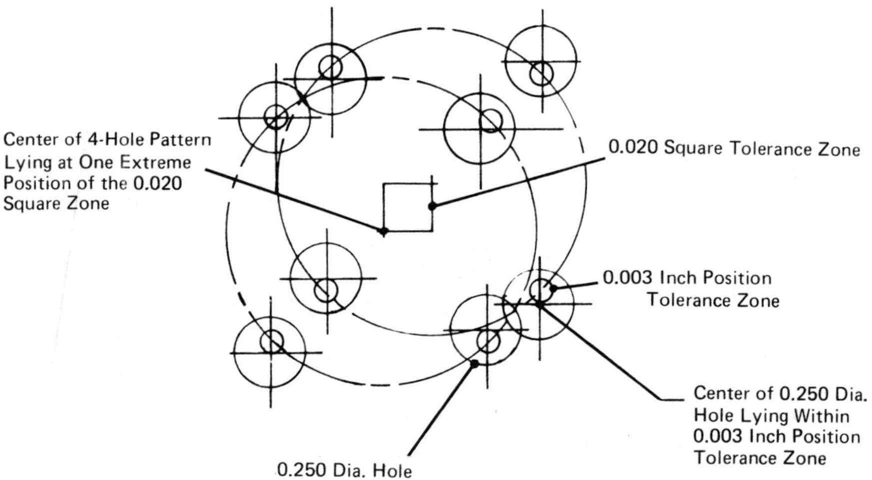

Figure 3.78 Several extreme positions of the four-hole bolt circle pattern.

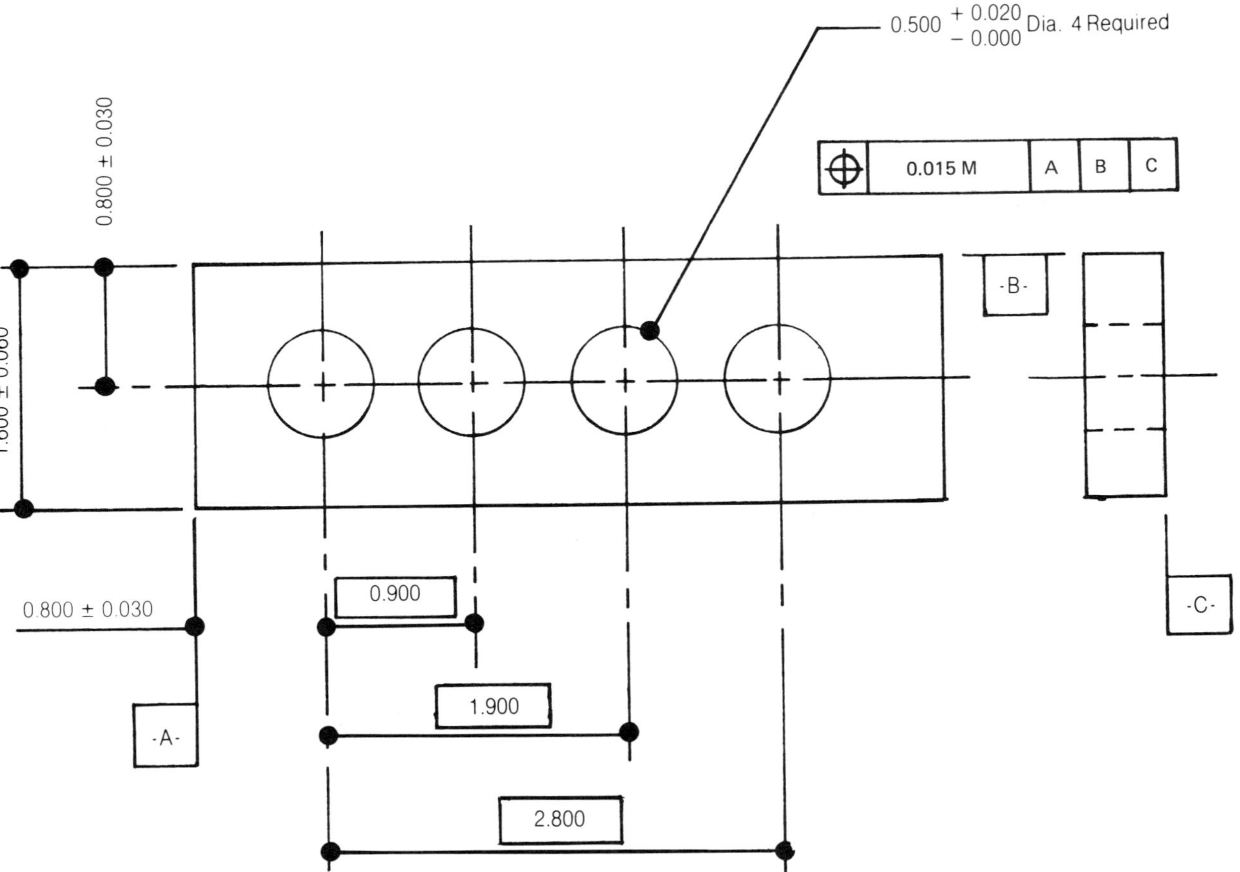

Figure 3.79 Coordinate and position tolerancing.

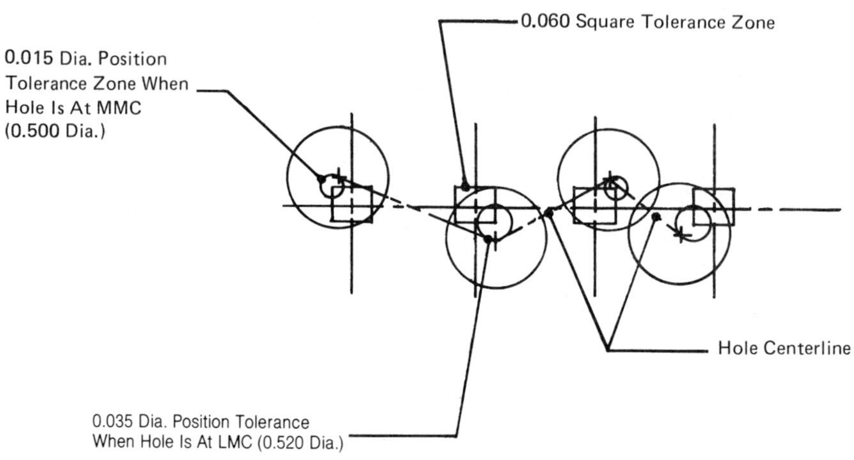

Figure 3.80 The possible deviation of holes with coordinate and position tolerancing.

the four-hole pattern which are possible with this rectangular coordinate tolerancing of ±0.01.

In Figure 3.76, rectangular tolerances are applied to the nominal location dimensions of 0.78 and 1.65, thus, establishing a bilateral type of tolerance zone (±0.010), while with position tolerancing there is no need for such a bilateral tolerance zone. With position tolerancing, the tolerance zone would be specified with the particular feature for example, the four 0.250 diameter holes have a 0.003 diameter position tolerance associated with them and the center of each 0.750 diameter hole must lie within its 0.003 diameter position tolerance zone. Therefore, the 1.50 diameter dimension used for locating these four holes along their bolt circle does not need an associated tolerance, therefore, a box is placed around this dimension to indicate that condition. Figure 3.78 shows several possible arrangements of these four 0.250 holes.

Figure 3.79 shows a part with four aligned holes with coordinate and position tolerance. The coordinate tolerance zones locating the pattern of the four holes are 0.060 square zones. The cylindrical position tolerance zones of 0.015 with a 0.500 diameter hole and 0.035 with a 0.520 diameter hole. Each of the hole's axes must lie within, or on, the periphery of the position tolerance zone and the position tolerance zone's axis must lie within, or on, the perimeter of the square tolerance zone. The placement of the holes on the part could deviate considerably as shown in Figure 3.80.

3.5.7. Composite Position Tolerancing

Composite position tolerancing is the recommended method used for the location of a pattern of features from datum surfaces where the location of the pattern is less important than the accuracy required for locating the features with respect to each other within the pattern. The advantage in using composite position tolerancing is the complete utilization of the features' pattern as well as the individual feature interrelationship within the pattern.

Two position tolerances are used with a single position tolerance symbol during composite tolerancing. The applicable datums and the pattern locating position tolerance are placed in the upper box with the applicable datum, and the feature relating position tolerance is placed in the lower box. Figure 3.81 shows such an arrangement.

In Figure 3.82, the center of the pattern of the six holes is located by the 2.563 and 4.121 basic dimensions and is related to datums A, B, and C. This center is the axis for the 1.862 basic diameter. The centers of the 0.250 diameter holes are determined by the intersection of the 60 degree basic angles at the 1.862 basic diameter. This pattern of six holes, taken as a unit, is to be located within a position tolerance of 0.02 inches when the holes are at MMC with respect to datums A, B, and C. Tables 3.1 and 3.2 list the additional tolerance for each hole equal to the amount of departure from MMC toward LMC. Remember, this additional tolerance is added to the position tolerance for locating the holes' pattern. The pattern locating position tolerance zone would be 0.02 diameter at the holes' MMC and 0.026 diameter at the holes' LMC. The feature relating position tolerance zones would be 0.005 in. in diameter at the holes' MMC (0.250) and 0.011 in. in diameter at the holes' LMC (0.256).

Table 3.1 Additional Hole Pattern Locating Tolerance

Hole measurement	Hole pattern locating tolerance
0.250	0.020
0.251	0.021
0.252	0.022
0.253	0.023
0.254	0.024
0.255	0.025
0.256	0.026

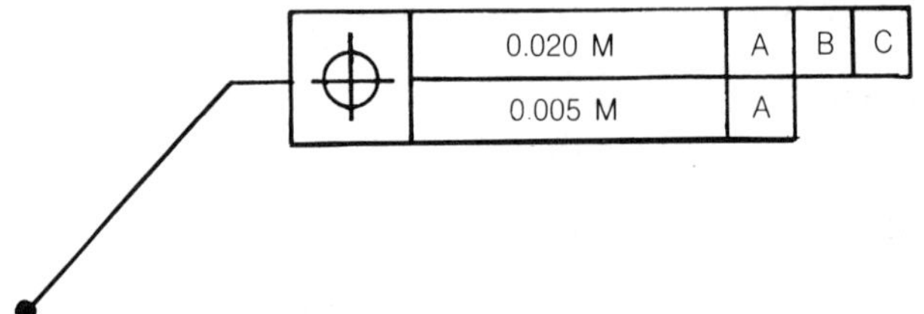

Figure 3.81 Composite position tolerancing geometric box.

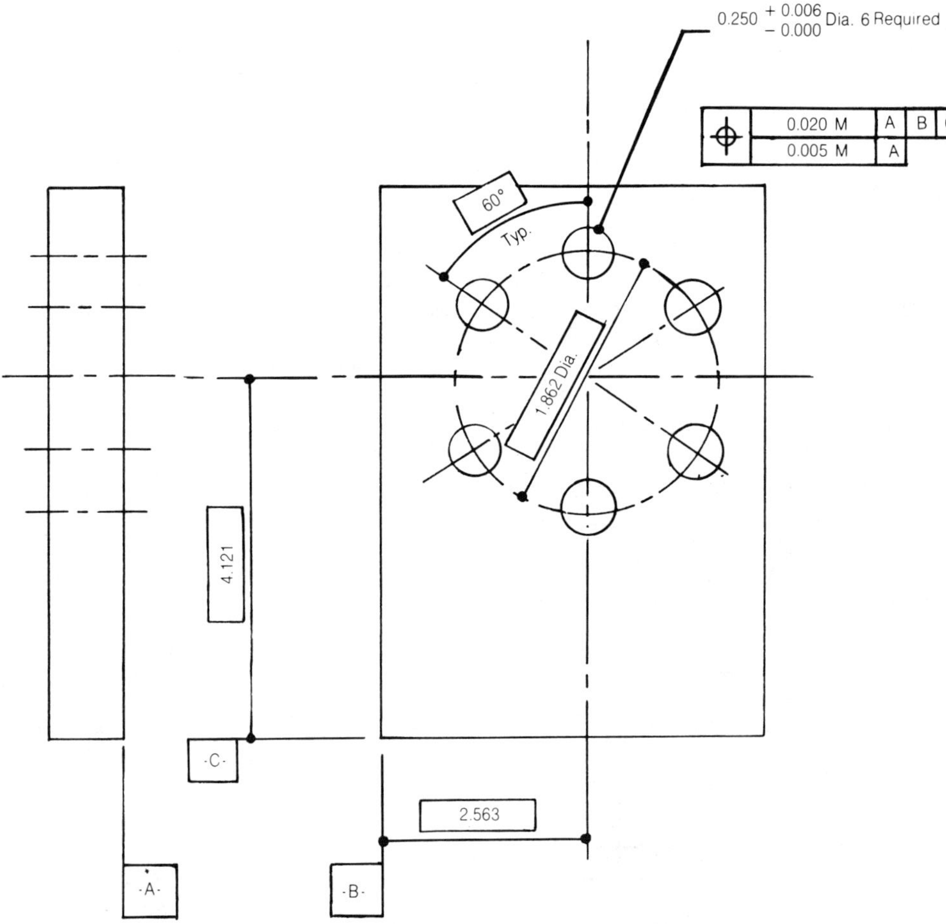

Figure 3.82 A part dimensioned and toleranced using composite position tolerancing.

Table 3.2 Additional Hole Tolerance

Hole Measurement	Hole Tolerance
0.250	0.005
0.251	0.006
0.252	0.007
0.253	0.008
0.254	0.009
0.255	0.010
0.256	0.011

Figure 3.83 shows a cross section of one of the six holes with the axes of the large (pattern) and small (feature) tolerance zones parallel to each other. The perpendicularity of each hole must lie within the 0.005-in. diameter tolerance zone, which is the same tolerance zone for locating the hole within the pattern.

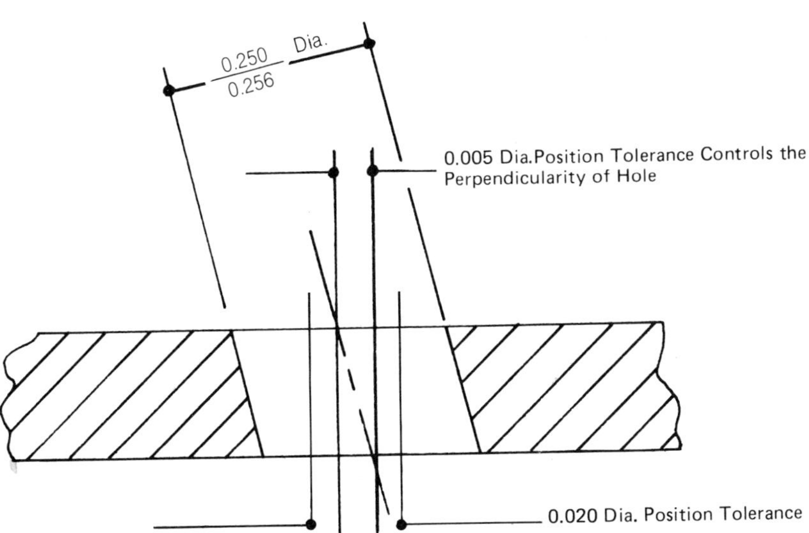

Figure 3.83 Position tolerance controlling perpendicularity of a hole.

3.5.8. Dimensioning a Pattern of Features Relative to a Datum Feature

Frequently, a pattern of features (i.e., a hole pattern) is located by using another feature (i.e., a hole) as a datum and dimensioning with respect to it. In Figure 3.84, the four 0.500 diameter holes are dimensioned from the 0.750-in. diameter hole. If the location of the 0.750-in. diameter hole moves, the position of the four 0.500-in. diameter holes move accordingly (Fig. 3.85). The letter M is used in the geometric box to indicate a maximum material condition (MMC).

The 0.750 diameter hole has a position tolerance diameter of 0.007 in. at MMC and is located from datum surfaces A, B, and C. The position tolerance of 0.007 in. diameter with respect to datums A, B, and C is greater than the perpendicularity tolerance of 0.002 in. with respect to datum A. The perpendicularity tolerance must be more accurate than the position tolerance; otherwise, the perpendicularity tolerance would control the position tolerance. The 0.750-in. diameter hole is designated as datum D at the MMC and the four 0.500-in. diameter holes are located with respect to it. This means if the 0.750-in. diameter hole gets bigger by 0.002 of an inch, the position and perpendicularity tolerances will increase an equal amount (Table 3.3). This 0.002 value will affect the location of the four 0.500 in. holes, since they are located from this datum (-D-) hole. Likewise, the position tolerance for each of the four 0.500-in. diameter holes is 0.003 in. at MMC. The least material condition for these holes would be 0.505 in. with a corresponding position tolerance of 0.008 in. (Table 3.4).

If the position tolerance for the 0.750-in. diameter hole did not depend on maximum material conditions and remained constant between the upper and lower limits of the 0.750-in. diameter hole, then an S modifier would replace the M symbol in the geometric box (Fig. 3.86). The S symbol represents the statement "regardless of feature size" or RFS.

Table 3.3 Additional Position and Perpendicularity Tolerance

MEASUREMENT OF OF 0.750 DIA. HOLE	ASSOCIATED POSITION TOLERANCE	ASSOCIATED PERPENDICULARITY TOLERANCE
0.750	0.007	0.003
0.751	0.008	0.004
0.752	0.009	0.005

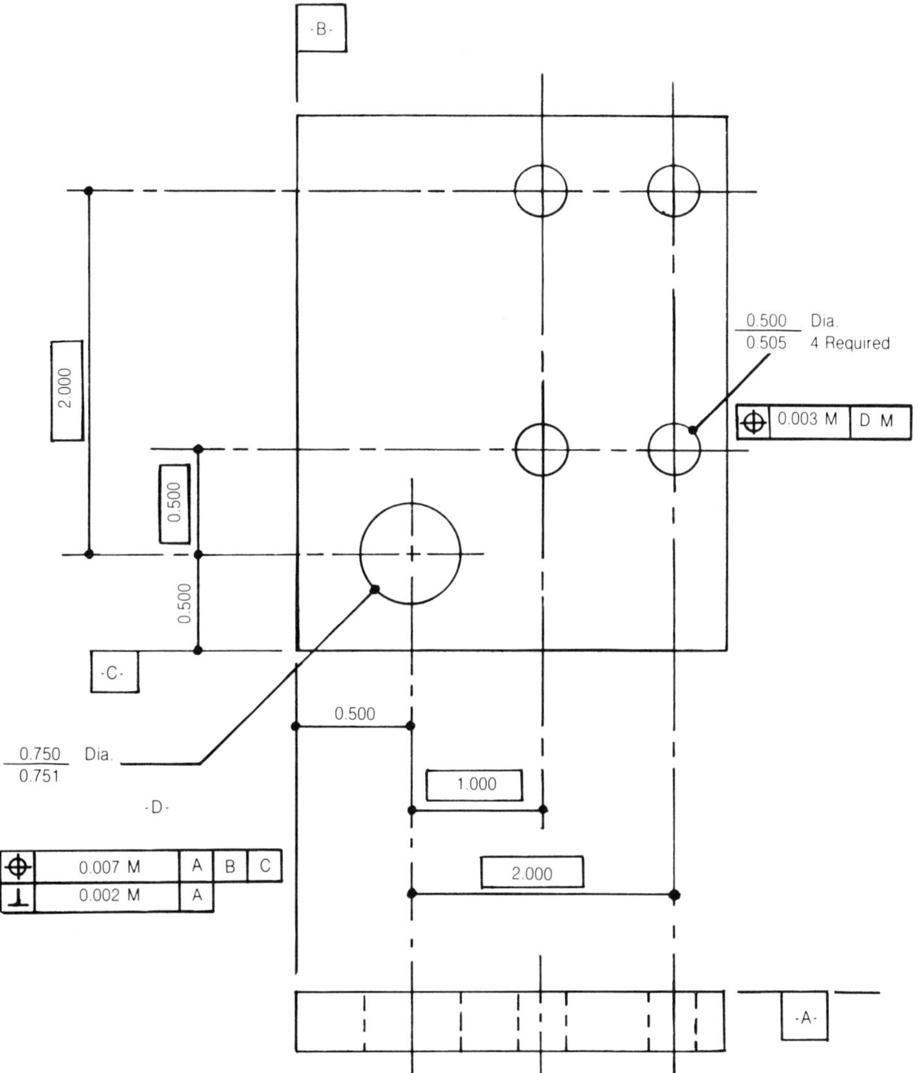

Figure 3.84 Four holes dimensioned and toleranced from a datum hole.

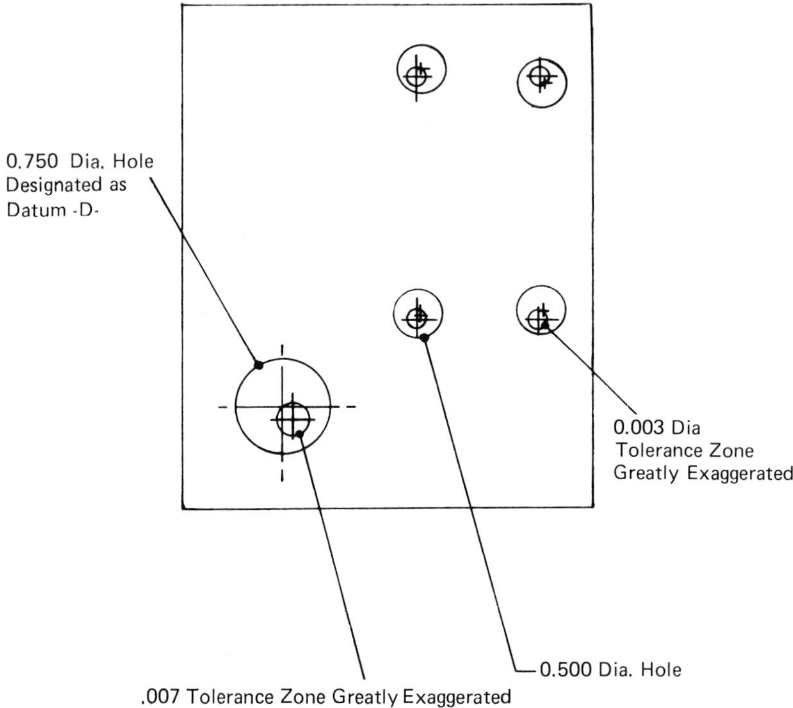

Figure 3.85 Possible extreme positions of the holes.

Table 3.4 Additional Position Tolerance

MEASUREMENT OF OF 0.500 DIA. HOLE	ASSOCIATED POSITION TOLERANCE
0.500	0.003
0.501	0.004
0.502	0.005
0.503	0.006
0.504	0.007
0.505	0.008

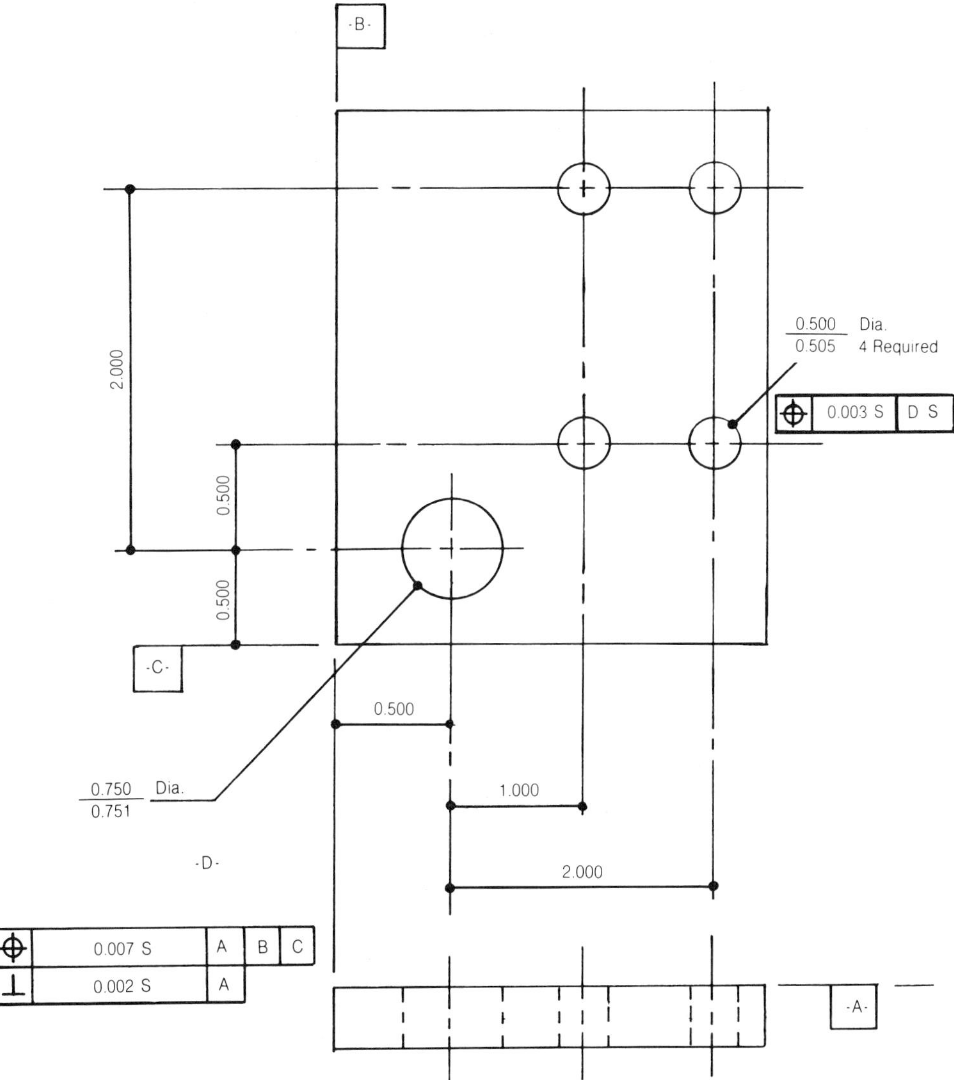

Figure 3.86 Replacing the maximum material condition of the holes with a regardless of feature size symbol.

Now the four 0.500 holes are located from the exact center of the datum hole D at whatever size the datum hole is produced (RFS). Again, the four 0.500 hole pattern varies with the movement which would be 0.007 of the datum D hole; since, their location depends on the location of the datum D. In other words, the variation in size of the datum D hole (0.750-0.752 in.) will have no effect on the four hole pattern relationship. In this instance, a closer relationship is maintained between the four holes and their datums. It must be noted that a RFS modifier S must be placed by datum D in the geometric box associated with the 0.500 holes to ensure the correct interpretation.

3.5.9. Projected Tolerance Zone

Position tolerancing is used to ensure the interchangeability of mating parts in assembly. Not only can this tolerance be used for locational control of features on a part, but it can be used for controlling the

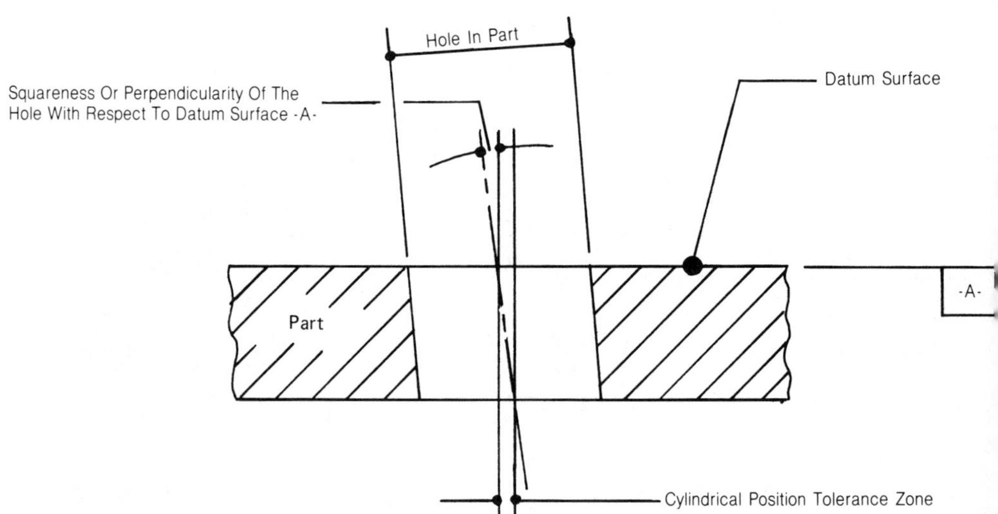

Figure 3.87 Controlling the squareness of a hole with a projected tolerance zone.

squareness of the feature related to a datum or other features on the part (Fig. 3.87). Further restrictions can be made on this squareness tolerance by projecting the position tolerance the depth of the mating part (Fig. 3.88). For any application where squareness is critical, a perpendicular tolerance is included with the position tolerance. This perpendicular tolerance must be smaller in diameter than the position tolerance if it is to control the attitude (perpendicularity) of the hole. In Figure 3.89, the axis of the 0.88-in. diameter hole must lie within the perpendicularity tolerance of 0.005-diameter, which in turn, must lie within the position tolerance zone of 0.010 diameter. The axis of the 0.250-20UNC threaded hole in Figure 3.90 must lie within a 0.002-in. diameter cylindrical zone which is theoretically 90 degrees with respect to the surface datum designated by the letter A.

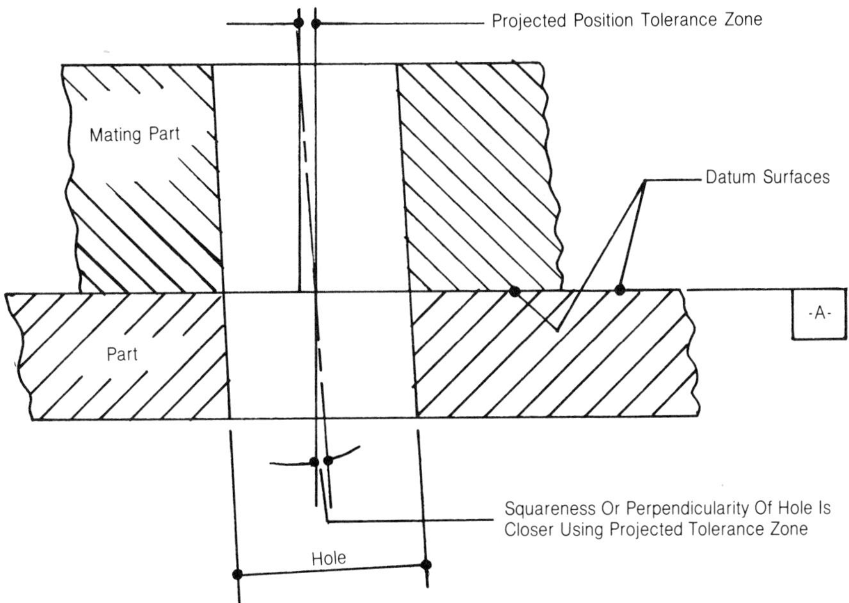

Figure 3.88 Projected tolerance zone in the mating part.

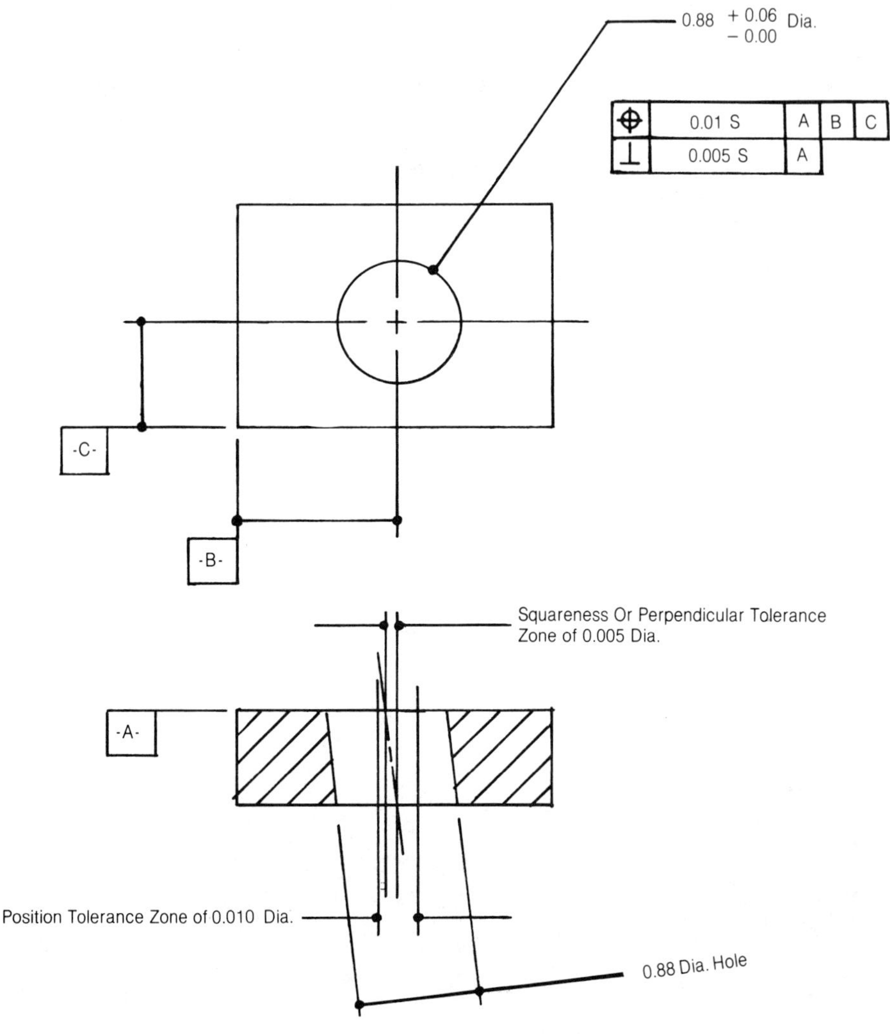

Figure 3.89 A hole having a separate position tolerance and perpendicularity tolerance.

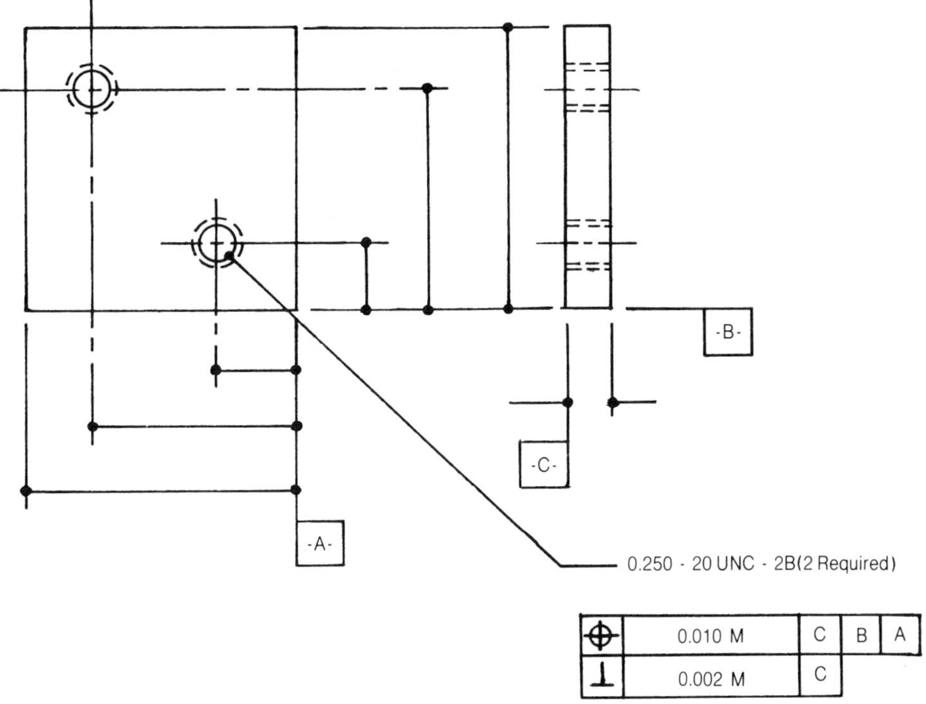

Figure 3.90 A part showing separate position tolerancing and perpendicularity tolerancing.

3.6. POSITION TOLERANCING OF NONCYLINDRICAL FEATURES

Usually position tolerances are cylindrical zones applied to cylindrical features within which the centerline of that feature must lie. For noncylindrical features such as slots and projections, the position tolerance is a zone bounded by two parallel planes within which the centerplane of the feature must lie (Fig. 3.91). The distance between these two parallel planes is the position tolerance and is expressed as a width. As shown in Figure 3.92, it is possible that the centerplane of the slot or projections will not necessarily be perpendicular to the surface of the part. Since the centerplane must lie within the position tolerance zone, it can be said that the position tolerance controls position as well as perpendicularity. The maximum material and regardless of feature size conditions can be used.

In Figure 3.93, part 1 is shown with a slot that has a width of S and a tolerance of $^{+0.005}_{-0.000}$. The position tolerance is 0.007 inches wide when

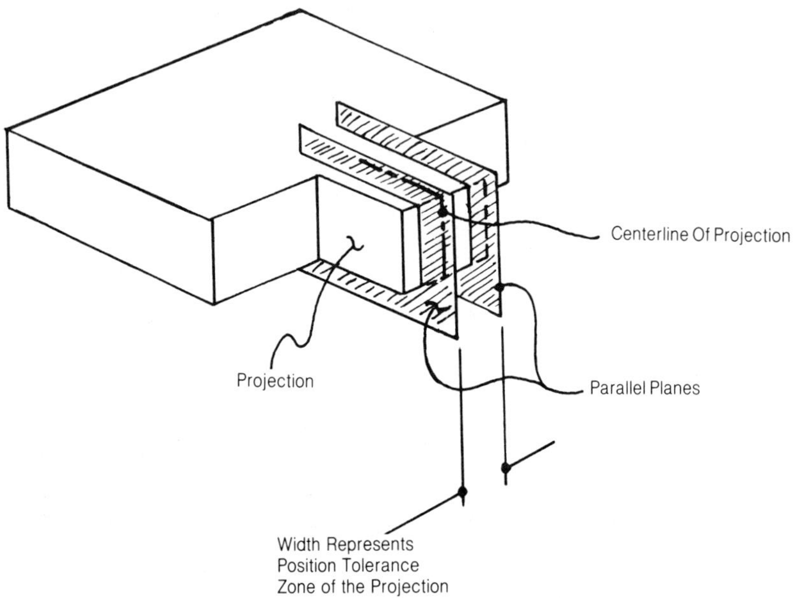

Figure 3.91 Position tolerancing of a projection.

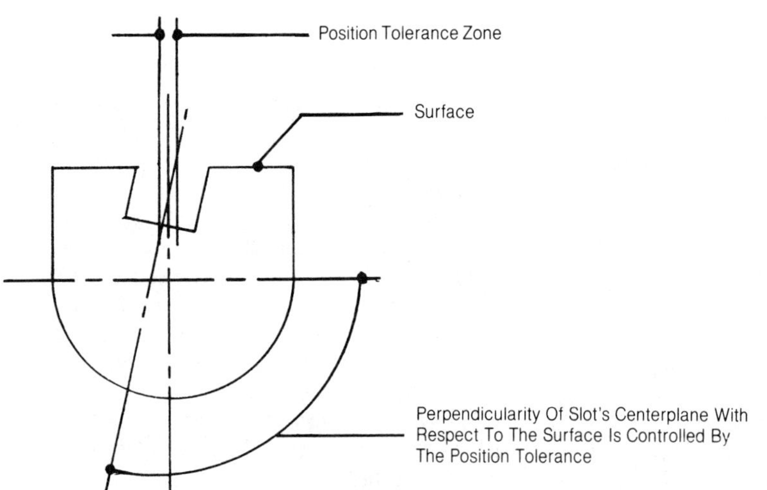

Figure 3.92 The position tolerance zone of a slot.

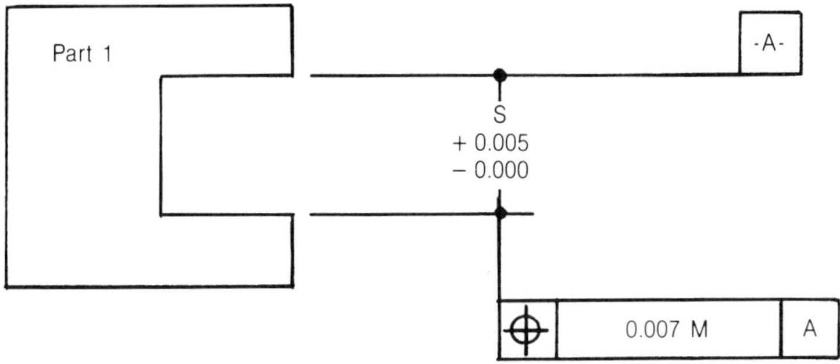

Figure 3.93 Part 1.

the slot is at its MMC size and has a depth equal to the thickness of the part.

Part 2, which is the mating part to part 1, is shown in Figure 3.94. It is necessary to calculate the width of the minimum slot that can be assembled with the projection under the worst condition of assembly. The formula $S = P_s + P_t + T$ can be used for slot and projection assemblies where S = minimum (MMC) slot width, T = maximum (MMC) projection width, P_s = position tolerance width of the slot at MMC, and P_t = position tolerance width of the projection at MMC.

Using the formula for slot and projection assemblies, the minimum slot width for part 1 to be properly assembled with part 2 would be S = 0.745 + 0.007 + 0.010 or S equals 0.762 inches.

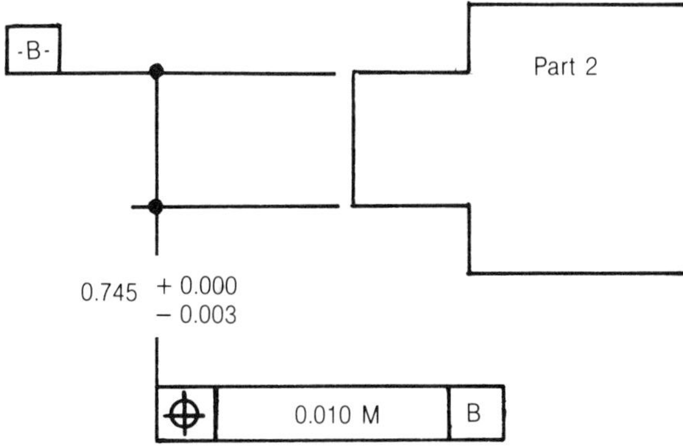

Figure 3.94 Part 2.

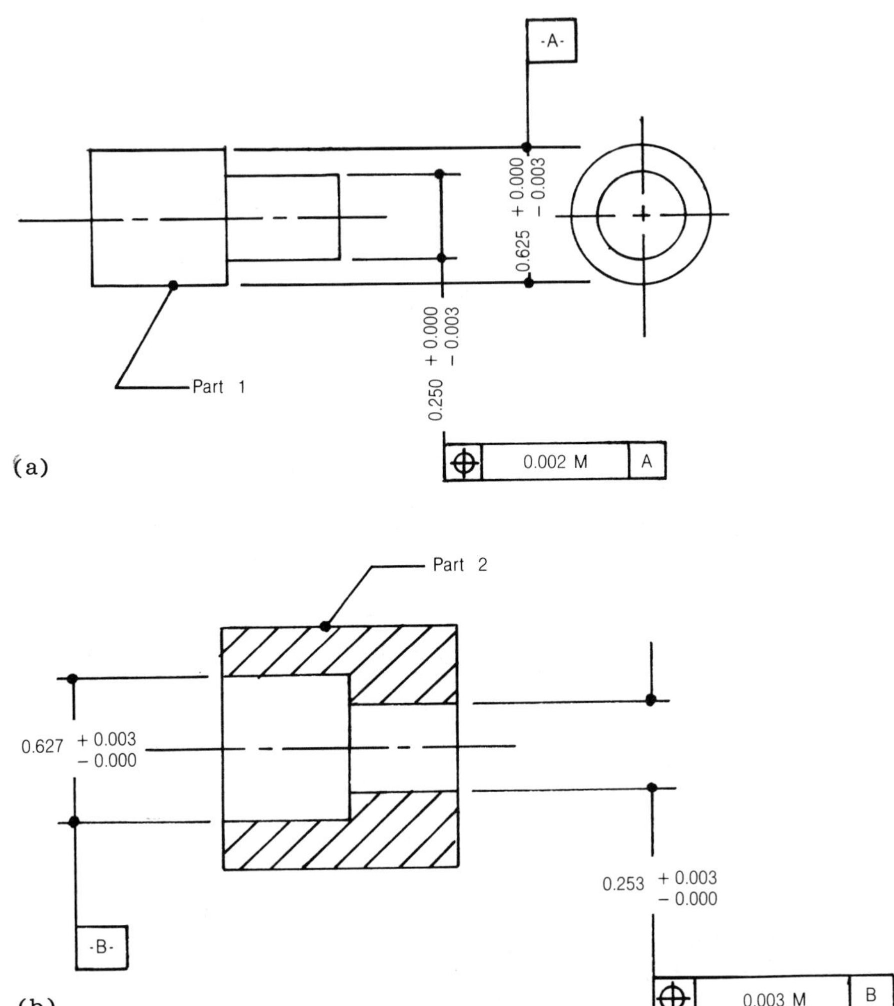

Figure 3.95 Part 1 (a) and part 2 (b) shown with their respective position tolerancing.

3.7. POSITION TOLERANCING OF COAXIAL FEATURES

Position tolerancing of coaxial features is used if there is a need to control the total cylindrical or profile surface and its axes relative to the datum axis on a MMC basis.

The position tolerances for parts 1 and 2 are to be calculated. The MMC size hole of Part 2 is 0.253 and the MMC size shaft of part 1 is 0.250 inches. When the shaft mates the the hole, the maximum clearance would be equal to 0.253 – 0.250 = 0.003. The MMC size of the *datum* hole of part 2 is 0.627 and the MMC size of the datum shaft of part 1 is 0.625 inches. When the shaft mates with the hole, the maximum clearance of 0.002 would result. The total tolerance is the sum of the coaxial clearances or 0.003 + 0.002 = 0.005. The sum of the position tolerances on part 1 and 2 must total 0.005; therefore, select 0.002 for the position tolerance on part 1 and select 0.003 for the position tolerance on part 2. These position tolerances are shown on parts 1 and 2 appearing in Figure 3.95.

The 0.250-in. diameter of the shaft on part 1 has a 0.002-in. position tolerance zone at its MMC size of 0.250. This position tolerance of 0.002 inches increases an amount equal to the departure from MMC size of the shaft (difference between the shaft's upper and lower limits) which can be as much as 0.250 - 0.246 = 0.004 of an inch. An additional amount equal to the departure from MMC size of the *datum* shaft up to 0.625 - 0.620 = 0.005 of an inch can also occur. This sum of 0.004 + 0.005 = 0.009 can be added directly to the shaft's 0.002 position tolerance zone to yield 0.011. Now, this 0.011 will be the permissible position tolerance at the least material conditions of the shaft and its *datum* shaft on part 1.

Likewise, a similar calculation can be performed on the hole of part 2 which has a position tolerance of 0.003 inches. This position tolerance of 0.003 increases an amount equal to the departure from MMC size of the hole (difference between the hole's upper and lower limits) which can be as much as 0.259 - 0.253 = 0.006 of an inch. An additional amount equal to the departure from MMC size of the *datum* hole up to 0.632 - 0.627 = 0.005 of an inch can occur. This sum of 0.006 + 0.005 = 0.011 can be added directly to the hole's 0.003 position tolerance to yield 0.014. Now, this 0.014 can be the permissible position tolerance at the LMC of the hole and its *datum* hole on part 2.

3.8. THE INTERRELATIONSHIP BETWEEN FORM AND POSITION TOLERANCING

Again, it is very difficult to visualize what happens when form and position tolerances are combined on a part drawing for specifying a feature's shape and location. For instance, the part shown in Figure

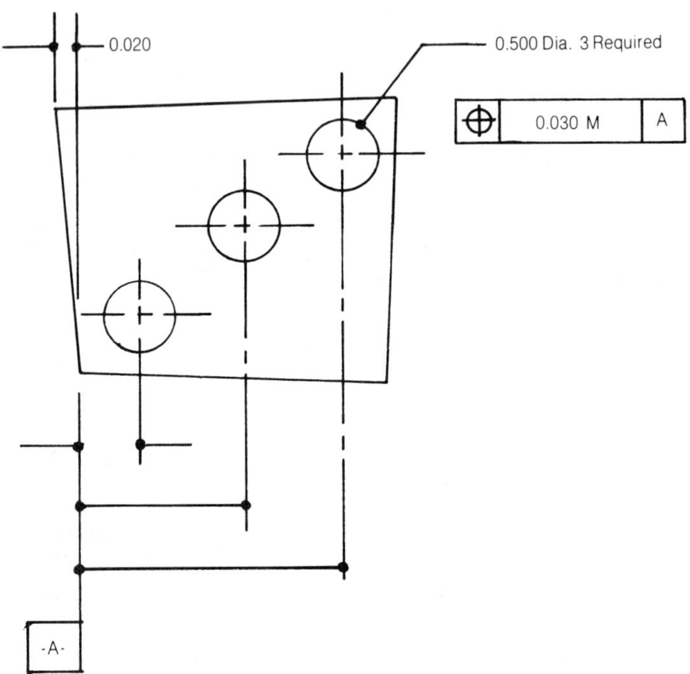

Figure 3.96 A part having form tolerancing and position tolerancing.

3.96 has three holes. The 0.500-in. diameter holes are located from a surface which is used as a datum.

Measurements taken from this datum surface may vary as much as 0.02 of an inch, depending where the measurement is taken from the datum surface marked A. Since the position tolerance for each 0.500 diameter hole is 0.03 inches, this surface variation will influence the position tolerance zone. To separate the effects form tolerancing has on position tolerancing, jigs, and fixtures are used in the shop. Jigs and fixtures are production tools used to accurately manufacture duplicate parts.

A jig is a special device which holds and supports a part which is to be machined. It is a production tool made so that it not only locates and holds the workpiece, but also guides the cutting tool as the machining operation is performed. Jigs are usually fitted with hardened steel bushings for guiding drills or other cutting tools.

A fixture is a production tool which locates, holds, and supports the work securely so the required machining operations can be performed. A fixture should be securely fastened to the table of the machine upon which the work is done.

3.9. COMPUTER GRAPHICS APPLICATIONS

The cam, rocker arm, and push rod assembly (Figs. 3.97 and 3.98) used in Chap. 2 Section 2.9, will be used as a position and geometric tolerancing example on a computer graphics system. The detail drawings of the six parts must be changed from rectangular coordinate dimensioning and tolerancing to position dimensioning and tolerancing (Figs. 3.99—3.104).

The position tolerances can be determined by two approaches. In either case, one must take a look at the square tolerance zone shown in Figure 3.105. The position tolerance zone is circular in shape and its diameter could be the length of the square's sides or the square's

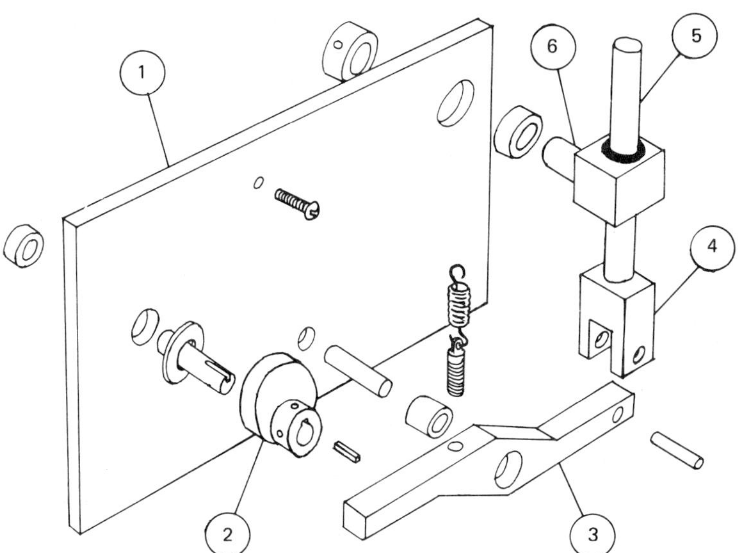

Figure 3.97 Cam, rocker, and push rod system.

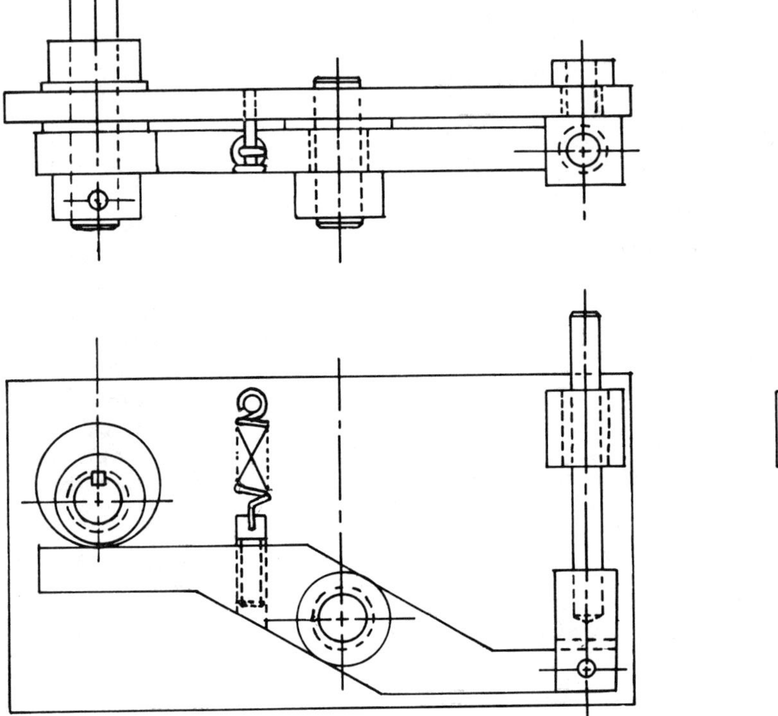

Figure 3.98 Cam, rocker, and push rod assembly.

diagonal, depending on how the square tolerance was derived. For instance, if the designer calculated the zone by using its diagonal for interchangeability purposes, the diagonal would be the diameter of the position tolerance. If the sides of the zone are used for interchangeability purposes, then the sides would be used as the diameter of the position tolerance.

Again, lets begin with the plate by placing it on the display screen of the computer graphics system. In this case, a 0.06-in. diameter position tolerance zone has been decided upon. Since the extreme position of the center of the holes in the plate would lie on the circumference of the position tolerance zone, it is easy to imagine how computer graphics could rotate a feature or a part along the circumference of this circular zone. But one must be careful that the orientation of the feature or part does not change during this rotation.

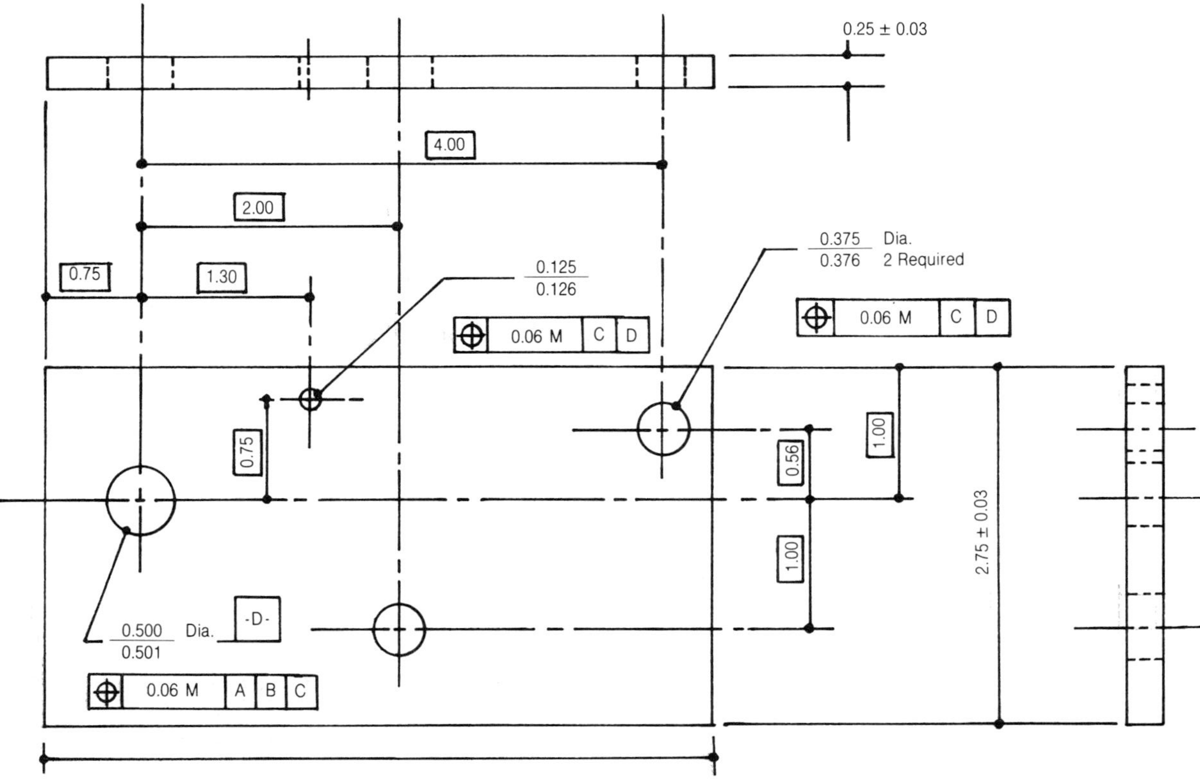

Figure 3.99 Plate detail.

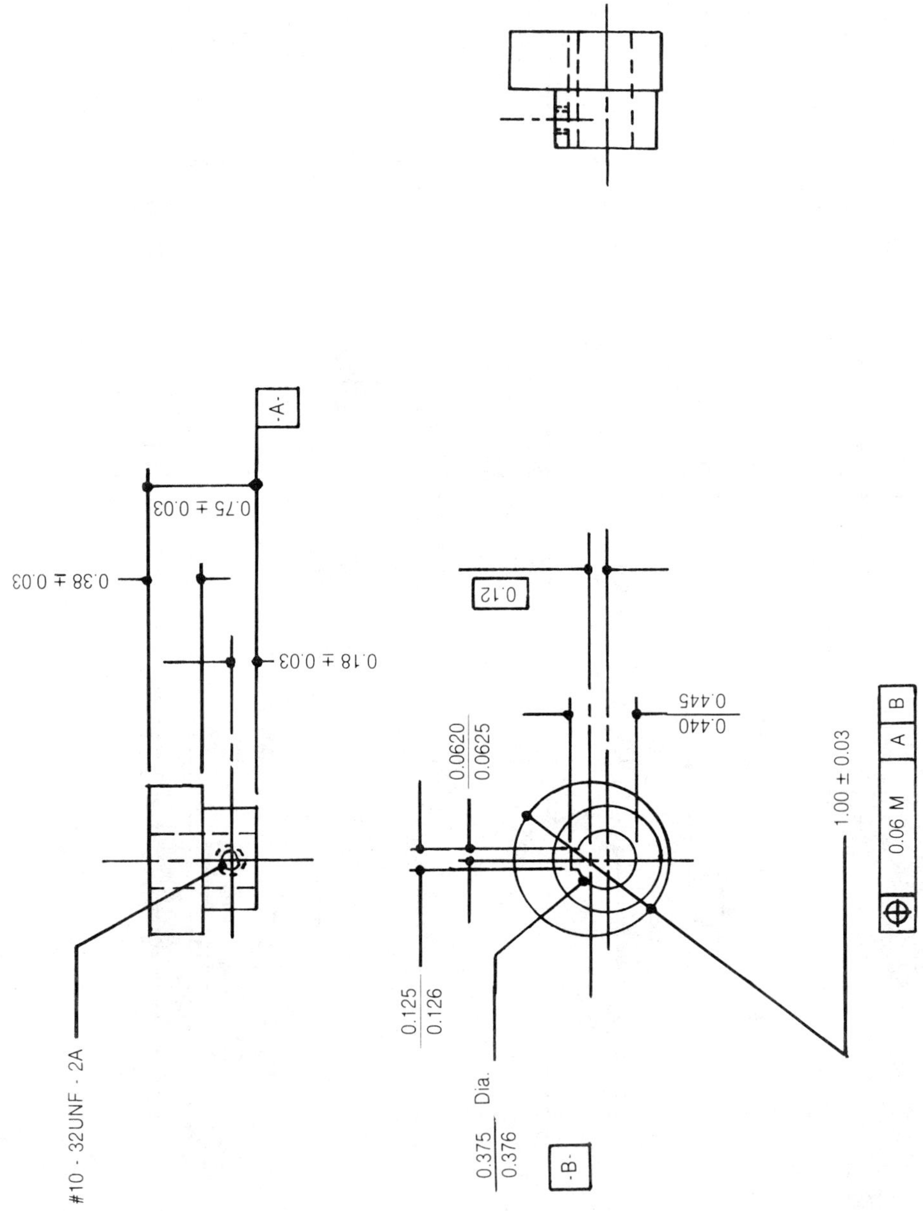

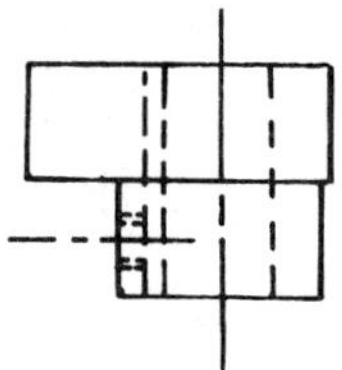

Figure 3.100 Cam detail.

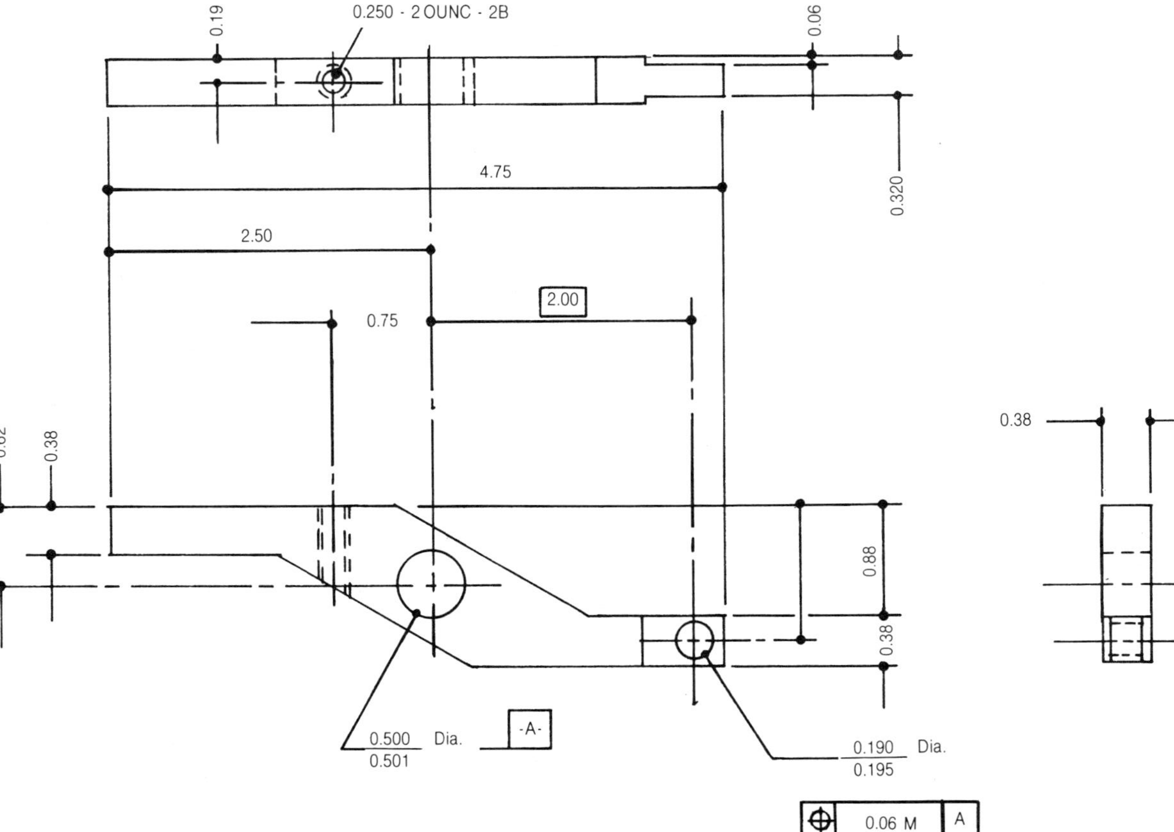

Figure 3.101 Rocker detail.

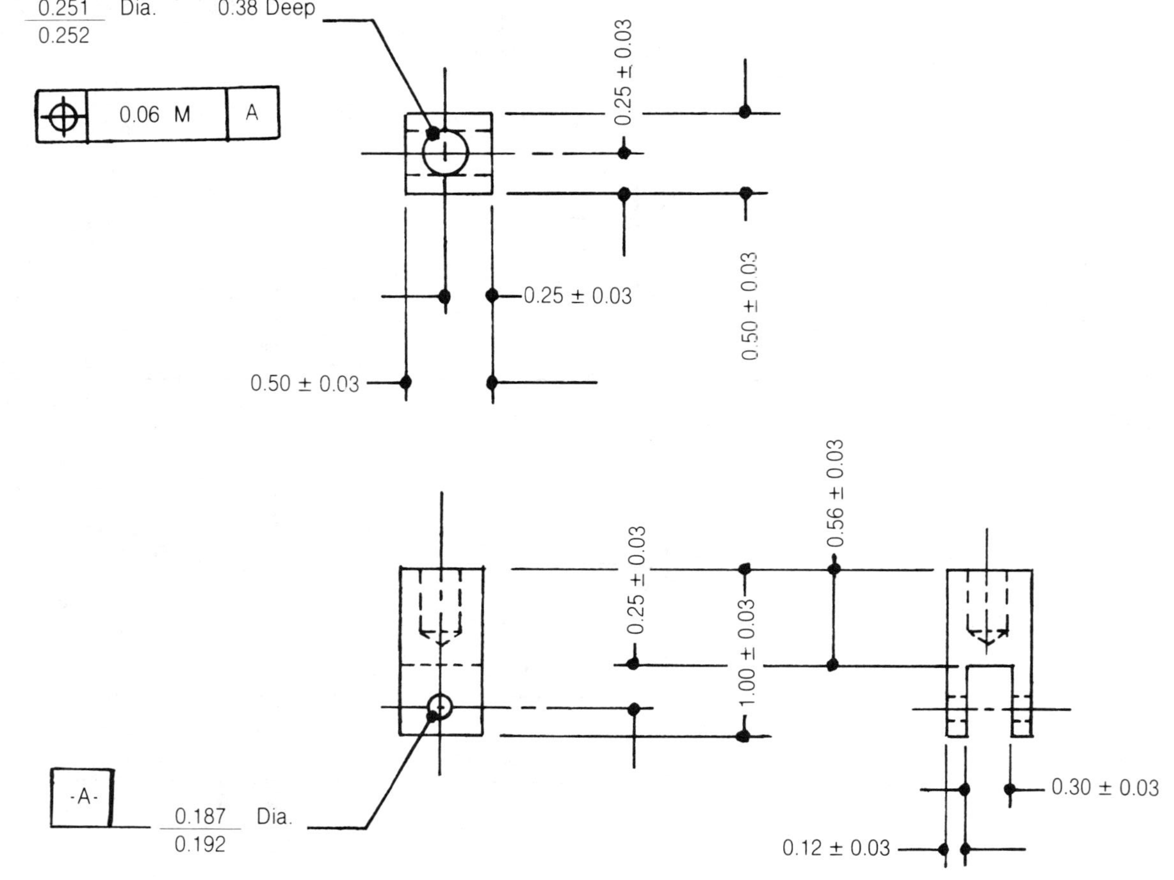

Figure 3.102 Clevis detail.

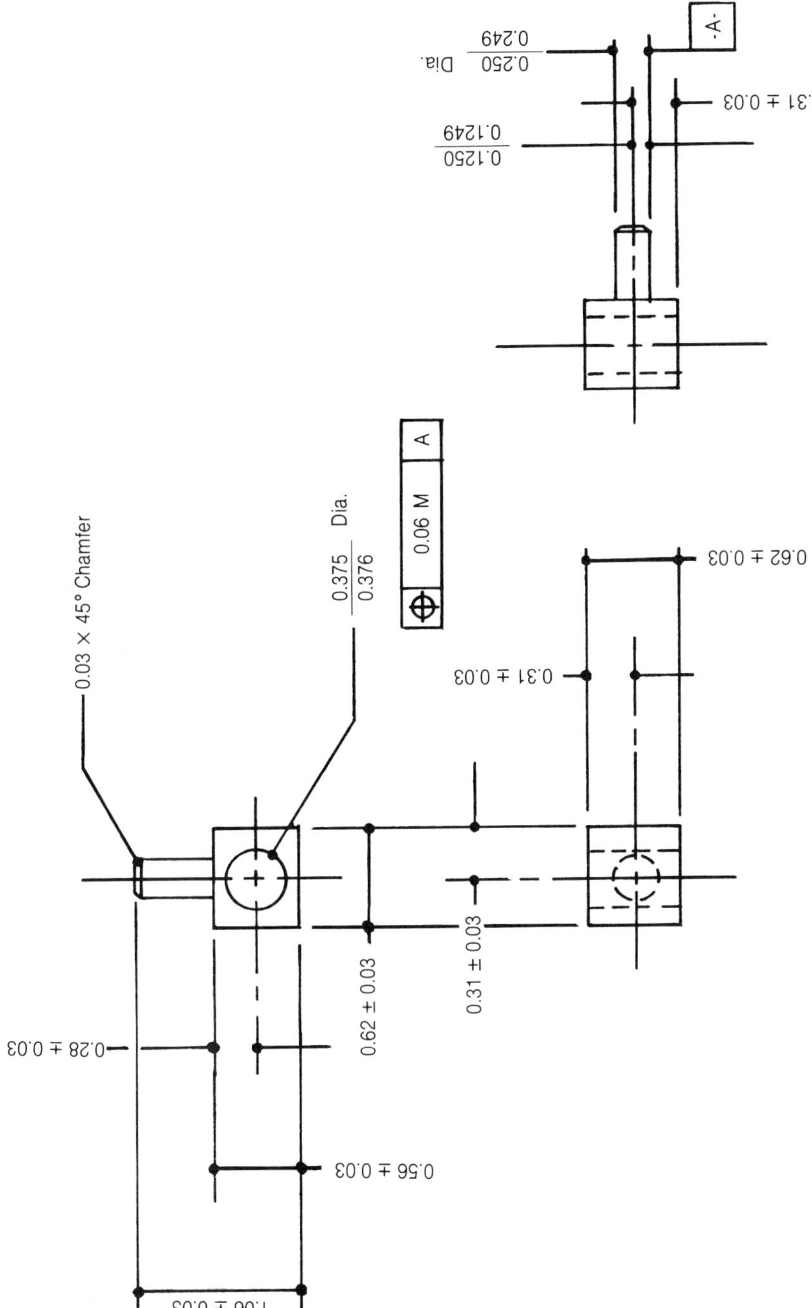

Figure 3.103 Slider block detail.

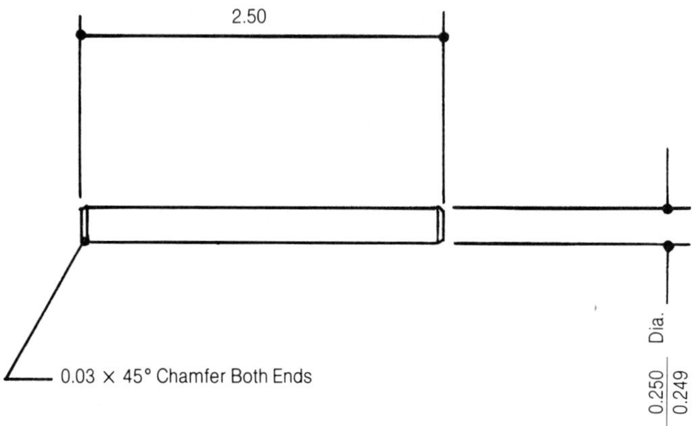

Figure 3.104 Push rod detail.

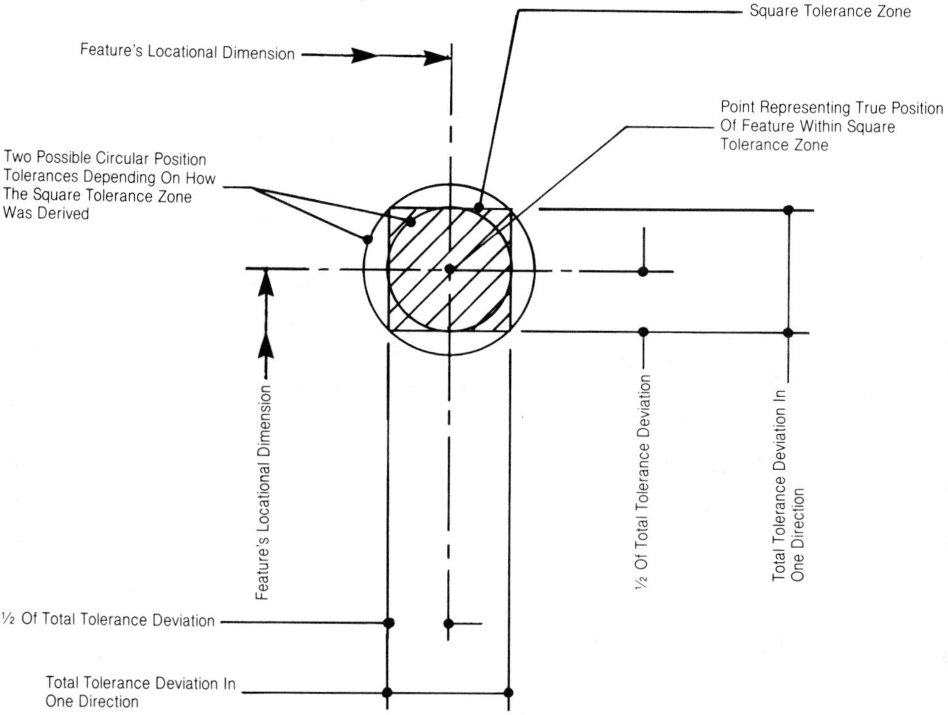

Figure 3.105 Comparing the square tolerance zone with the position tolerance zone.

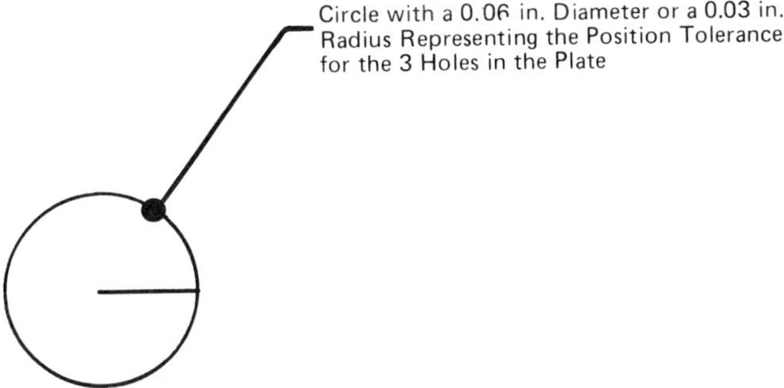

Figure 3.106 0.060 diameter circle representing the position tolerance zone of the three holes.

Assuming the position tolerance for the three holes in the plate is 0.060 inches in diameter, then these position tolerances can be represented by a circle with a 0.030 radius originating from the center of each of these circular zones (Fig. 3.106). The commands used to place these radii within their respective position tolerance is PLACE LINE HORIZONTAL LONG 0.030 / INPUT XOYO D1; INPUT X2Y-1 D_2; INPUT X4Y0.5 D_3 CR and Figure 3.107 shows how digitizers D_1, D_2, and D_3 are placed on the display screen in order to specify the proper direction of these horizontal lines. The results are shown in Figure 3.108. Since the radii are very small, it is difficult to see them on the display screen. Therefore, the operator must enlarge the viewing area on the display screen where each position tolerance is located. This is done by using the commands ENLARGE 10 / UNIT WINDOW D_1 D_2 CR (Fig. 3.109). Everything inside the window which is defined by the diagonal of the two digitizers D_1 and D_2 is enlarged ten times the size that it appears on the display screen. The result is shown in Figure 3.110.

A point is placed on the end of each radius by using the command PLACE POINT / INPUT END,D_1 CR. These points will be used to locate the various parts when they are placed on the display screen. For instance, if we take the cam which has been constructed using the maximum material conditions and place it on the display screen at one of its extreme positions along the position tolerance circumference. This cam is stored under its file name of MMC cam. It is placed on the display screen by using the command PLACE PART MMC CAM ENLARGE 10 / INPUT POINT D_1 CR (Fig. 3.111). The part will appear at the location of the digitized point on the display screen and it will appear ten times its size, since everything has been enlarged by a factor of ten. To place this cam exactly at the end of the radius, which represents a location on the position tolerance's circumference, one must

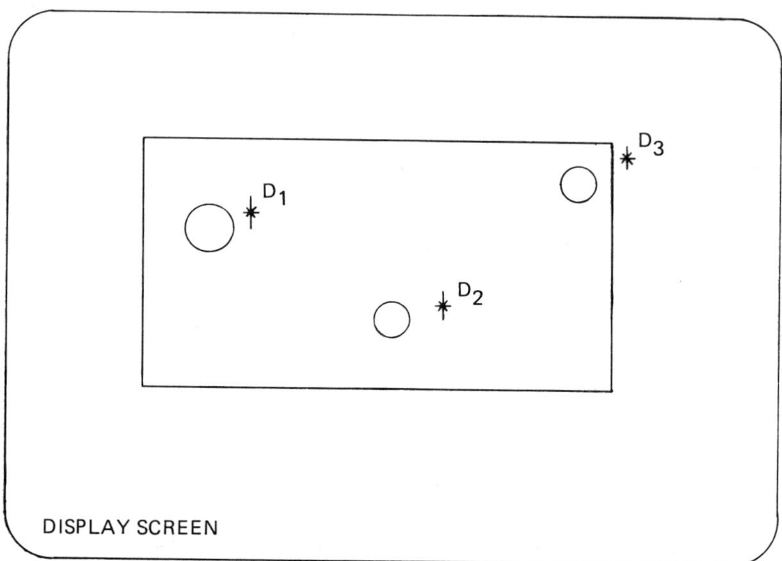

Figure 3.107 Plate detail shown on the display screen.

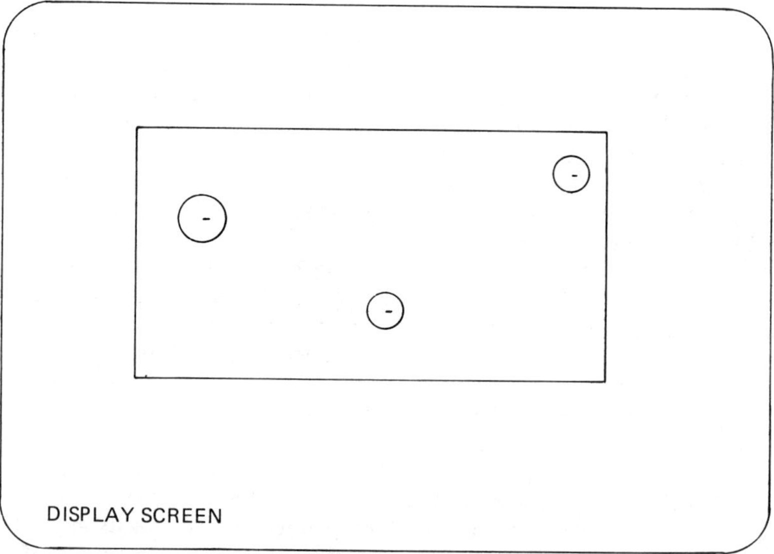

Figure 3.108 Position tolerances with a radius of 0.030 is represented by the three very small horizontal lines.

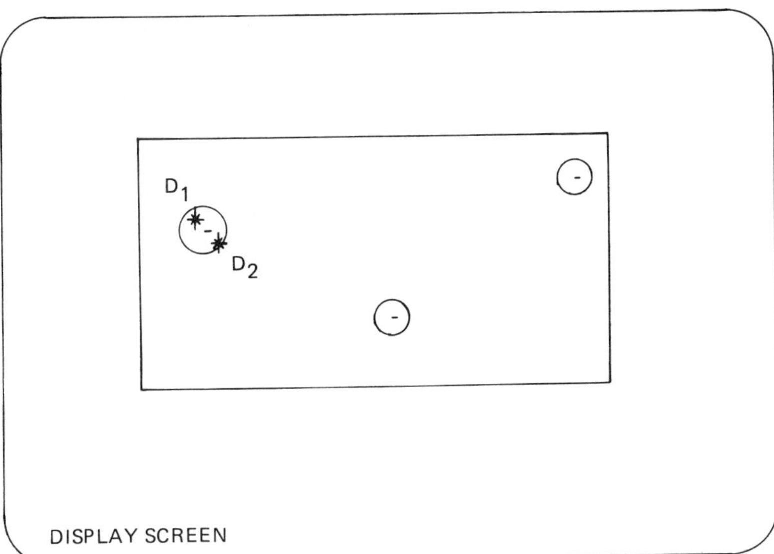

Figure 3.109 Digitizing a window on the display screen for enlarging one of the small horizontal lines.

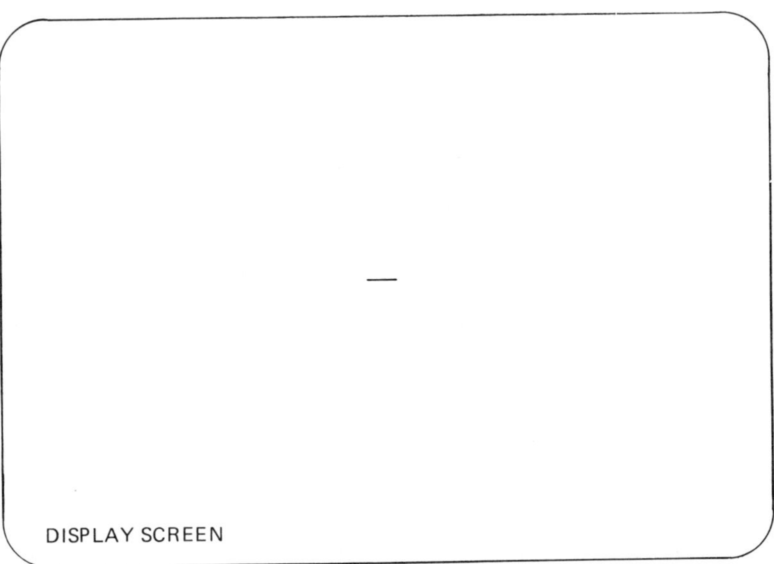

Figure 3.110 One of the horizontal lines enlarged ten times its actual size.

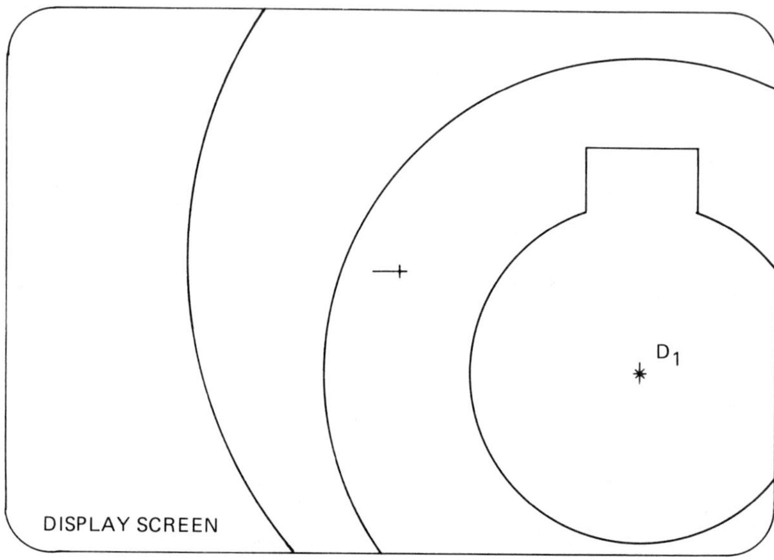

Figure 3.111 The MMC cam is placed on the display screen ten times its actual size.

use the command MOVE / UNIT POINT D_1 D_2 CR (Fig. 3.112). Figure 3.113 shows the cam moved to the exact point lying on the end of the radius.

The cam can be moved to another position along the circumference of the position tolerance zone by using the command ROTATE ANGLE 45 / UNIT D_1 D_2 INPUT END D_3 CR (Fig. 3.114). This command will move the line and its end point in a counterclockwise direction by 45 degrees about the center of the position tolerance zone (Fig. 3.115). The cam, itself, will remain at the previous location, since it was not digitized and therefore was not included in the rotation command. This procedure was purposely done, since the rotation command would have changed the orientation of the cam.

If the radius has been rotated to its new location (Fig. 3.115) without the cam, then the cam can be moved to the end point of the radius by using the move command, again, MOVE / UNIT POINT D_1 D_2 CR (Fig. 3.116). The result is shown in Figures 3.117 and 3.118 showing everything on the display screen reduced ten times by using the command REDUCE 10 EVERYTHING CR. This procedure can be repeated, thereby, moving the cam along the circumference of the position tolerance zone in increments of 45 degrees. Any angle of rotation can be used for the rotation, since this angle is specified in the rotate command.

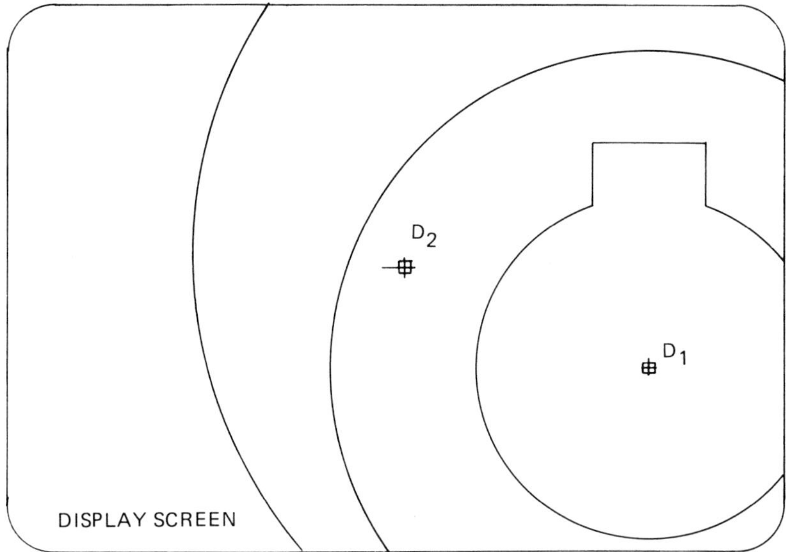

Figure 3.112 The MMC cam is moved to the end of the horizontal line representing the position tolerance zone.

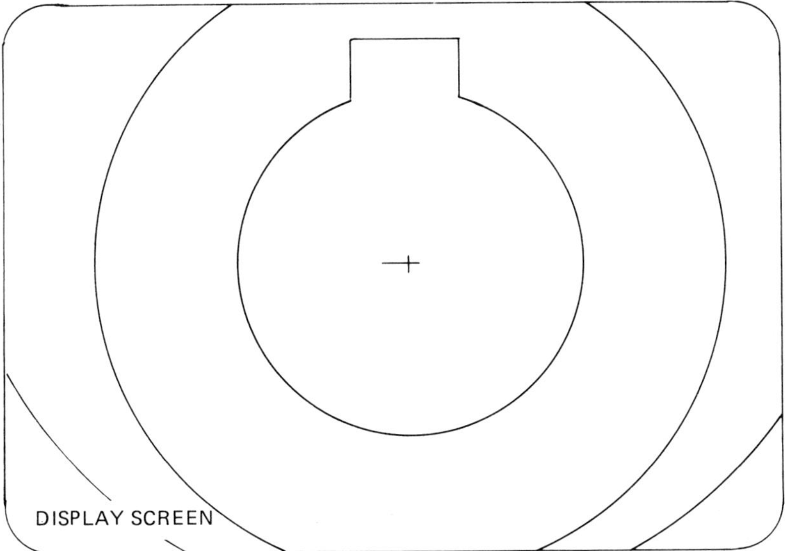

Figure 3.113 The MMC cam located at the end of the horizontal line.

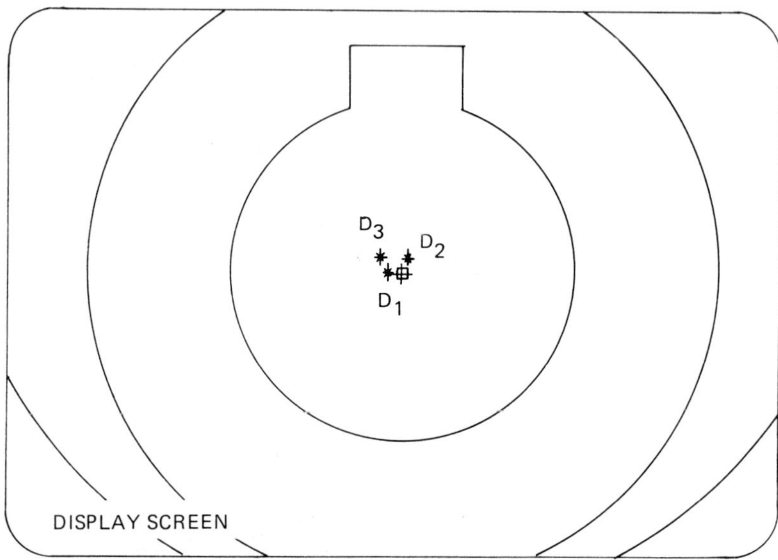

Figure 3.114 The horizontal line is rotated 45 degrees in a counterclockwise direction.

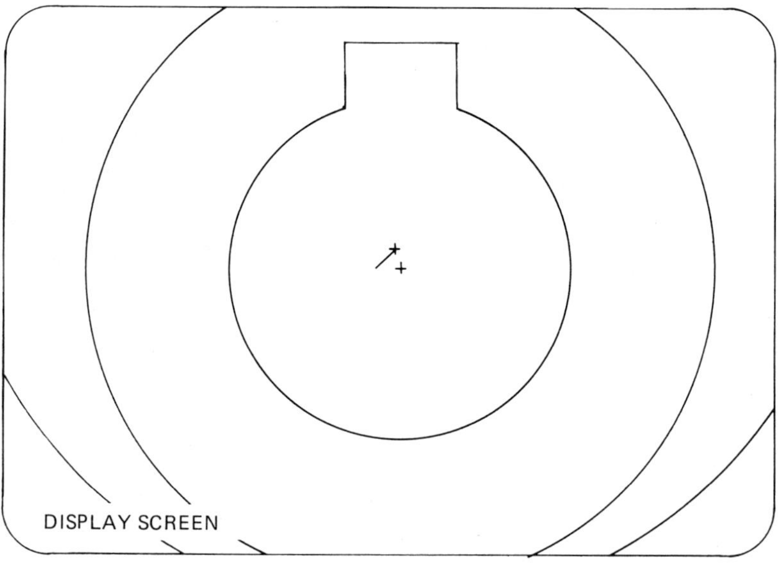

Figure 3.115 The rotated line represents an extreme position lying on the circumference of the position tolerance zone.

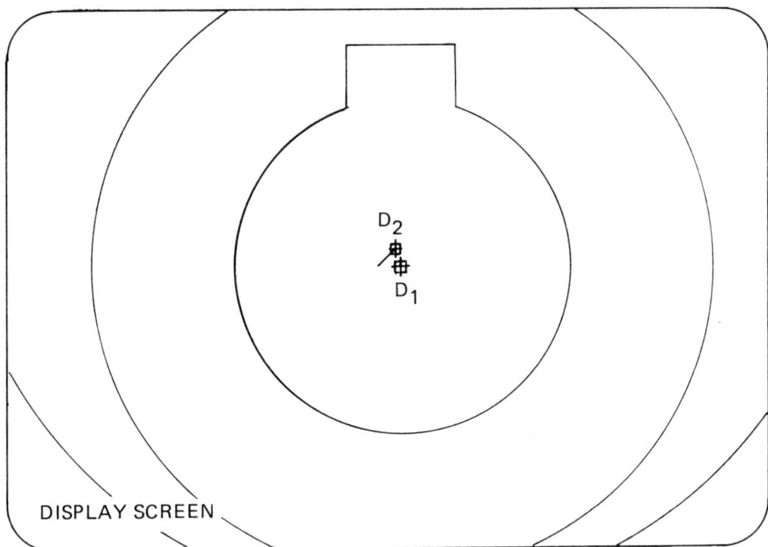

Figure 3.116 The enlarged MMC cam is moved to the end of the rotated line.

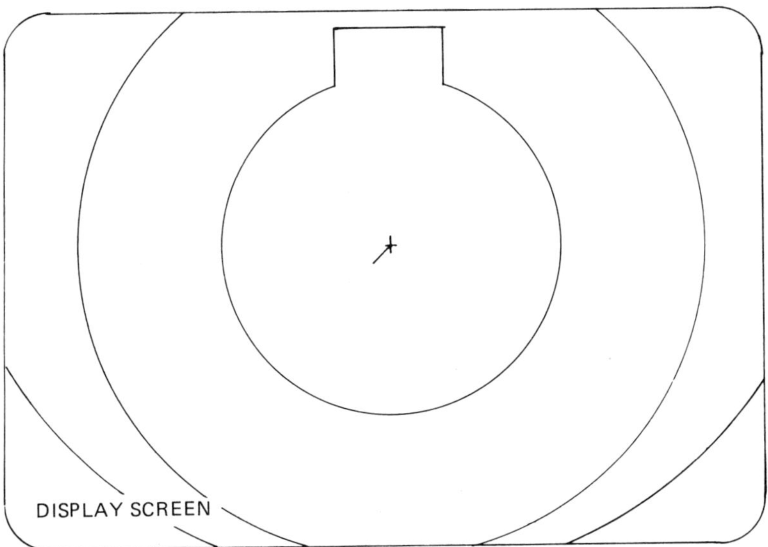

Figure 3.117 The MMC cam is shown at the end of the rotated line.

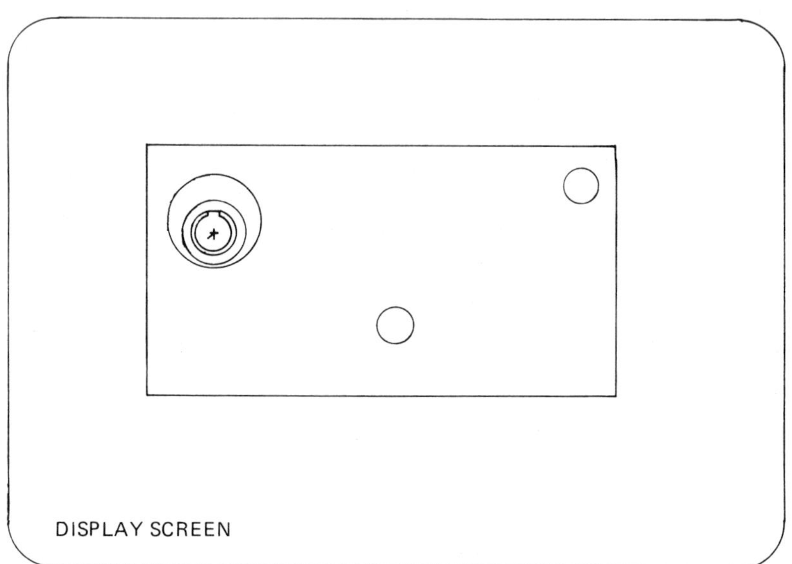

Figure 3.118 The plate and cam are reduced by a factor of ten.

One can imagine performing the same procedure on the rocker arm, push rod, and slider block, but it isn't quite that simple. One must remember that there are interrelationships which must be kept between mating parts through this procedure. For example, the rocker arm must always touch the cam at a tangent point, no matter how the cam and rocker arm are moved and/or rotated.

One's ingenuity is helpful in simplifying this task. For instance, the rocker arm can be simplified into a kinematic linkage composed of two circles with two tangent lines (Fig. 3.119) representing the mating linkages between the cam and the push rod. This simplified kinematic linkage is easily moved and rotated about its position tolerance zone and can be related to the cam and push rod very easily. Also, the holes in the plate and cam are erased with the cam's keyseat, since they do no serve any purpose in this analysis. This type of part geometry simplification eliminates drawing time plus memory space inside the computer. In addition, the drawing appearing on the display screen is easier to understand.

By using the same procedure, the rocker arm can be placed about a point on the circumference of its position tolerance zone. One must remember that the size of the display screen on a computer graphics system is limited; therefore, the operator must enlarge these zones

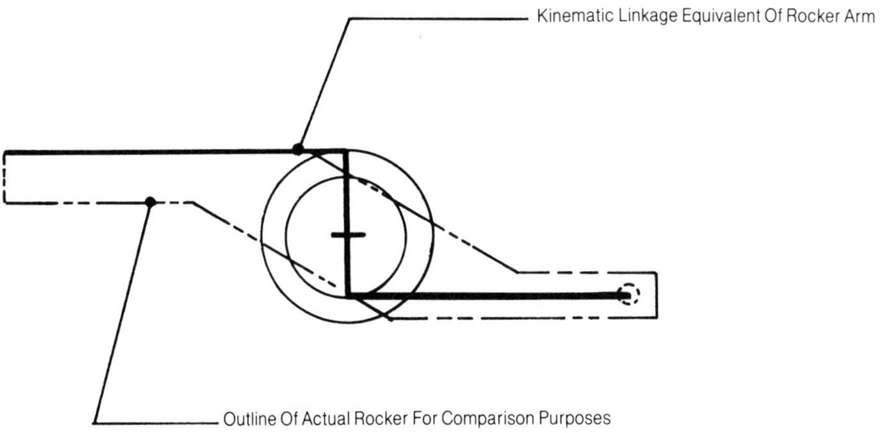

Figure 3.119 Equivalent kinematic linkage of the rocker arm.

and perform the necessary tasks and then reduce them again and repeat this procedure for each zone that is being worked on. It is assumed that the rocker will be simplified into two circles with diameters of 7/8 and 1-1/4 in. placed about a point lying on the circumference of its position tolerance zone by using the previous commands for locating the cam about its point or by constructing the 7/8 and 1-1/4-inch diameter circles about the point.

If the screen is filled with the cam and the circles representing the rocker arm, Figure 3.120, would be the result. Now, a line is drawn from the cam's surface tangent to the circle by using the command PLACE LINE TANGENT TO / UNIT D_1 D_2 CR (Fig. 3.121). This tangent line represents one of the arms attached to the rocker. The push rod is connected to the other arm of the rocker. This arm can be obtained by first, placing a perpendicular line from the end of the previous tangent line. This perpendicular line would pass through the center of the circle, and have as its length the diameter of the circle. The command used for placing this perpendicular line would be PLACE LINE PERPENDICULAR LONG 1.062 / UNIT D_1 INPUT END D_2 D_3 CR. A line perpendicular to this line or parallel to the previous tangent line with a length of 2 inches, which is connected to the push rod, would complete the kinematic drawing of the rocker arm (Fig. 3.122). The command used to perform this task would be similar to the previous command. One must remember that this kinematic linkage of the rocker arm must be constructed whenever the cam is rotated to a new position because of its interrelationship (mating) with the cam.

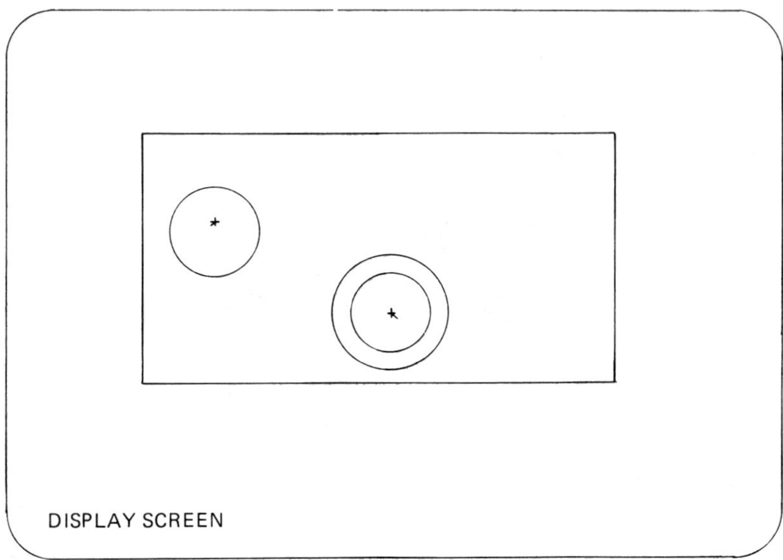

Figure 3.120 Placing the cam and the two circles representing the two arms of the rocker.

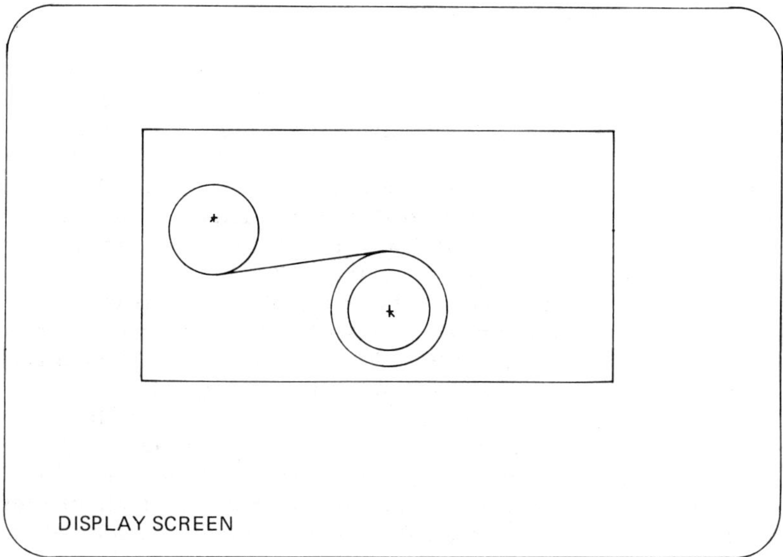

Figure 3.121 Placing a tangent line between the cam and one of the rocker arm's circle.

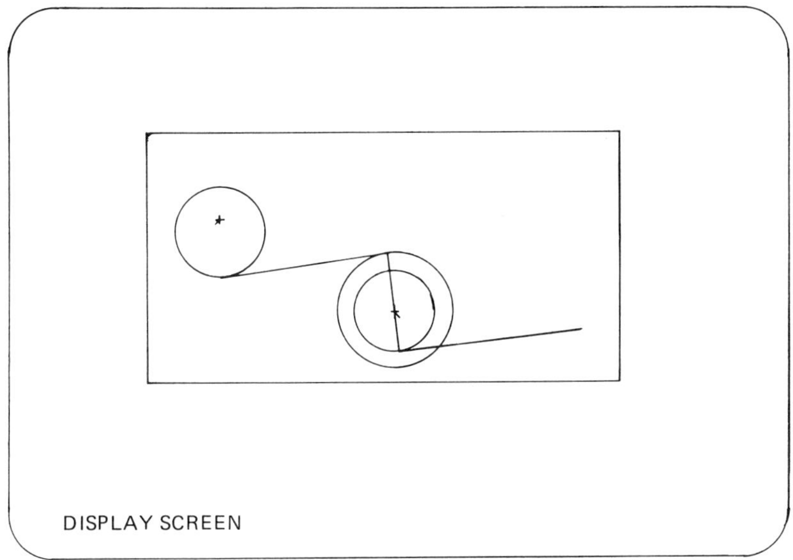

Figure 3.122 Placing the other arm of the rocker from its other circle.

The kinematic linkage drawing of the push rod is a straight line. This 3-inch line is attached to the end of the rocker's arm and would pass through the pivot point of the slider block. This pivot point could lie on the circumference of a circular positional tolerance zone, too. But in this case, it will be placed at the center of the zone. The command used for doing this task is PLACE LINE LONG 3 / INPUT END D_1 INPUT D_2 CR. The result is shown in Figure 3.123. This straight line would have to be constructed, since it is dependent on where the end of the rocker's arm is located. This would be considered the interrelationship (mating) between the rocker arm and the push rod.

The variation of movement of the push rod can be obtained by first establishing a point on the push rod before the cam and rocker arm are placed on the circumferences of their position tolerance zones and before the cam is rotated 180 degrees. This can be done by using the command PLACE POINT / INPUT END D_1 CR (Fig. 3.124). Then the cam and rocker arm are placed on the circumference of their position tolerance zone and the cam is rotated 180 degrees following the previously mentioned procedure and another point is placed on the end of the push rod (Fig. 3.125). The distance between these two points can be considered the output of the push rod. The distance between these two points can be measured by using the command MEASURE DISTANCE / UNIT POINTS D_1 D_2 CR (Fig. 3.126) and the computer

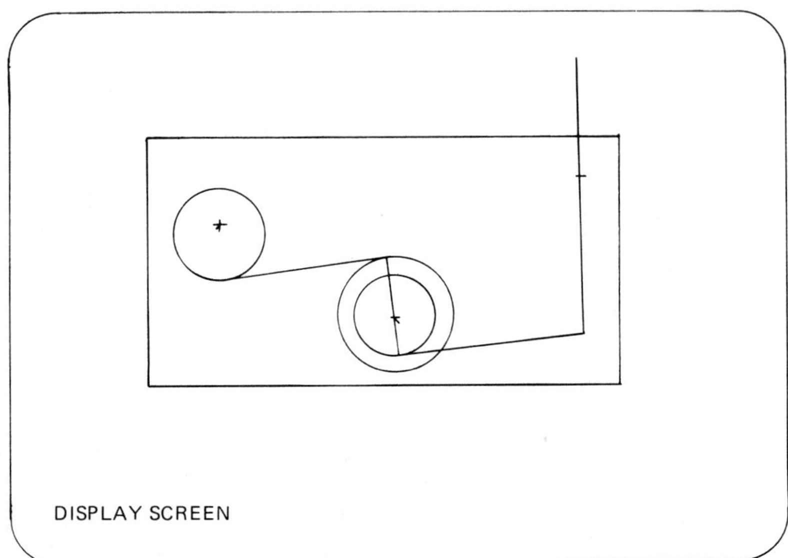

Figure 3.123 Attaching the push rod to the arm of the rocker.

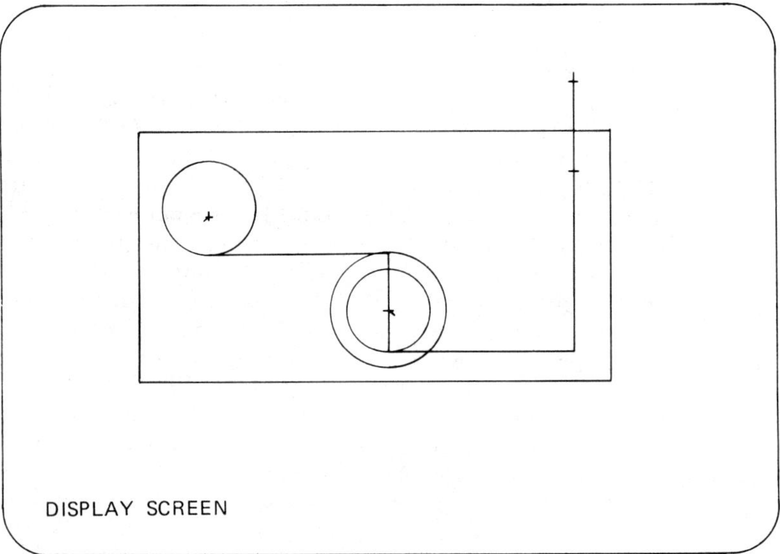

Figure 3.124 Establishing a point on the push rod before the cam and rocker are rotated.

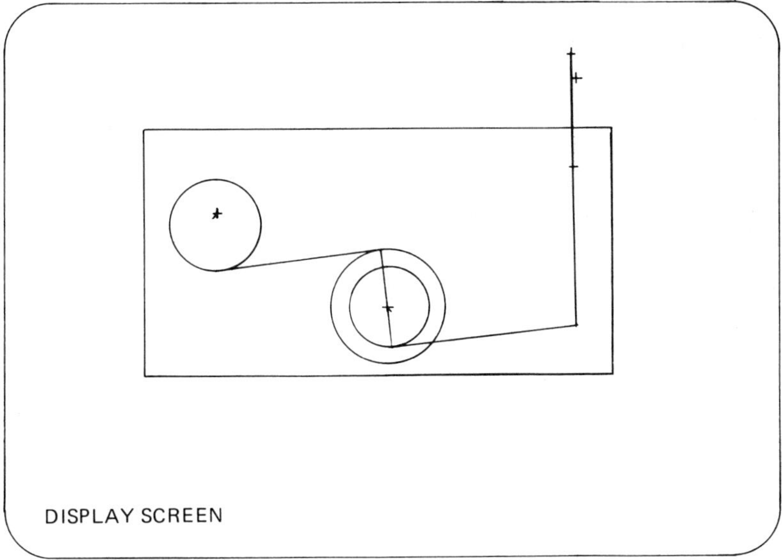

Figure 3.125 Establishing a point on the push rod after the cam and rocker are rotated.

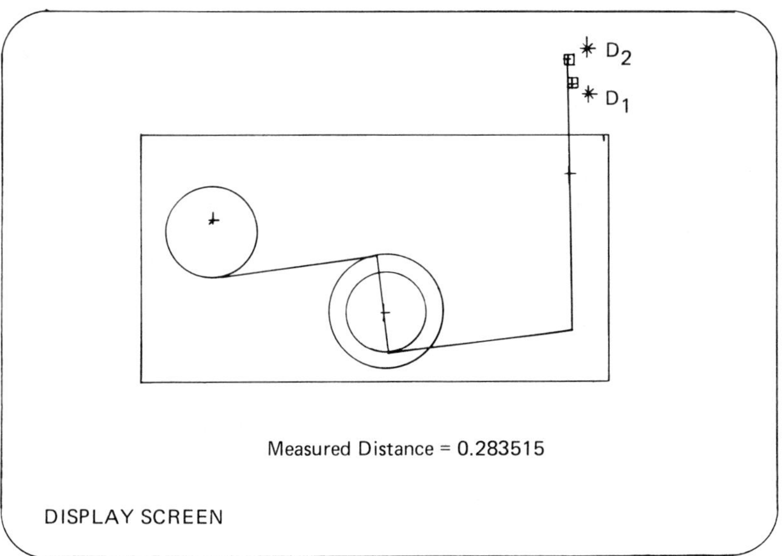

Figure 3.126 Measuring the distance between the two points to obtain the push rod's output.

will measure the distance between these two points with an accuracy of a millionth of an inch, in this case, 0.283515 inches.

This entire process would be repeated for a new set of position tolerances and/or material conditions. The output movement of the push rod for each set of specifications can now be compared in order to determine the greatest possible movement variation of the push rod due to the position tolerances or material conditions or both.

This variation could be critical, especially in an engine where the clearance between the piston and valves is controlled by position tolerances and material conditions of the cams, push rods, and rocker arms. Today it is possible to determine the variations which will occur when mass-produced parts are assembled. Computer graphics is the tool which has made this possible.

Most computer graphics systems have software that can be preprogrammed for analyzing tolerance zones exclusively. For instance, let's imagine the designer would like to be able to analyze a simple tolerance stackup of four parts that are to be assembled inside an enclosure (frame). The four parts mate with each other as shown in Figure 3.127 and parts A and D mate with the enclosure. It is desirable to know the minimum and maximum clearances between parts A, B, C, and D with the enclosure. The manufacturing tolerance is ±0.001 for the parts and ±0.002 for the enclosure. For simplicity, let's assume each part has a nominal dimension of 1.000 inches and the enclosure has a 4.062 nominal dimension in the direction of the stackup. The maximum and minimum clearance occurring between the parts and the enclosure can be easily obtained by using computer graphics.

The maximum (MMC) and minimum or least material conditions (LMC) of each part can be represented by boundaries about each part. Most computer graphics systems have color, so it is very easy to have the maximum material boundary one color (i.e., red) and the minimum material boundary another color (i.e., green). Each part would have red and green as their MMC and LMC boundaries, respectively, enabling the operator to distinguish these boundaries easier on the display

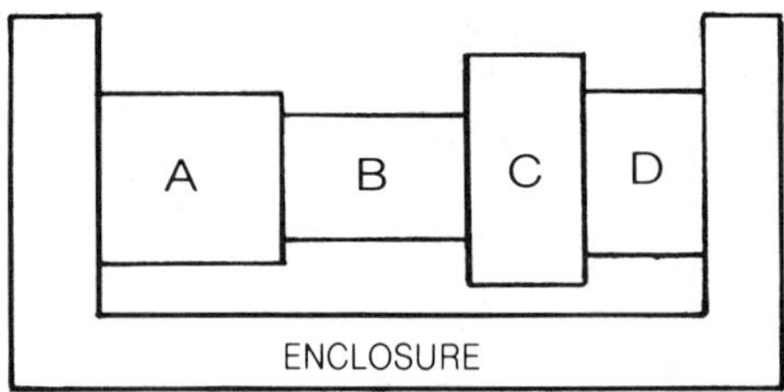

Figure 3.127 Four parts mating with each other and an enclosure.

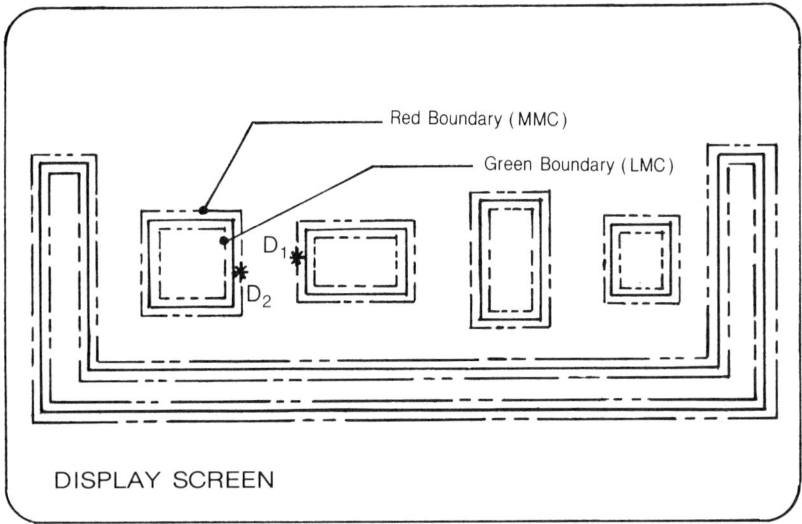

Figure 3.128 The four parts and enclosure appearing on a computer graphics' display screen.

screen. Figure 3.128 shows how these four parts would appear on the display screen with their respective colored MMC and LMC boundaries.

For manipulating parts on the display screen, a touch command would be utilized. Such a command would move parts about on the display screen automatically using a digitizing scheme. For example, if one would like to move part B so that its maximum material boundary touches (mates) with the maximum material boundard (red) of part A, the operator only needs to use the following command TOUCH ANGLE 180 / UNIT D_1 D_2 CR. The entire part with its boundaries will move horizontally so that the two digitized surfaces will occupy the same space or, in other words, touch. The direction of movement of the part is determined by having 0 degrees located at the 3 o'clock position with the angle increasing in the counterclockwise direction. Along this axis, the part will move until its digitized boundary occupies the same space as the second digitized boundary. In this case, part B was moved to the left; therefore, the angle was 180 degrees. The result is shown in Fig. 3.129. The same task is performed on parts C, D, and the enclosure. Now, it is a simple task of measuring the distance between the red boundaries of part D and the enclosure as shown in Fig. 3.130 to obtain the minimum clearance with respect to the maximum size of the enclosure.

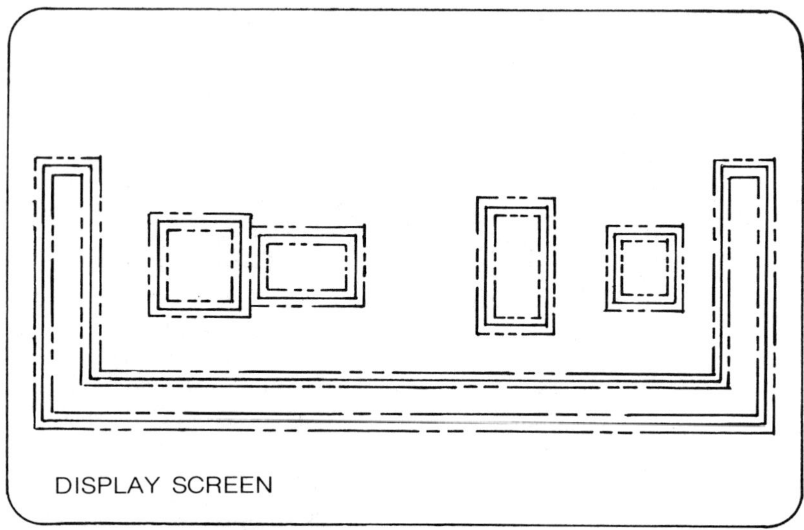

Figure 3.129 Mating two parts.

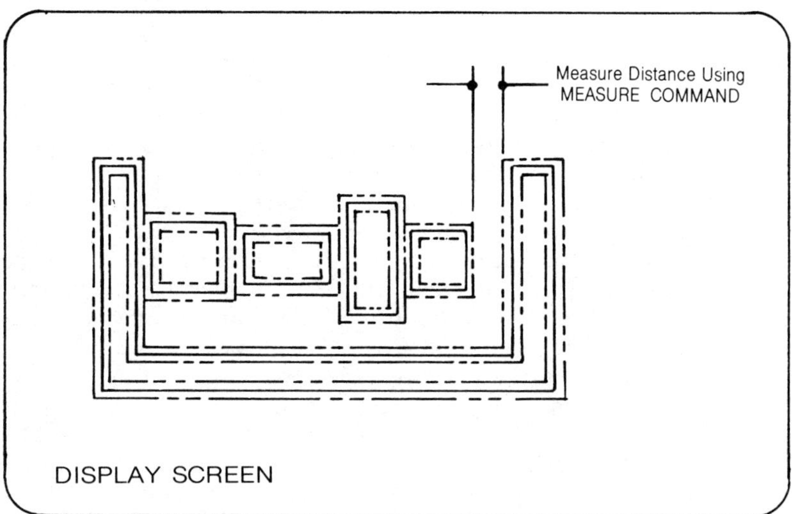

Figure 3.130 Mating four parts and measuring minimum clearance.

Similarly, the minimum clearance between these four parts and the enclosure can be obtained by moving parts A, B, C, D and the enclosure so they touch each other at the green boundaries (LMC). Then measure the distance between the green boundaries of part D and the enclosure.

This information is very useful in determining interchangeability among mating parts being assembled. If the maximum and minimum clearances or interferences are predetermined (engineering allowances) and the manufacturing tolerances known, computer graphics becomes a very powerful tool in determining the dimensions and tolerances of parts so when these parts are assembled they will satisfy function and interchangeability specifications (design integrity).

3.10. WORK PROBLEMS

1. Make working drawings of the parts of the rotary joint shown in Figure 3.131. Dimension according to the form and position dimensioning and tolerancing system, using the appropriate place decimals, datums, etc. Change all fractions to decimals with one place decimals equal to ±0.1, two place decimals equal to ±0.01, and three place decimals equal to ±0.001 inches. Use tables from Chapter 1 for the fits.

2. Make working drawings of the details of the bench threading machine shown in Figure 3.132. Dimension according to the form and position dimensioning and tolerancing system, using the appropriate place decimals, datums, etc. Change all fractions to decimals with one place decimals equal to ±0.1, two place decimals equal to ±0.01, and three place decimals equal to ±0.001 inches. Use tables from Chapter 1 for the fits.

3. Make working drawings of the parts of the double-action valve shown in Figure 3.133. Dimension according to the form and position dimensioning and tolerancing system, using the appropriate place decimals, datums, etc. Change all fractions to decimals with one place decimals equal to ±0.1, two place decimals equal to ±0.01, and three place decimals equal to ±0.001 inches. Use tables from Chapter 1 for the fits.

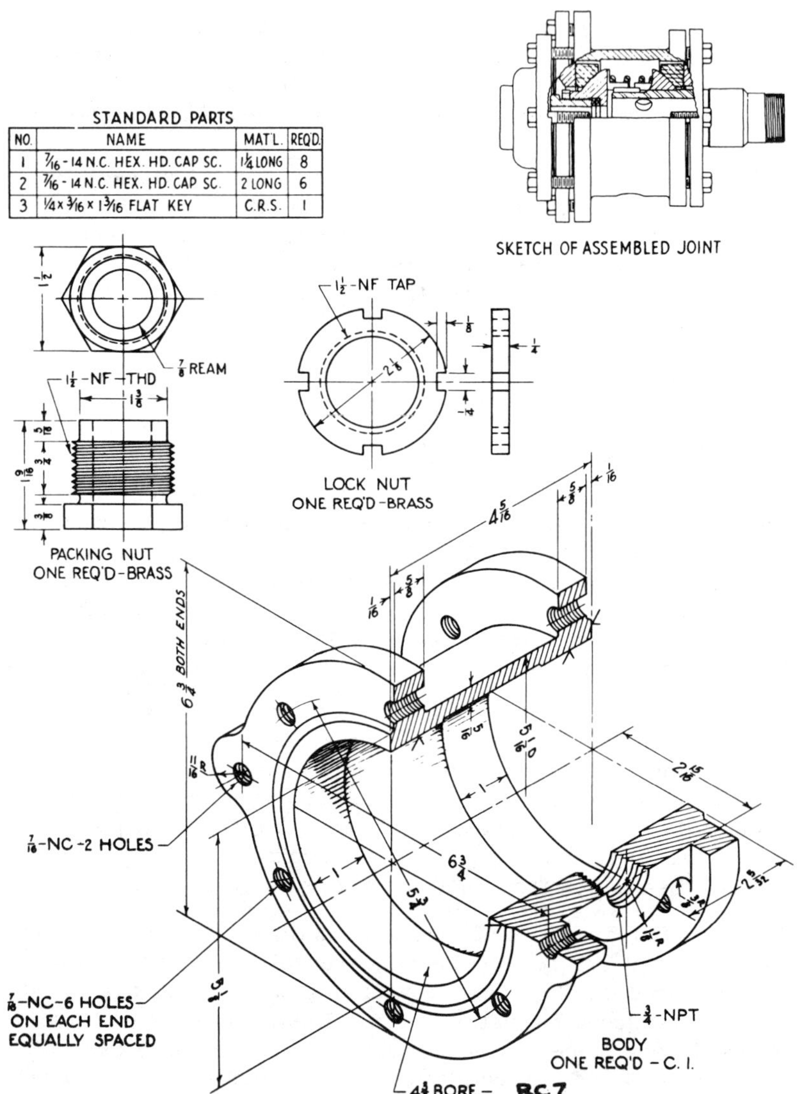

Figure 3.131 Rotary valve. (From H. D. Orth, R. R. Worsencroft, and H. B. Doke, *Theory and Practice of Engineering Drawing,* William C. Brown, Dubuque, Iowa, 1959.)

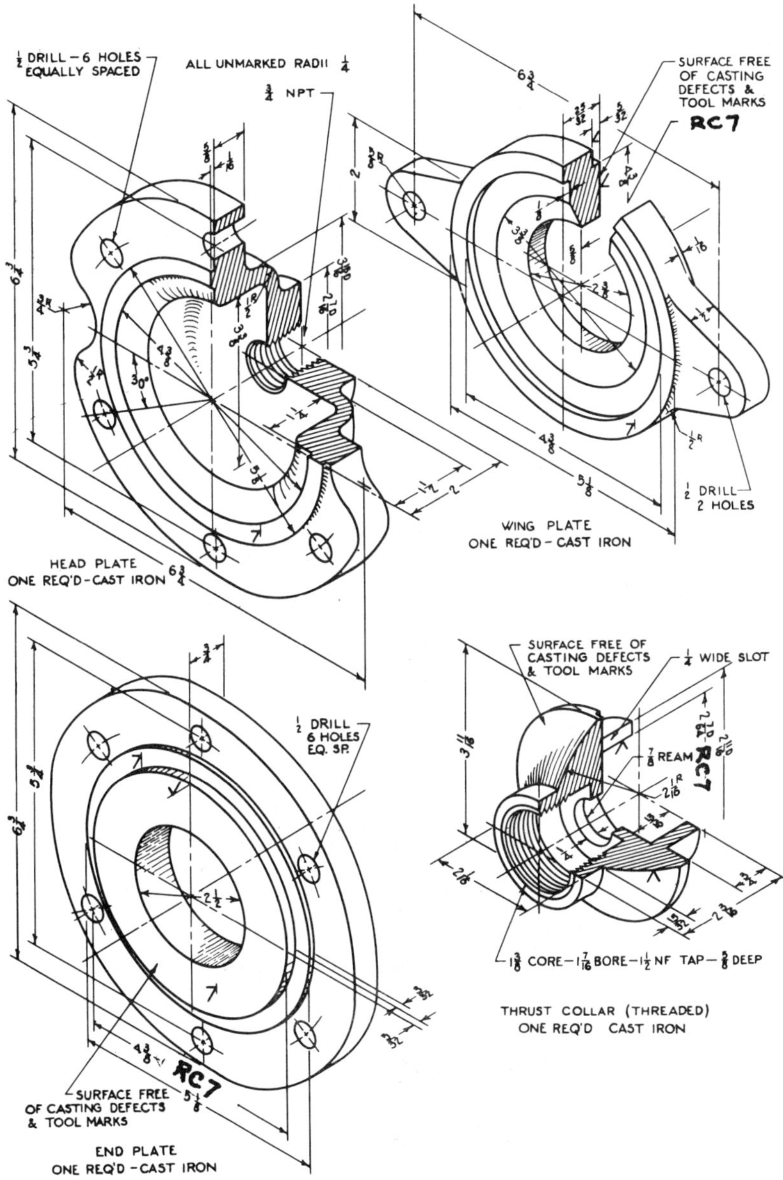

Figure 3.131 (Continued)

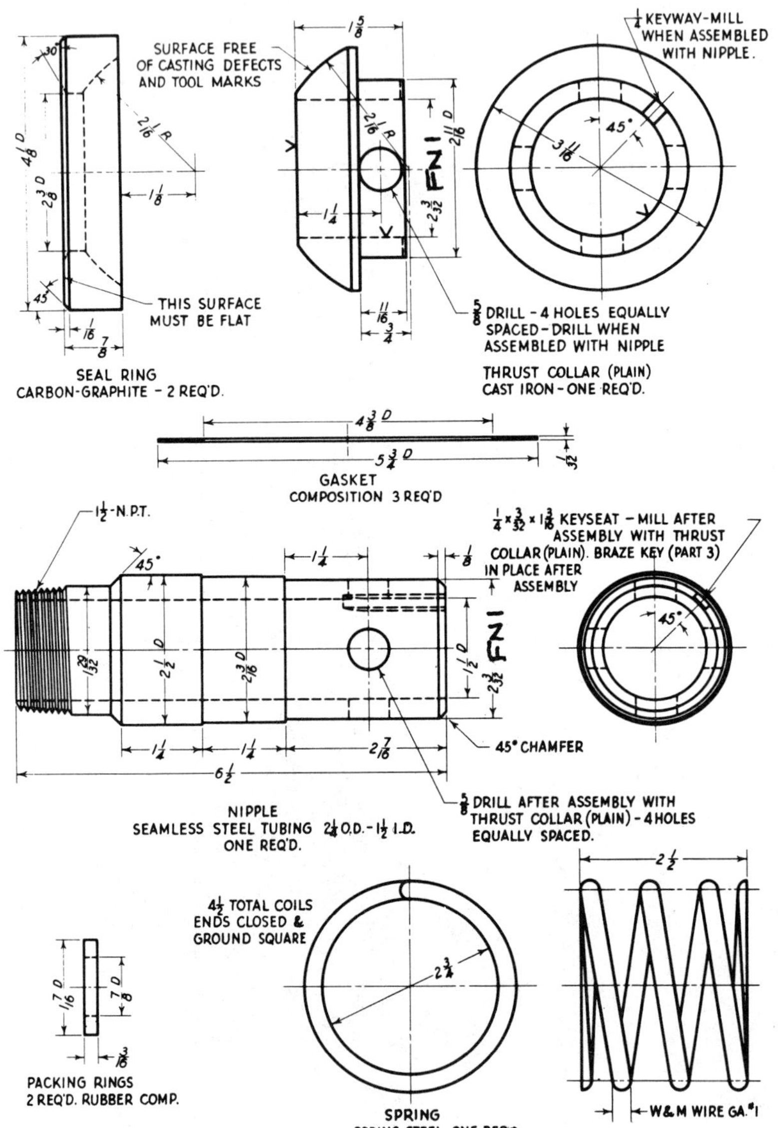

Figure 3.131 (Continued)

NOTES:

1. **The orthographic views shown are not necessarily the proper views to draw.**
2. **Mating surfaces are keyed to assist in assembling. Thus surface A on part No. 1 mates with surface A on part No. 2.**
3. **Provide proper thread notations and limit dimensions from ASA tables where called for.**
4. **Missing dimensions on various details may be found by reference to the part which mates with it.**
5. **Clearance has been provided between Pulleys (Part 8) and Clutch Disc (Part 10) to allow for reversing direction of rotation of the Spindle (Part 7).**

LIST OF PARTS

NO.	NAME	REQ'D	MAT'L	NOTES
1	Base	1	C.I.	
2	Spindle Bear. Bracket	1	C.I.	
3	Allen Socket Cap Sc.	4	Std.	1/2-13NC-1" Long
4	Bushing	2	Bronze	
5	Thrust Bearing	2	Std.	#20BL-SAE
6	Oil Cup	2	Std.	1/8" Pipe thd.
7	Spindle	1	Tool St.	
8	V-Belt Pulley	2	C.I.	
9	Pulley Bushing	2	Bronze	
10	Clutch Disc	1	C.I.	
11	Clutch Lining	2	Therm.	Purchase
12	Lining Rivets	8	Copper	1/8" x 1" Long
13	Hdless Set Sc. Dog Pt.	1	Std.	1" Long
14	Work Holder Slide	1	C.I.	
15	Slide Gib	1	Steel	
16	1/4" Hdless Set Sc.	3	Std.	Flat Pt.-1" Long
17	3/8" Hex. Hd. Cap Sc.	1	Std.	3" Long
18	3/8" Hex. Lock Nut	1	Std.	
19	Linkage Rear Support	1	Steel	
20	Link	1	Steel	
21	Feed Arm	1	Steel	
22	Linkage Screw (1)	1	Steel	
23	Linkage Screw (2)	2	Steel	
24	Handle	1	Steel	
25	Pin	1	Std.	1/8" D.

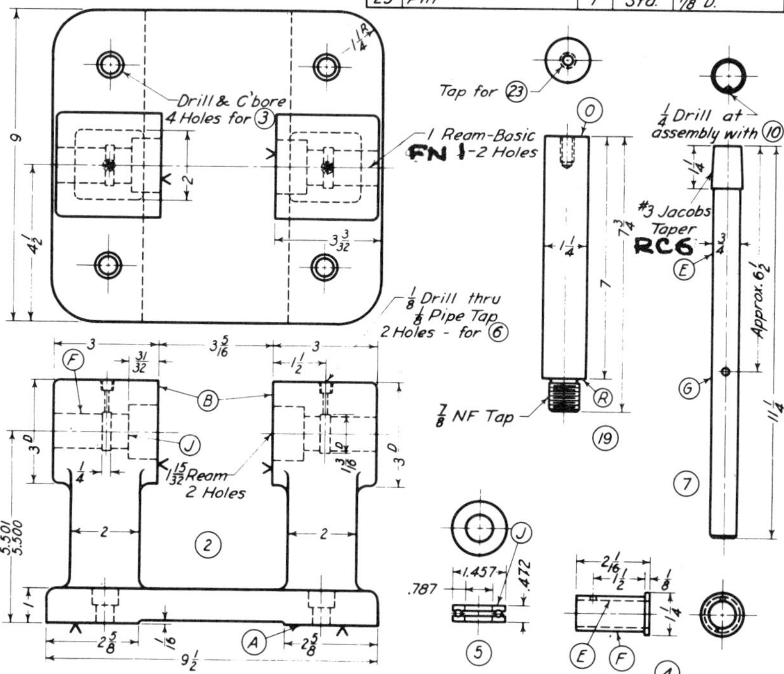

Figure 3.132 Bench threading machine. (From H. D. Orth, R. R. Worsencroft, and H. B. Doke, *Theory and Practice of Engineering Drawing*, William C. Brown, Dubuque, Iowa, 1959.)

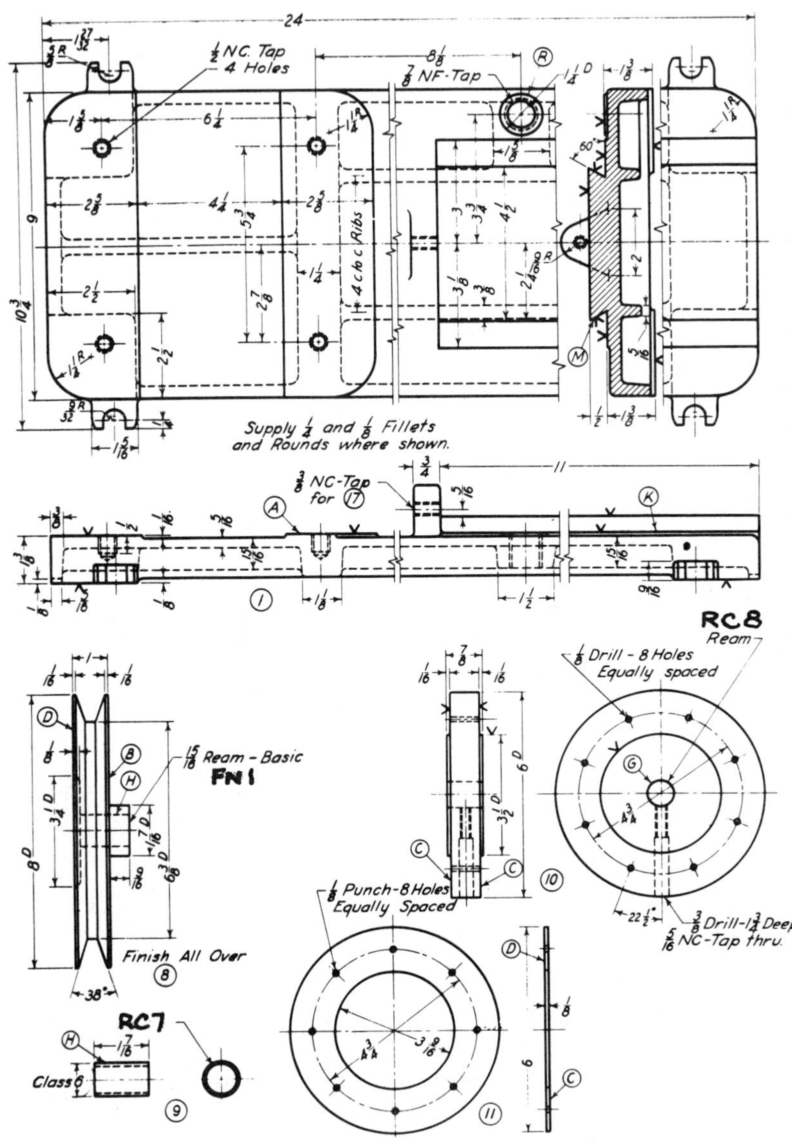

Figure 3.132 (Continued)

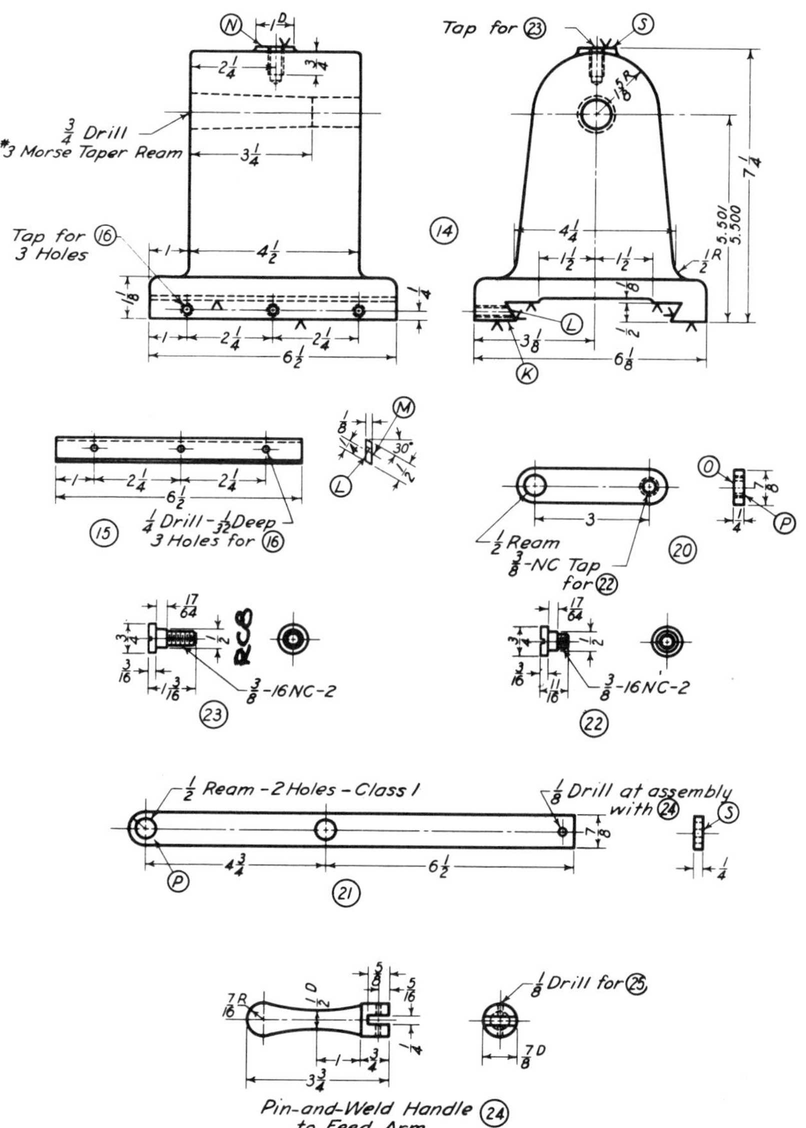

Figure 3.132 (Continued)

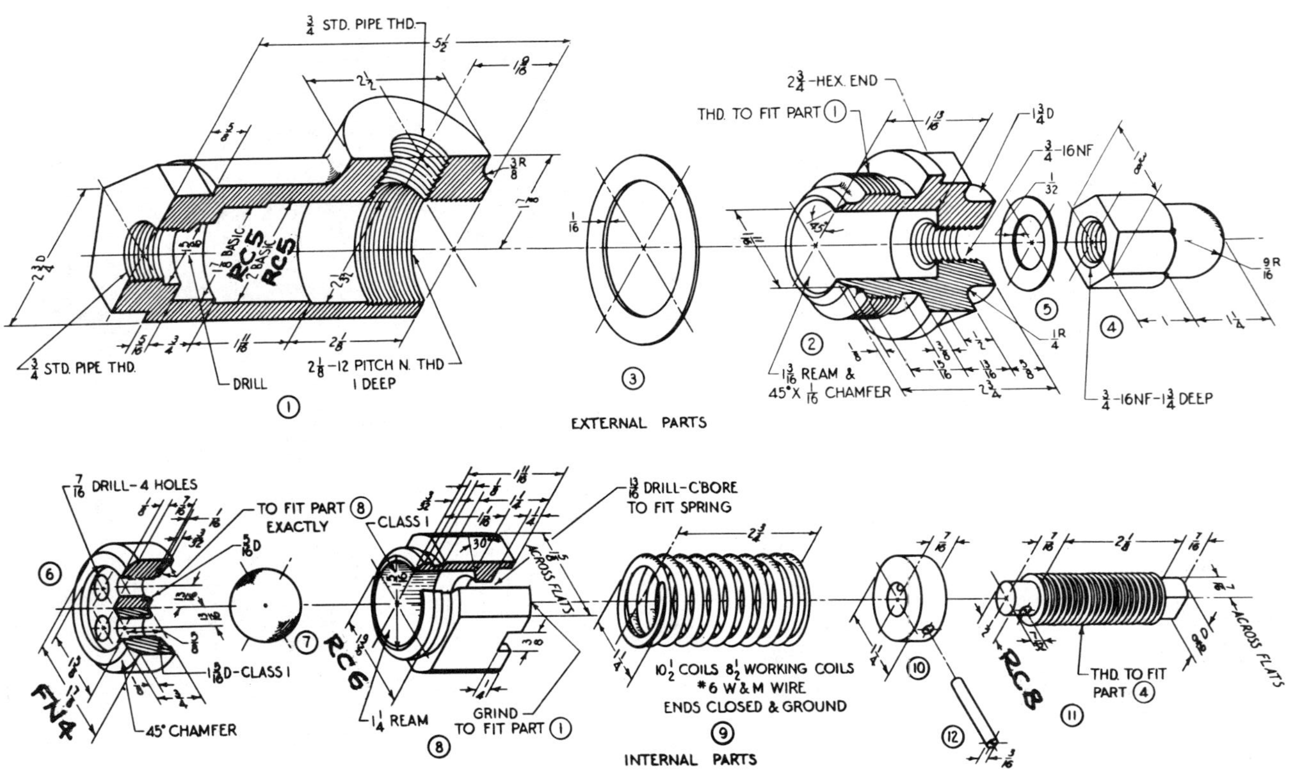

Prob. 16-23.
Make detail working drawings of the parts of the ¾″ Double Action Valve. Provide ASA Standard limit dimensions where fits are specified. Dimensions may not be completely or properly shown on all parts. The draftsman should supply and place them to conform to proper practice.

Prob. 16-24.
Make an assembly drawing of the ¾″ Double Action Valve, either from the details drawn for Prob. 16-23. or the given pictorial views.

NO.	NAME	MAT'L.	NO.	NAME	MAT'L.
1	VALVE BODY	CAST ST.	7	1″ BALL	STD.
2	VALVE CAP	STEEL	8	VALVE	STEEL
3	GASKET	COPPER	9	SPRING	SPR. ST.
4	LOCK NUT	STEEL	10	ADJUST. COLLAR	STEEL
5	GASKET	RUB. C.	11	ADJUST. SCREW	STEEL
6	VALVE SEAT	STEEL	12	PIN	STEEL

Figure 3.133 Double-action valve. (From H. D. Orth, R. R. Worsencroft and H. B. Doke, *Theory and Practice of Engineering Drawing*, William C. Brown, Dubuque, Iowa, 1959.)

4

Determining and Optimizing Dimensions and Their Tolerances

4.1. DIMENSIONING THEORY

For any machine or mechanical device to be commercially successful it must perform its function accurately and satisfactorily. The components, therefore, must operate properly and wear well. In addition, the parts must be manufactured as economically as possible if the device is to be sold and used profitably.

Meeting these requirements is largely the objective of dimensioning. To function properly in the machine, parts must be dimensioned correctly, that is, they must have the necessary dimensions placed on them in the proper positions. To be manufactured economically, however, the numerical values of these dimensions must be given in such a way as to permit as much latitude in their construction as is compatible with the functioning of the parts. For example, the outside contours of a machine require far less numerical accuracy in dimensioning than does the size of a hole in which a shaft revolves at high speed.

4.2. ALLOWABLE VARIATION IN DIMENSIONING (TOLERANCING)

Since no measurement can ever be made exactly to the dimension value that gives its size, the problem arises as to how we can allow for as great a variation as possible from the dimension value during the manufacturing process and yet limit this error where necessary so that parts will mate and function properly in the finished machine or product. This is accomplished by the use of tolerances.

The development of production processes for large-volume manufacturing at low cost has been largely dependent upon interchangeability of component parts. Thus, the designer must determine both the proper tolerances for the individual parts and the correct amount of clearance or interference to permit assembly with mating parts. The designer uses dimensional and tolerance stackups to achieve this purpose.

4.3. DIMENSIONAL AND TOLERANCE STACKUPS

Dimensional and tolerance stackups are methods of calculating clearances or interferences, between mating parts when they are assembled either graphically or mathematically, or both.

The simplest stackup of an assembly of parts is an arithmetic sum of the parts' dimensions and their respective tolerances. The grinder head shown in Figure 4.1a will be used as an example. Suppose the designer chose to check the placement of the shaft seal (drawing no. 400-01) with respect to the spindle's diameter (drawing no. 300-00) to ensure proper sealing. The stackup would consist of the following parts: grinder housing (drawing no. 200-00, Fig. 4.1b), front cap (drawing no. 200-01, Fig. 4.1c), seal (drawing no. 400-01), and spindle (drawing no. 300-00, Fig. 4.1d). The dimensions and their associated tolerances for each part to be used in the stackup are 5.000 ±0.002, 1.500 ±0.005, $0.500^{+0.000}_{-0.020}$, and 6.500 ±0.005, respectively. This stackup is shown graphically in Figure 4.2.

A similar stackup can be done for the locknut (drawing no. 100-03) and front cap (drawing no. 200-01) to determine if they will interfere with each other in assembly. The graphic representation of this stackup is shown in Figure 4.3. This type of stackup depends on the previous stackup which is usually the case of complicated assemblies (assemblies comprised of many parts which are interrelated). It is important to have the same datums used in calculating stackups throughout the assembly; otherwise, it becomes an almost impossible task when calculating stackups of such assemblies. In this case, the left counterbore of the housing (drawing no. 200-00) which locates the rear bearing (drawing no. 100-02) is used as a datum while the centerline shared by the spindle, housing, caps, pulley, etc. is the other datum. These datums establish the x, y, and z directions of the various stackups that will be performed on the assembly.

Usually, stackups are performed by a designer before the part drawings of the individual parts of the assembly are drawn by the drafter. These stackups ensure proper assembly and function of the parts when they are placed together as a unit and the dimensions and tolerances on the part drawing reflect these stackups. At this stage, there is

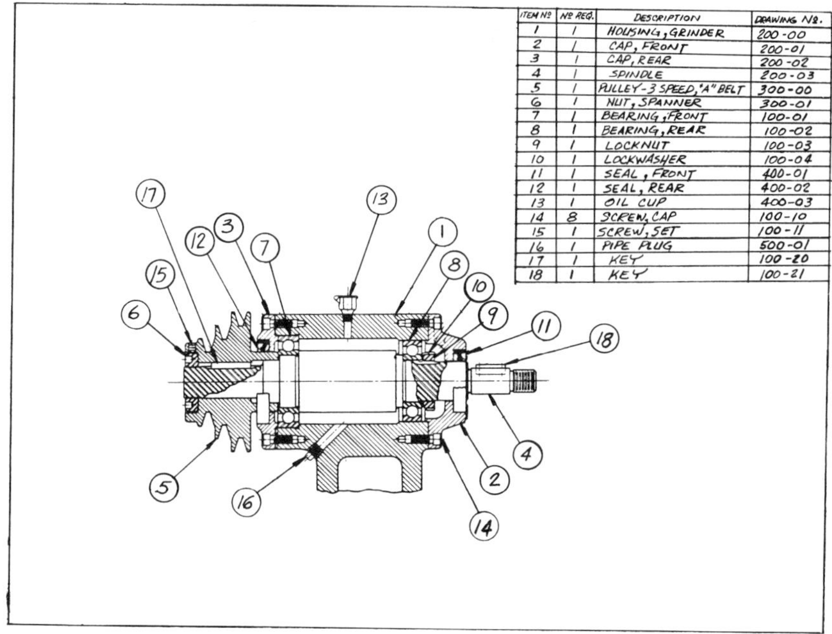

ITEM №	№ REQ.	DESCRIPTION	DRAWING №.
1	1	HOUSING, GRINDER	200-00
2	1	CAP, FRONT	200-01
3	1	CAP, REAR	200-02
4	1	SPINDLE	200-03
5	1	PULLEY-3 SPEED, "A" BELT	300-00
6	1	NUT, SPANNER	300-01
7	1	BEARING, FRONT	100-01
8	1	BEARING, REAR	100-02
9	1	LOCKNUT	100-03
10	1	LOCKWASHER	100-04
11	1	SEAL, FRONT	400-01
12	1	SEAL, REAR	400-02
13	1	OIL CUP	400-03
14	8	SCREW, CAP	100-10
15	1	SCREW, SET	100-11
16	1	PIPE PLUG	500-01
17	1	KEY	100-20
18	1	KEY	100-21

Figure 4.1a Grinder head.

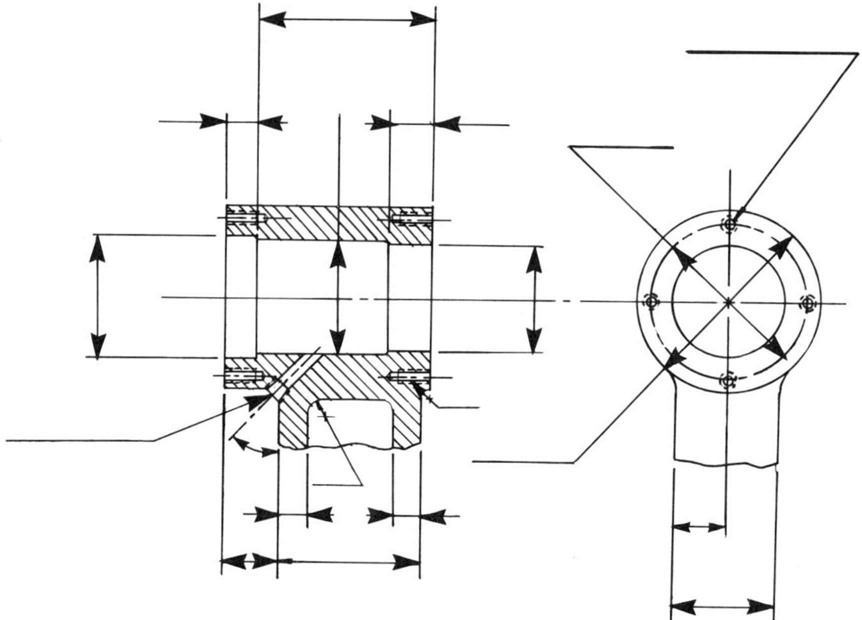

Figure 4.1b Grinder housing (drawing no. 200-00).

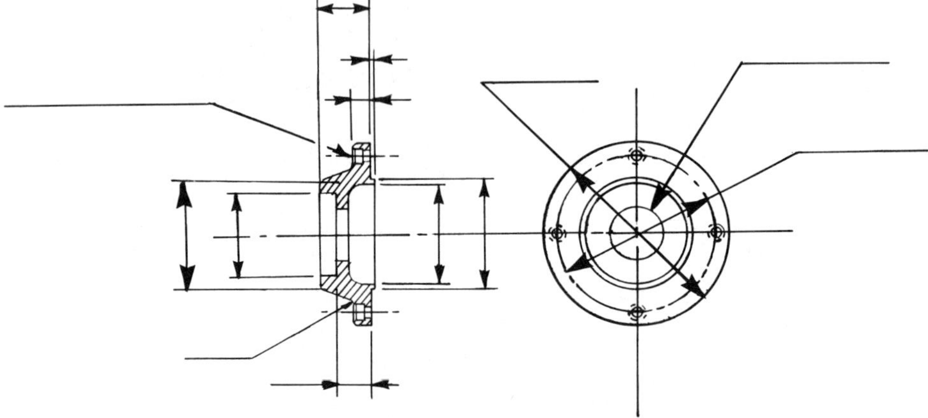

Figure 4.1c Front cap (drawing no. 200-01).

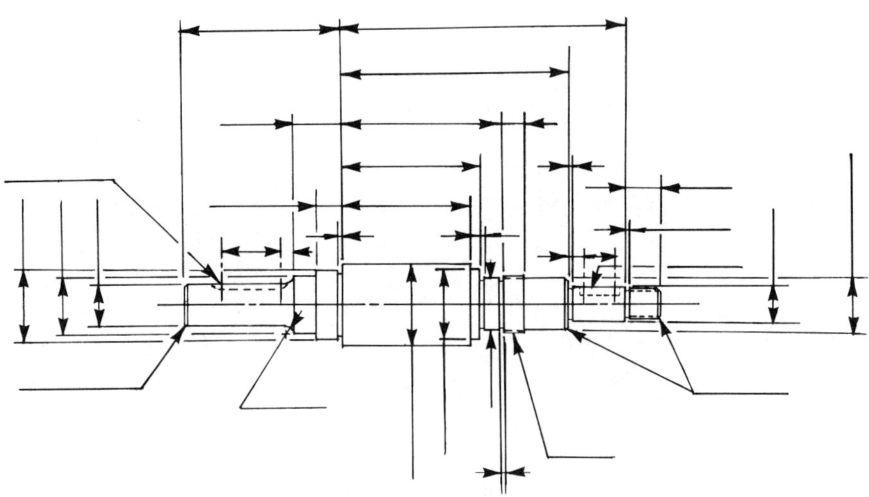

Figure 4.1d Spindle (drawing no. 300-00).

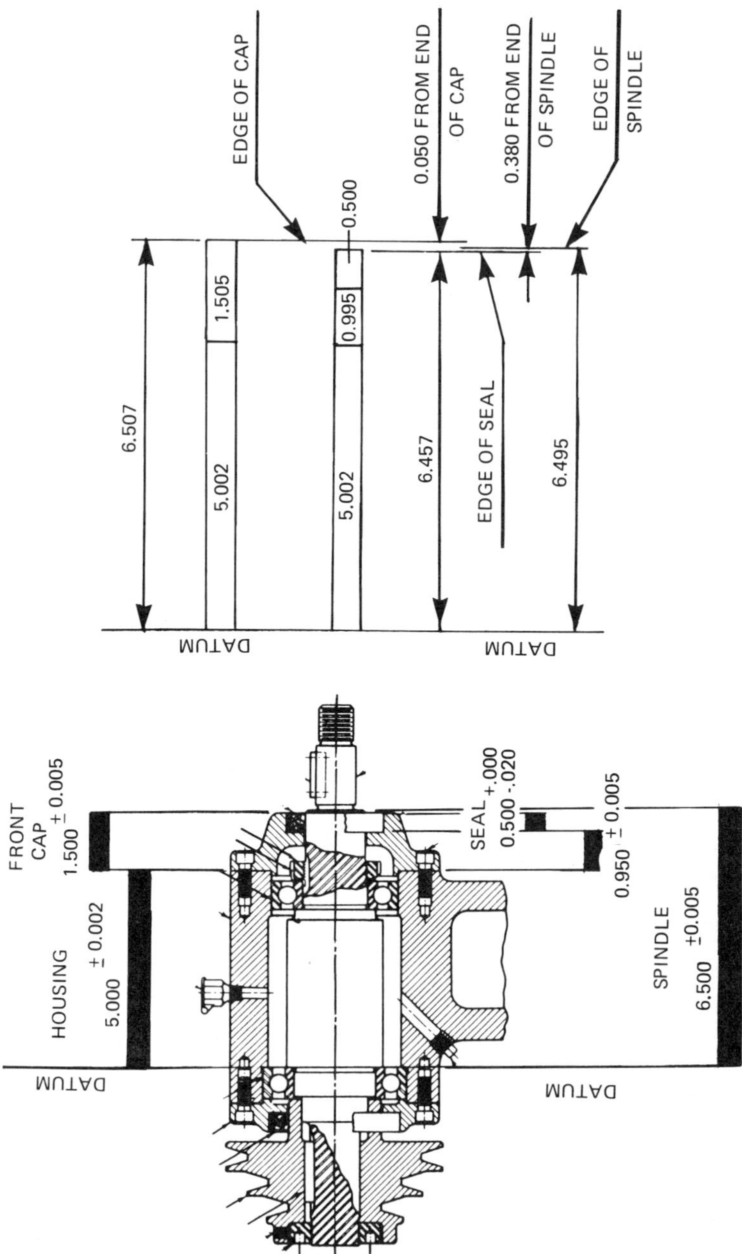

Figure 4.2 Stackup.

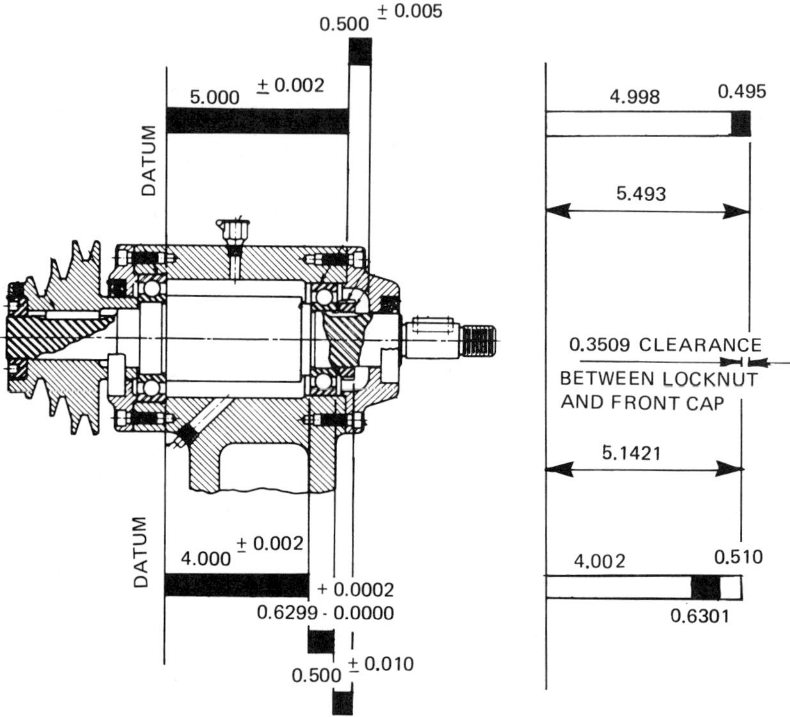

Figure 4.3 Stackup.

little information or data available concerning the distribution of the parts' dimensions and tolerances due to manufacturing. This information must be gathered by manufacturing people in the shop and relayed to the designer via engineering and manufacturing standards.

These engineering and manufacturing standards are usually established by using some form of statistical analysis of each manufacturing process in the particular shop. The following section discusses an approach to optimizing dimensions and their associated tolerances by using a normal distribution.

4.4. OPTIMIZING DIMENSIONS AND THEIR TOLERANCES

The laws of chance apply to the size variations encountered when parts are made in mass production. Measurements of parts from a random sample will usually establish a pattern which will be typical of the entire production run. Statistically, this pattern is known as a normal distribution whose graphical representation is a bell-shaped curve, it is then possible to anticipate the number of parts that will be acceptable and the number of parts to be scrapped or reworked.

Therefore, the dimensions and tolerances placed on part drawings by the designer must be such that they can be obtained by the manufacturing process involved in producing the particular part. When equipment cannot meet specified tolerances, cost is added in the form of scrap, rework, or both.

The equation for the normal distribution or normal curve is given as

$$f(z) = \left(\frac{1}{\sqrt{2\pi\, e}}\right)^{-z^2/2} = 0.3989\ e^{-z^2/2}$$

where $\pi = 3.14159$, $e = 2.71828$, $z = (x - \bar{x})/\sigma$, x is the actual measurement form the random sample, $\bar{x}$ is the mean of the normal distribution, and σ is the standard deviation of the normal distribution. The variable z is called the standardized normal value and the upper z is equal to the upper tolerance limit minus $\bar{x}$ of the sample divided by the sample's standard deviation. The lower z is equal to the lower tolerance limit minus $\bar{x}$ of the sample divided by the sample's standard deviation. Figure 4.4 shows the normal curve in its standard form, which is with a mean of zero and a standard deviation of 1. It is noted that the curve is asymptotic at $z = -3$ and $z = +3$.

The area under the curve is equal to 1.00 or 100%, and therefore, can be easily used for probability calculations. Since the area under the curve between various points is a very useful statistical tool, a normal area table is provided (Table 4.1).

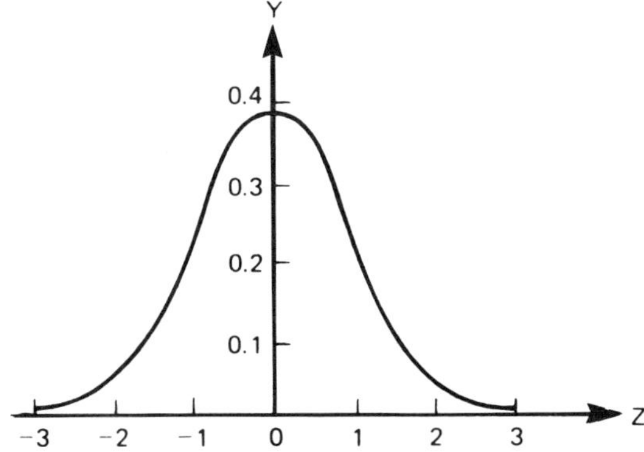

Figure 4.4 A normal curve with six standard deviations.

Table 4.1 Area Under the Normal Curve of 6 Standard Deviations

$\frac{X_i - \bar{X}'}{\sigma'}$	0.00	0.01	0.02	0.03	0.04	0.05	0.06	0.07	0.08	0.09
+0.0	0.5000	0.5040	0.5080	0.5120	0.5160	0.5199	0.5239	0.5279	0.5319	0.5359
+0.1	0.5398	0.5438	0.5478	0.5517	0.5557	0.5596	0.5636	0.5675	0.5714	0.5753
+0.2	0.5793	0.5832	0.5871	0.5910	0.5948	0.5987	0.6026	0.6064	0.6103	0.6141
+0.3	0.6179	0.6217	0.6255	0.6293	0.6331	0.6368	0.6406	0.6443	0.6480	0.6517
+0.4	0.6554	0.6591	0.6628	0.6664	0.6700	0.6736	0.6772	0.6808	0.6844	0.6879
+0.5	0.6915	0.6950	0.6985	0.7019	0.7054	0.7088	0.7123	0.7157	0.7190	0.7224
+0.6	0.7257	0.7291	0.7324	0.7357	0.7389	0.7422	0.7454	0.7486	0.7517	0.7549
+0.7	0.7580	0.7611	0.7642	0.7673	0.7704	0.7734	0.7764	0.7794	0.7823	0.7852
+0.8	0.7881	0.7910	0.7939	0.7967	0.7995	0.8023	0.8051	0.8079	0.8106	0.8133
+0.9	0.8159	0.8186	0.8212	0.8238	0.8264	0.8289	0.8315	0.8340	0.8365	0.8389
+1.0	0.8413	0.8438	0.8461	0.8485	0.8508	0.8531	0.8554	0.8577	0.8599	0.8621
+1.1	0.8643	0.8665	0.8686	0.8708	0.8729	0.8749	0.8770	0.8790	0.8810	0.8830
+1.2	0.8849	0.8869	0.8888	0.8907	0.8925	0.8944	0.8962	0.8980	0.8997	0.9015
+1.3	0.9032	0.9049	0.9066	0.9082	0.9099	0.9115	0.9131	0.9147	0.9162	0.9177
+1.4	0.9192	0.9207	0.9222	0.9236	0.9251	0.9265	0.9279	0.9292	0.9306	0.9319
+1.5	0.9332	0.9345	0.9357	0.9370	0.9382	0.9394	0.9406	0.9418	0.9429	0.9441
+1.6	0.9452	0.9463	0.9474	0.9484	0.9495	0.9505	0.9515	0.9525	0.9535	0.9545
+1.7	0.9554	0.9564	0.9573	0.9582	0.9591	0.9599	0.9608	0.9616	0.9625	0.9633
+1.8	0.9641	0.9649	0.9656	0.9664	0.9671	0.9678	0.9686	0.9693	0.9699	0.9706
+1.9	0.9713	0.9719	0.9726	0.9732	0.9738	0.9744	0.9750	0.9756	0.9761	0.9767
+2.0	0.9773	0.9778	0.9783	0.9788	0.9793	0.9798	0.9803	0.9808	0.9812	0.9817
+2.1	0.9821	0.9826	0.9830	0.9834	0.9838	0.9842	0.9846	0.9850	0.9854	0.9857
+2.2	0.9861	0.9864	0.9868	0.9871	0.9875	0.9878	0.9881	0.9884	0.9887	0.9890
+2.3	0.9893	0.9896	0.9898	0.9901	0.9904	0.9906	0.9909	0.9911	0.9913	0.9916
+2.4	0.9918	0.9920	0.9922	0.9925	0.9927	0.9929	0.9931	0.9932	0.9934	0.9936
+2.5	0.9938	0.9940	0.9941	0.9943	0.9945	0.9946	0.9948	0.9949	0.9951	0.9952
+2.6	0.9953	0.9955	0.9956	0.9957	0.9959	0.9960	0.9961	0.9962	0.9963	0.9964
+2.7	0.9965	0.9966	0.9967	0.9968	0.9969	0.9970	0.9971	0.9972	0.9973	0.9974
+2.8	0.9974	0.9975	0.9976	0.9977	0.9977	0.9978	0.9979	0.9979	0.9980	0.9981
+2.9	0.9981	0.9982	0.9983	0.9983	0.9984	0.9984	0.9985	0.9985	0.9986	0.9986
+3.0	0.99865	0.99869	0.99874	0.99878	0.99882	0.99886	0.99889	0.99893	0.99896	0.99900
+3.1	0.99903	0.99906	0.99910	0.99913	0.99915	0.99918	0.99921	0.99924	0.99926	0.99929
+3.2	0.99931	0.99934	0.99936	0.99938	0.99940	0.99942	0.99944	0.99946	0.99948	0.99950
+3.3	0.99952	0.99953	0.99955	0.99957	0.99958	0.99960	0.99961	0.99962	0.99964	0.99965
+3.4	0.99966	0.99967	0.99969	0.99970	0.99971	0.99972	0.99973	0.99974	0.99975	0.99976
+3.5	0.99977	0.99978	0.99978	0.99979	0.99980	0.99981	0.99981	0.99982	0.99983	0.99983

An example of a production process is the best way to show a statistical analysis of dimensions and tolerances. Assume a bushing is to be press fitted into a hole in a waterpump housing. The impellor shaft is held in place by this bushing. Each bushing is to be force fitted into a hole with a class of fit of 2. The limits of interference vary from −0.007 to −0.0019 of an inch. The negative sign indicates an interference fit while a positive sign indicates a clearance fit. The

Table 4.1 (Continued)

$\frac{X_i-\bar{X}'}{\sigma'}$	0.09	0.08	0.07	0.06	0.05	0.04	0.03	0.02	0.01	0.00
−3.5	0.00017	0.00017	0.00018	0.00019	0.00019	0.00020	0.00021	0.00022	0.00022	0.00023
−3.4	0.00024	0.00025	0.00026	0.00027	0.00028	0.00029	0.00030	0.00031	0.00033	0.00034
−3.3	0.00035	0.00036	0.00038	0.00039	0.00040	0.00042	0.00043	0.00045	0.00047	0.00048
−3.2	0.00050	0.00052	0.00054	0.00056	0.00058	0.00060	0.00062	0.00064	0.00066	0.00069
−3.1	0.00071	0.00074	0.00076	0.00079	0.00082	0.00085	0.00087	0.00090	0 00094	0.00097
−3.0	0.00100	0.00104	0.00107	0.00111	0.00114	0.00118	0.00122	0.00126	0.00131	0.00135
−2.9	0.0014	0.0014	0.0015	0.0015	0.0016	0.0016	0.0017	0.0017	0.0018	0.0019
−2.8	0.0019	0.0020	0.0021	0.0021	0.0022	0.0023	0.0023	0.0024	0.0025	0.0026
−2.7	0.0026	0.0027	0.0028	0.0029	0.0030	0.0031	0.0032	0.0033	0.0034	0.0035
−2.6	0.0036	0.0037	0.0038	0.0039	0.0040	0.0041	0.0043	0.0044	0.0045	0.0047
−2.5	0.0048	0.0049	0.0051	0.0052	0.0054	0.0055	0.0057	0.0059	0.0060	0.0062
−2.4	0.0064	0.0066	0.0068	0.0069	0.0071	0.0073	0.0075	0.0078	0.0080	0.0082
−2.3	0.0084	0.0087	0.0089	0.0091	0.0094	0.0096	0.0099	0.0102	0.0104	0.0107
−2.2	0.0110	0.0113	0.0116	0.0119	0.0122	0.0125	0.0129	0.0132	0.0136	0.0139
−2.1	0.0143	0.0146	0.0150	0.0154	0.0158	0.0162	0.0166	0.0170	0.0174	0.0179
−2.0	0.0183	0.0188	0.0192	0.0197	0.0202	0.0207	0.0212	0.0217	0.0222	0.0228
−1.9	0.0233	0.0239	0.0244	0.0250	0.0256	0.0262	0.0268	0.0274	0.0281	0.0287
−1.8	0.0294	0.0301	0.0307	0.0314	0.0322	0.0329	0.0336	0.0344	0.0351	0.0359
−1.7	0.0367	0.0375	0.0384	0.0392	0.0401	0.0409	0.0418	0.0427	0.0436	0.0446
−1.6	0.0455	0.0465	0.0475	0.0485	0.0495	0.0505	0.0516	0.0526	0.0537	0.0548
−1.5	0.0559	0.0571	0.0582	0.0594	0.0606	0.0618	0.0630	0.0643	0.0655	0.0668
−1.4	0.0681	0.0694	0.0708	0.0721	0.0735	0.0749	0.0764	0.0778	0.0793	0.0808
−1.3	0.0823	0.0838	0.0853	0.0869	0.0885	0.0901	0.0918	0.0934	0.0951	0.0968
−1.2	0.0895	0.1003	0.1020	0.1038	0.1057	0.1075	0.1093	0.1112	0.1131	0.1151
−1.1	0.1170	0.1190	0.1210	0.1230	0.1251	0.1271	0.1292	0.1314	0.1335	0.1357
−1.0	0.1379	0.1401	0.1423	0.1446	0.1469	0.1492	0.1515	0.1539	0.1562	0.1587
−0.9	0.1611	0.1635	0.1660	0.1685	0.1711	0.1736	0.1762	0.1788	0.1814	0.1841
−0.8	0.1867	0.1894	0.1922	0.1949	0.1977	0.2005	0.2033	0.2061	0.2090	0.2119
−0.7	0.2148	0.2177	0.2207	0.2236	0.2266	0.2297	0.2327	0.2358	0.2389	0.2420
−0.6	0.2451	0.2483	0.2514	0.2546	0.2578	0.2611	0.2643	0.2676	0.2709	0.2743
−0.5	0.2776	0.2810	0.2843	0.2877	0.2912	0.2946	0.2981	0.3015	0.3050	0.3085
−0.4	0.3121	0.3156	0.3192	0.3228	0.3264	0.3300	0.3336	0.3372	0.3409	0.3446
−0.3	0.3483	0.3520	0.3557	0.3594	0.3632	0.3669	0.3707	0.3745	0.3783	0.3821
−0.2	0.3859	0.3897	0.3936	0.3974	0.4013	0.4052	0.4090	0.4129	0.4168	0.4207
−0.1	0.4247	0.4286	0.4325	0.4364	0.4404	0.4443	0.4483	0.4522	0.4562	0.4602
−0.0	0.4641	0.4681	0.4721	0.4761	0.4801	0.4840	0.4880	0.4920	0.4960	0.5000

bushing is used as the starting point (basic shaft system) in determining the tolerances.

The bushing is manufactured using a different process than the hole and therefore the bushing has a tolerance range of 0.0007 of an inch while the hole has a tolerance range of 0.0005 of an inch. The design size of the bushing is 0.7500 inches in diameter. Figure 4.5 shows this bushing and its hole with their respective upper and lower limits.

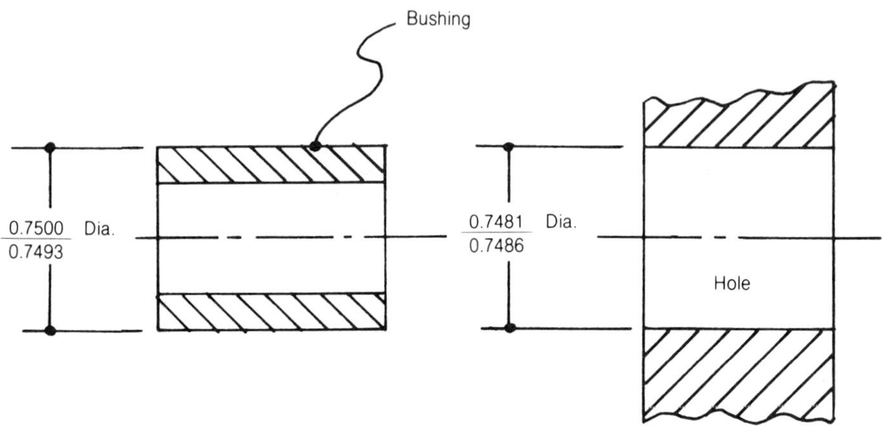

Figure 4.5 A bushing and a hole.

If a lot of 500 bushings are produced, the immediate question is what probable percentage of scrap or rework will arise from manufacturing. A random sample of 50 bushings are taken from the production run and each bushing's outside diameter is measured. These measurements are shown in Table 4.2. This random sample will be used to represent the entire production run and the manufacturing capabilities of the machine used in machining these bushings.

If the production process followed the laws of chance, the dimensions would show a normal distribution curve. The random sample measurements are plotted on the x axis and the number of parts per measure-

Table 4.2 List of the Measurements of 50 Randomly Picked Bushings from a Production Run of 500 Bushings

.7496	.7498	.7495	.7496	.7500
.7499	.7494	.7499	.7494	.7496
.7497	.7496	.7498	.7496	.7495
.7495	.7498	.7497	.7497	.7495
.7492	.7496	.7497	.7493	.7499
.7497	.7494	.7493	.7495	.7496
.7494	.7496	.7495	.7498	.7494
.7496	.7498	.7497	.7494	.7493
.7497	.7496	.7492	.7498	.7497

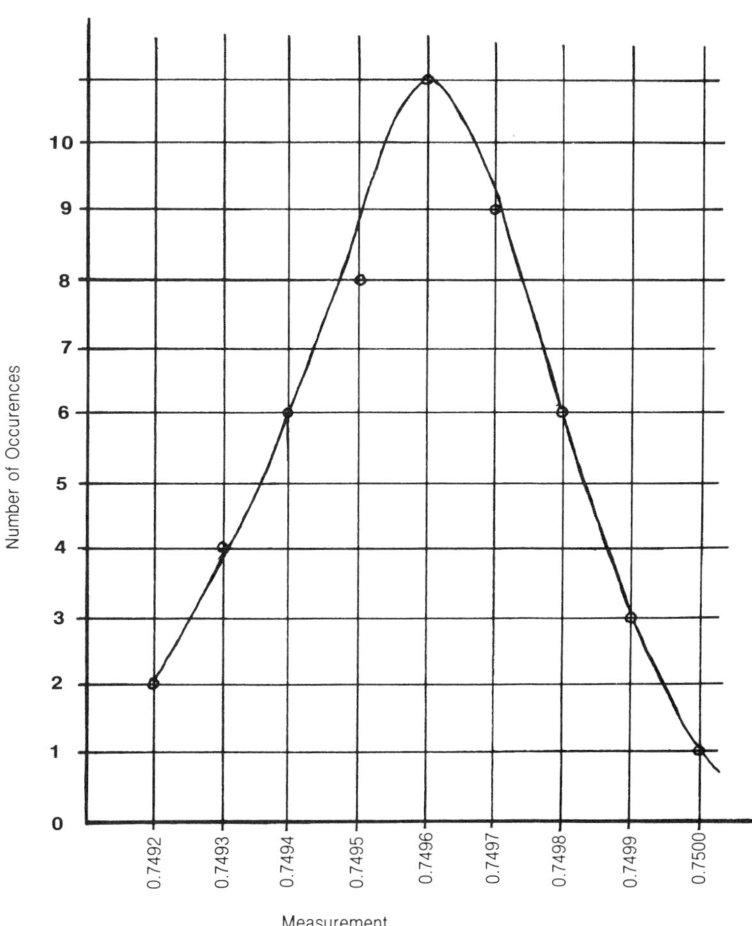

Figure 4.6 Normal curve associated with the production of 500 bushings.

ment are plotted along the y axis. This bell-shaped curve (Fig 4.6) is known as the normal curve associated with this production run. The mean $\bar{x}$ and standard deviation σ of this normal curve are calculated. Table 4.3 contains the information required to calculate the mean and standard deviation for the measurements appearing in Table 4.2.

Table 4.3 Calculating the Mean and Standard Deviation of the 50 Randomly Picked Bushings

Measurement (m)	Frequency (f)	m·f	x-$\bar{x}$	$(x-\bar{x})^2$	$m\cdot(x-\bar{x})^2$
0.7492	2	1.4984	−0.0004	16×10^{-8}	32×10^{-8}
0.7493	4	2.9972	−0.0003	9×10^{-8}	36×10^{-8}
0.7494	6	4.4964	−0.0002	4×10^{-8}	24×10^{-8}
0.7495	8	5.9960	−0.0001	1×10^{-8}	8×10^{-8}
0.7496	11	8.2456	−0.0000	0	0
0.7497	9	6.7473	+0.0001	1×10^{-8}	9×10^{-8}
0.7498	6	4.4988	+0.0002	4×10^{-8}	24×10^{-8}
0.7499	3	2.2497	+0.0003	9×10^{-8}	
0.7500	1	0.7500	+0.0004	16×10^{-8}	16×10^{-8}

$$\sum f = 50 \qquad \sum m\,(x - \bar{x})^2 = 176 \times 10^{-8}$$

$$\sum m \times f = 37.4794 \qquad \sigma = \sqrt{\frac{\Sigma[m \times (x - \bar{x})^2]}{(\Sigma f)}}$$

$$\bar{x} = \frac{\Sigma m \times f}{\Sigma f} = \frac{37.4794}{50} \qquad \sigma = \sqrt{\frac{176 \times 10^{-8}}{50}}$$

$$\bar{x} = 0.7496 \qquad \sigma = 0.00018762 \text{ in.}$$

The terms in Table 4.3 are defined as follows:

m = measurement
f = frequency

$$\bar{x} = \text{mean} = \frac{\Sigma\, m \times f}{\Sigma f}$$

$$= \text{standard deviation} = \sqrt{\frac{\Sigma m \cdot (x - \bar{x})^2}{\Sigma f}}$$

and the mean is equal to 37.4794/50 or 0.7496 inches. The standard deviation is equal to $\sqrt{176 \times 10^{-8}/50}$ or 0.0001876 inches. This normal curve with its standard deviation is shown in Figure 4.7.

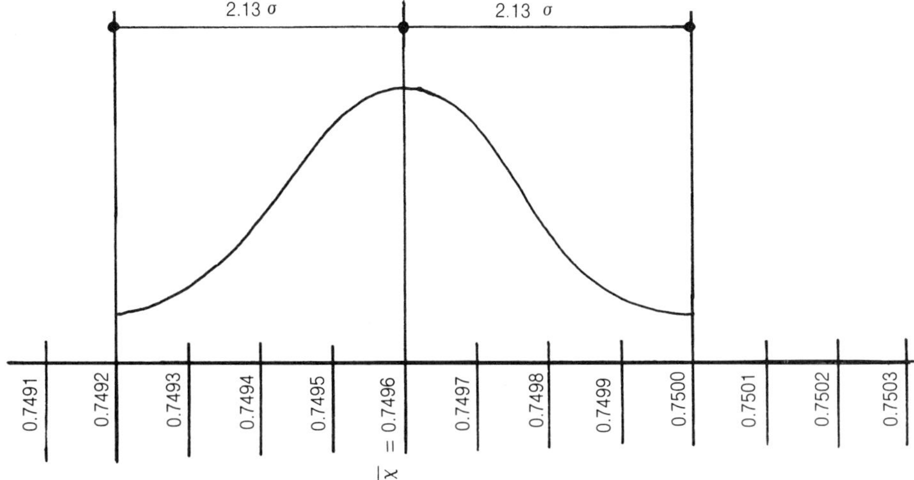

$\overline{X}$ = 0.7496 inches

σ = 0.00018762 inches

Figure 4.7 Mean and standard deviation associated with the production run of bushings.

The same procedure is followed in determining the mean and standard deviation of the holes. The measurement corresponding to the 50 holes in the random sampling is shown in Table 4.4.

Table 4.4 List of the Measurements of 50 Randomly Picked Holes from a Production Run of 500 Holes

.7484	.7484	.7485	.7483	.7484
.7484	.7482	.7484	.7484	.7485
.7483	.7483	.7484	.7484	.7482
.7484	.7483	.7481	.7482	.7483
.7482	.7483	.7483	.7486	.7484
.7486	.7484	.7481	.7481	.7486
.7485	.7485	.7486	.7482	.7483
.7481	.7486	.7482	.7486	.7484
.7485	.7482	.7484	.7485	.7485
.7483	.7483	.7483	.7481	.7482

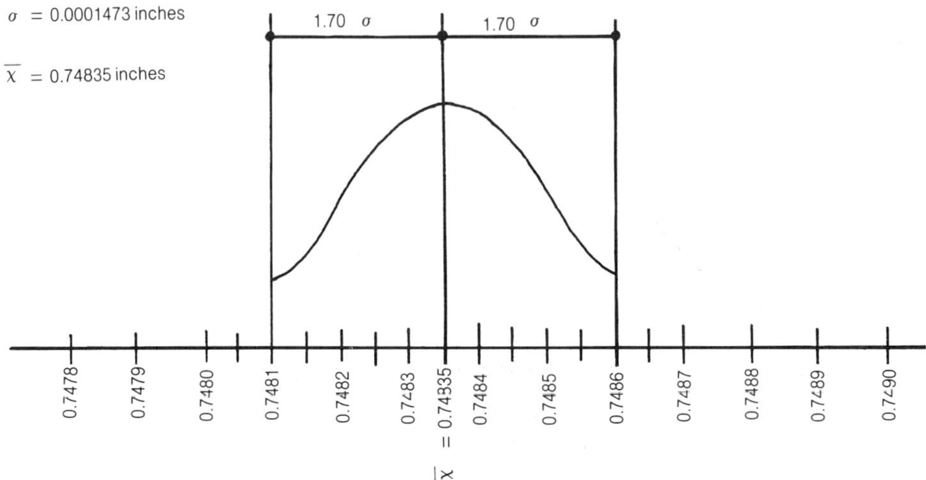

Figure 4.8 Mean and standard deviation associated with the production run of holes.

The mean and standard deviation of this random sample of hole sizes are calculated using Table 4.4. The mean is 0.74835 inches and the standard deviation is 0.0001473 inches and they are shown on their normal curve in Figure 4.8.

The normal curve for the bushings and the normal curve for the holes are both placed on the same coordinate axis for comparison purposes (Fig. 4.9). From Figure 4.9, one can obtain the normal curve associated with the fits that occur between the 50 bushings and the 50 holes of the sample. The mean of this normal curve is the mean press fit or 0.00125 inches and the standard deviation of the fits is equal to the square root of the sum of the square of the standard deviation of the bushing sample and the square of the standard deviation of the hole sample or,

$$\sigma_{fit} = \sqrt{\sigma^2_{bushing} + \sigma^2_{hole}}$$

$$\sigma_{fit} = \sqrt{(0.0001876)^2 + (0.0001473)^2}$$

$$\sigma_{fit} = 0.0002383 \text{ in.}$$

The normal curve of the fits between the 50 bushings and the 50 holes is shown in Figure 4.10 with its standard deviation.

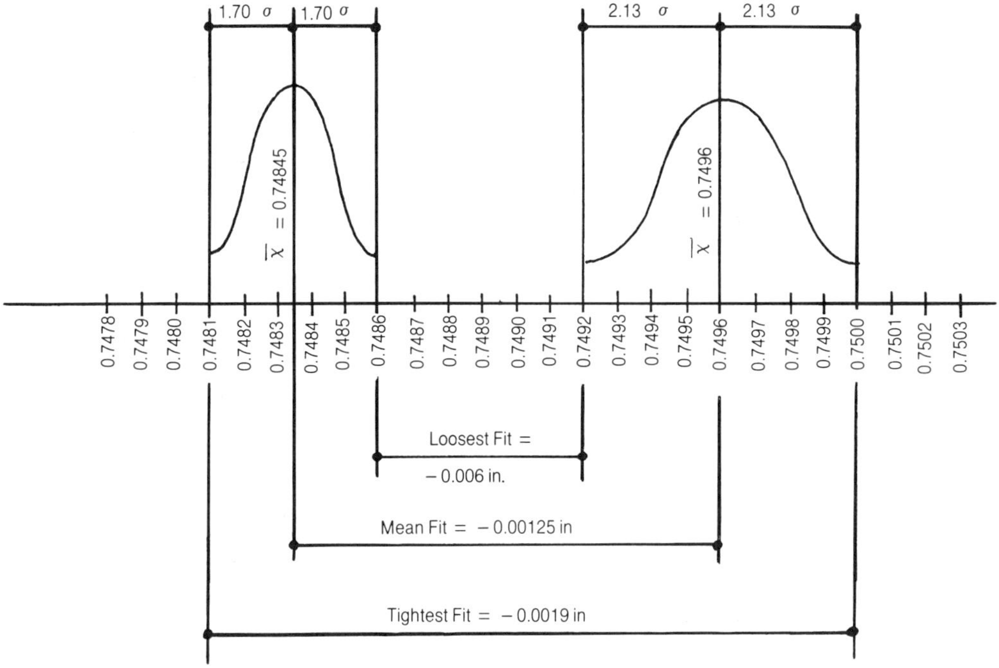

1.70 σ
1.70 σ
2.13 σ
2.13 σ
X̄ = 0.74845
X̄ = 0.7496
0.7478
0.7479
0.7480
0.7481
0.7482
0.7483
0.7484
0.7485
0.7486
0.7487
0.7488
0.7489
0.7490
0.7491
0.7492
0.7493
0.7494
0.7495
0.7496
0.7497
0.7498
0.7499
0.7500
0.7501
0.7502
0.7503
Loosest Fit =
– 0.006 in.
Mean Fit = – 0.00125 in
Tightest Fit = – 0.0019 in

Figure 4.9 Comparing the normal curve of the bushings with the normal curve of the holes.

2.73 σ
2.73 σ
– 0.0006 in
– 0.00125 in
– 0.0019 in

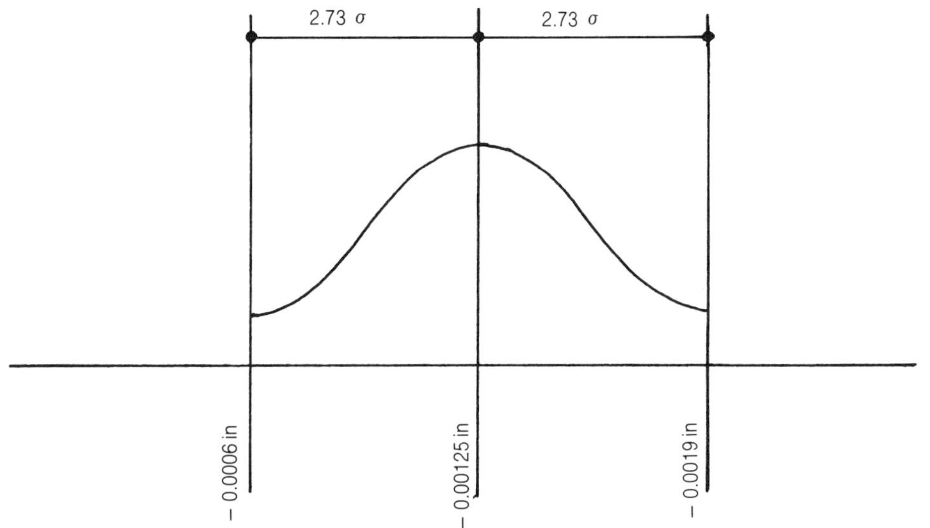

Figure 4.10 Normal curve of the fits between the 50 bushings and the 50 holes.

The theoretical percentages of acceptable bushings and holes can be calculated from their respective curves (Figs. 4.7 and 4.8). Similarly, the percentage of the bushings that will mate correctly with the holes can be calculated from the normal curve shown in Figure 4.10.

The percentage of acceptable bushings is acquired by using the bushing's normal curve shown in Figure 4.7. The area between z = −2.13 and z = +2.13 is 97.7% using Table 4.1. Remember that $z = (x - \bar{x})/\sigma$ when reading Table 4.1 and for the value of z = −2.13, one would get the area of 0.0166 using Table 4.1. The z value is a minus amount in this region and the table represents the area below the normal curve from a minus z value to a plus z value. For the value of z = −2.13, one would get the corresponding area of 0.0166 $in.^2$ under the normal curve as shown in Figure 4.11.

For the value of z = +2.13, one would get the corresponding area of 0.9834 $in.^2$ under the normal curve as shown in Figure 4.12. The actual area under the normal curve lies between the limits of z = −2.13 and z = +2.13 or one must substract 0.0166 from 0.9834 to obtain 0.9768. This value represents the theoretical amount of acceptable bushings out of a population of one or unity. If this value is multiplied by 100%, the resulting percentage of 97.68% would represent the amount of acceptable bushings out of this lot of 500 bushings. Also, 1.15% of the bushings will be considered scrap (undersized) and 1.15% of the bushings will be considered rework (oversized) from this lot of 500 bushings.

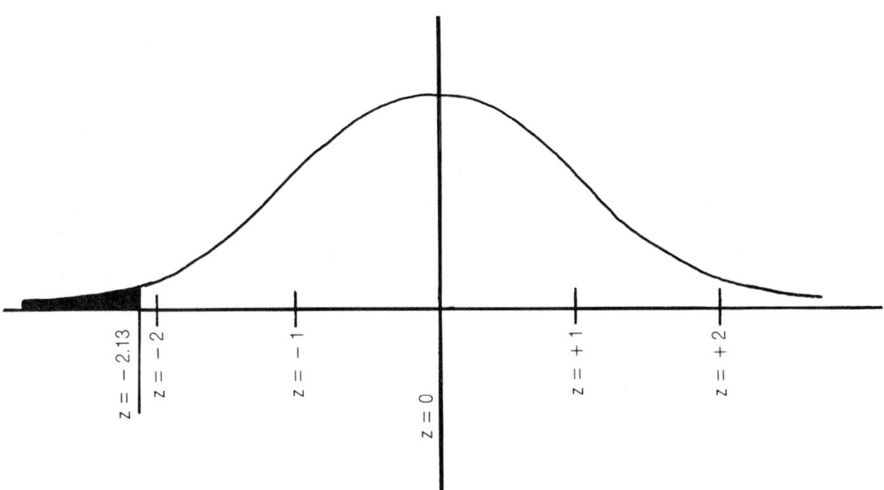

Figure 4.11 Area under the normal curve for a value of z = - 2.13.

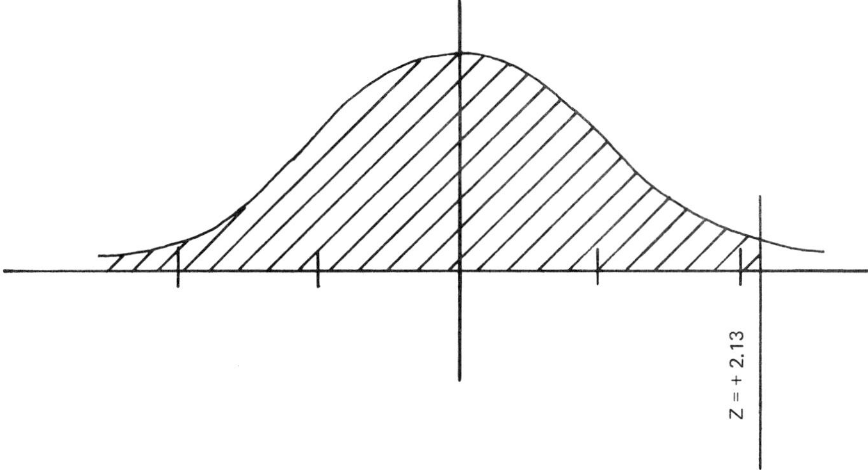

Figure 4.12 Area under the normal curve for a value of z = + 2.13.

The percentage of acceptable holes is acquired by using the hole's normal curve shown in Figure 4.8. The area between z = −1.70 and z = +1.70 is 0.9554 minus 0.0367 or 0.9187 by using Table 4.3. This means that 91.9% of the hole sizes are acceptable while approximately 4% are scrap (oversized) and 4% have to be reworked (undersized).

Figure 4.10 shows the normal curve related to how the bushings and holes will mate while in assembly. The percentage of acceptable fits (matings) is 99.4% and is the area under this normal curve from z = −2.73 to z = +2.73. Similarly, 0.3% of the fits are too tight and 0.3% of the fits are too loose.

Therefore, the normal distribution of the parts due to the manufacturing process is a controlling factor in determining the dimensions and tolerances a designer places on his/her drawings. These dimensions and tolerances must be compatible to the manufacturing capabilities of the shop. To callout dimensions and tolerances on drawings that *cannot* be obtained in manufacturing is as poor practice as a machine that *does not* function correctly.

4.5. PROCESS CAPABILITY STUDY

The machine or process involved in making a quality product must be able to maintain the specified tolerances called out on the drawing. When equipment cannot meet these tolerances, cost is added in the form of scrap and/or rework.

First, let's define the word *process*. A *process* consists of three basic variables: people, machines, and materials. Therefore, a process includes raw materials, machines or equipment, the skill of the operator, the measuring instrument, the skill of the inspector, and any other variable that may be present during the process. Secondly, *process capability* is that which the process is capable of producing under normal, in-control conditions. Normal, in-control conditions are those conditions which yield parts having a pattern of variation that follows a predictable and normal distribution. In other words, the process must be in statistical control with no apparent assignable causes of variation present. The first step in determining the capability of a process is to determine if the process is in control.

Mathematically, process capability is defined as six standard deviation (σ) units. Remember, in a normal distribution or population the area bounded by plus and minus three standard deviations ($\pm\sigma$) from the average $\bar{x}$ contains 99.73% of all the readings (measurements). It can be said that the expected spread of the measurements generated in the process is obtained if the standard deviation of the process is calculated and this value is multiplied by six (6). The process is in control if, and only if, the spread of the process is the same as the capability of the process.

Standard deviation is the only number needed to determine the process capability, however, the first job of any quality control organization is to assure that the product meets the required specification. To determine this, the average of the set of numbers (measurements) must be known. By knowing the average and the standard deviation, the largest shaft diameter and the smallest shaft diameter can be obtained by merely adding three times the standard deviation to the average and subtracting three times the standard deviation from the average. If both of these values are within the tolerance limits of the dimension and the distribution is normal, it is said that all the shafts produced by this process are within specifications.

The mean $\bar{x}$ and standard deviation σ of the measurements listed in Table 4.5, appear in Table 4.6. The same method is used to calculate the mean and standard deviation of these measurements as covered in the previous section.

It will be assumed for the remainder of this section that the process is in-control. Normally, at least 50 readings (measurements) are used in calculating the capability of a process. Also, the order of the readings should be preserved so that, when necessary, remeasurement of any particular part can be accomplished easily. The measuring device should be capable of dividing the total tolerance band (plus and minus) into at least ten divisions and the capability into at least eight divisions. Lastly, the tolerance is bilateral or, in other words, equal tolerance values in the plus and minus directions.

Table 4.5 Measurements of 50 Shafts

.9210	.9220	.9210	.9215	.9210
.9205	.9205	.9215	.9210	.9210
.9225	.9210	.9205	.9210	.9205
.9200	.9205	.9210	.9220	.9210
.9205	.9215	.9205	.9205	.9220
.9215	.9200	.9205	.9210	.9215
.9200	.9215	.9215	.9210	.9210
.9210	.9220	.9210	.9200	.9210
.9205	.9215	.9210	.9210	.9215
.9220	.9210	.9215	.9215	.9200

In order to illustrate the calculation of a process capability, let's assume a manufacturing process. A shaft is turned to a specified diameter of 0.9210 ± 0.0030 in. on a lathe. It is known that the process is in control, so the measurement of 50 shafts were taken at random. The results of these measurements appear in Table 4.5.

The natural spread of this process is equal to 0.9225-0.9200 = 0.0025. The natural spread of this process does not exceed 75% of the tolerance spread (0.006 times 0.75 = 0.0045). Now the process capacity will be calculated.

The process capability is mathematically defined as six (6) standard deviations. To calculate the process capability of this particular process, the standard deviation is multiplied by 6.

$$6 \times 0.00059 = 0.00354 \text{ in.}$$

Table 4.6 The Mean and Standard Deviation of the Shaft Measurements

$$\sum m \times f = 46.052$$

$$\bar{x} = \frac{\Sigma\, m \times f}{\Sigma f} = \frac{46.052}{50} = 0.92104 \text{ in.}$$

$$\sum m\,(x - \bar{x})^2 = 0.00001742$$

$$\sigma = \sqrt{\frac{\Sigma m\,(x - \bar{x})^2}{\Sigma f}} = \sqrt{\frac{0.00001742}{50}}$$

$$\sigma = 0.00059 \text{ in.}$$

It would be expected that 99.73% of all the parts produced under these conditions would fall between $\bar{x} - 3$ and $\bar{x} + 3$. The upper and lower limits are:

(0.92104) − (3) (0.00059) = 0.9193 = the lower limit

(0.92104) + (3) (0.00059) = 0.9228 = the upper limit

One can see that the natural limits of the process are well within the tolerance limits; therefore, the process is capable of producing these shafts to the desired specification. To allow for movement of the average from the nominal, the capability or natural spread of the process should not exceed 75% of the tolerance spread. For example, the natural spread of this process is equal to 42% of the tolerance spread (0.0025/0.0060 times 100%).

4.6. OPERATIONS RESULTING IN SKEWED NORMAL CURVES

There are certain operations which will normally or usually yield frequency distributions that do not follow the normal curve. Unilateral tolerancing of a dimension produces a skewed normal curve. Some examples of unilateral tolerancing are: concentricity, straightness, flatness, roundness, parallelism, and perpendicularity. Since zero, or perfection, is a limit to one side of the distribution, the distribution will be skewed positively (Fig. 4.13) if the tolerance band is given as +.xxx and −.000 or skewed negatively (Fig. 4.14) if the tolerance band is given as +.000 and −.xxx.

The standard procedure cannot be used to find the process capability, but a slight modification of this procedure will yield the desired results. This modification is as follows:

1. Determine the mode which is the highest point on the distribution. (In this case *do not* use the average.)
2. Assume that the left half of the distribution is the half of a distribution having a certain standard deviation and the right half of the distribution is the half of the distribution having another standard deviation. This principle is illustrated in Figure 4.15.
3. By combining half of the process capability of the distribution represented by the left half of the original distribution represented by the right half of the original distribution, the result is approximately the process capability of the original distribution. This could be expressed as $3\sigma_1 + 3\sigma_2$.

In order to illustrate this procedure, assume that another specification on the shafts being studied is roundness. The specification calls for maximum out of roundness of 0.010 inches total indicator reading (TIR). These shafts are measured for this characteristic and the results of these measurements are given in Table 4.7 as a frequency distribution.

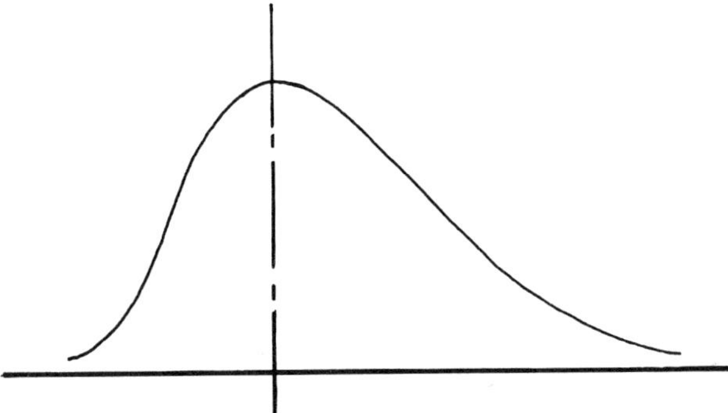

Figure 4.13 A positively skewed distribution.

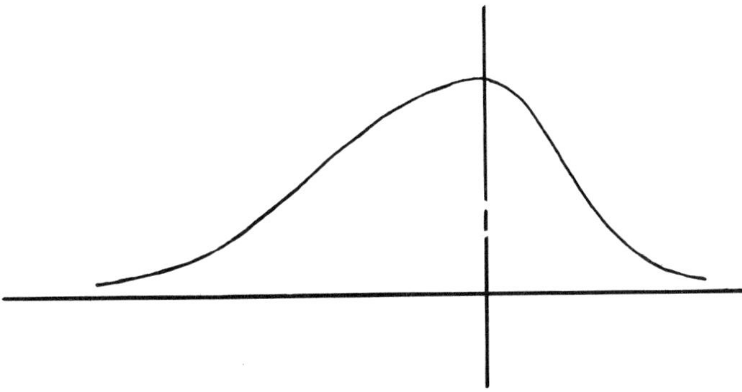

Figure 4.14 A negatively skewed distribution.

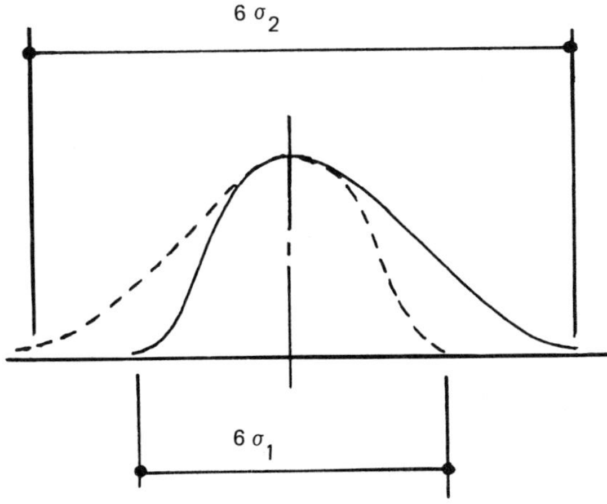

Figure 4.15 Adding a positive to a negative skewed distribution to obtain a normal distribution.

Table 4.7 Frequency Distribution

TIR	Frequency
.001	1
.002	6
.003	17
.004	9
.005	6
.006	5
.007	3
.008	2
.009	1

Now, the standard deviation of this set of measurements must be calculated. There is a simplified method for calculating standard deviation of grouped data. In order to calculate the average and standard deviation of grouped data, a number of coded values must first be calculated. The steps used in the calculation of these coded values are covered in the following discussion. Also, Table 4.8 which is used with these calculations in included.

1. Divide the readings (measurements) into groups called cells (there should be between 8 and 20 cells).
2. In the column noted as cell groups, record the lowest and highest reading in the cell.
3. In column 1 (cell midpoints) record the median (middle reading) of each cell.
4. In column 2 f record the number of measurements which fall at or between the lowest reading and the highest reading.
5. In column 3 d, place a zero opposite the cell in which you think the arithmetic mean of the 50 samples will lie.
6. Still in column 3, code each of the other cells in the following way:
   a. That cell which is one cell lower (in terms of the actual readings) is coded −1. That cell which is two cells lower is coded −2, three cells lower −3, and so on until the lowest number is reached.
   b. That cell which is higher than the cell coded as zero is coded +1. That cell which is two cells higher is coded +2, three cells higher +3, and so on until the highest cell is reached.
7. In column 4 fd, record the product of column 2 times column 3.
8. In column 5 fd^2 record the product of column 3 times column 4.

Table 4.8 Calculation of Process Capability

	COL. #1	COL. #2	COL. #3	COL. #4	COL. #5
		f	**d**	**fd**	**fd^2**
CELL GROUPS	CELL MIDPOINTS	FREQ. OF OCCURRENCE	CELL DEVIATION	COL. #2 × COL. #3	COL. #3 × COL. #4
		n =		fd =	fd^2 =

Ci = Class Interval = Difference Between Cell Midpoints

9. Calculate the coded value known as Σfd by adding column 4.
10. Calculate the coded value known as Σfd^2 by adding column 5.
11. The value n is the total number of readings or the total column 2.
12. The value known as Ci is the difference between the cell midpoints.

The formula used in the calculation of the average is given as:

$$\bar{x} = \bar{x}_a + \frac{\Sigma fd}{n} \times Ci$$

where

$\bar{x}$ = Average or arithmetic
$\bar{x}_a$ = The midpoint of the cell in which the average is assumed to be
fd = Total of column 4
n = Number of observations
Ci = Class interval

The formula used in the calculation of standard deviation is given as:

$$\sigma = Ci \cdot \sqrt{\frac{fd^2}{n} - \frac{\Sigma fd}{n^2}}$$

where:

σ = Standard deviation
Ci = Class interval
fd^2 = Total of column 5
fd = Total of column 4
n = Number of observations

This simplified method of determining standard deviation will be used on the frequency distribution listed in Table 4.7. The mode (measurement that occurs most frequently) is 0.003 in. in this case. It is this point (0.003) at which the distribution is divided into divisions to the left and to the right. Therefore, the frequency from 0.001 to 0.003 is taken directly from the frequency distribution representing the normal curve signified by $6\sigma_1$ in Figure 4.15. Now, assume the same frequency on the other side of the mode.

The same procedure is performed for the normal curve represented in Figure 4.8 by $6\sigma_2$ and the results are shown in Table 4.9.

Now, calculate the coded values of fd and fd^2 as previously discussed in Table 4.8. Since the distribution is symmetrical, the value of Σfd will always equal zero.

The formula for finding σ is equal to $\sqrt{Ci\ \Sigma fd^2/n - (\Sigma fd/n)^2}$ and since Σfd always equals zero in this kind of situation, and zero divided by any number equals zero, the second term can be dropped and the formula can be rewritten as:

$$Ci \times \sqrt{\frac{fd^2}{n}} = 0.001\sqrt{\frac{22}{32}} = 0.00083 \text{ in.}$$

Table 4.9 Calculating the Standard Deviation of the Left-Hand Side of the Distribution

CELL	f	d	fd	fd^2
.001	1	−2	−2	4
.002	7	−1	−7	7
.003	16	0	0	0
	7	+1	+7	7
	1	+2	+2	4
	n = 32		fd = 0	fd^2 = 22

Table 4.10 Calculating the Standard Deviation of the Right-Hand Side of the Distribution

CELL	f	d	fd	fd^2
	1	−6	−6	36
	2	−5	−10	50
	2	−4	−8	32
	4	−3	−12	36
	7	−2	−14	28
	10	−1	−10	10
.003	16	0	0	0
.004	10	+1	+10	10
.005	7	+2	+14	28
.006	4	+3	+12	36
.007	2	+4	+8	32
.008	2	+5	+10	50
.009	1	+6	+6	36
	n = 68		fd = 0	fd^2 = 384

Substituting the values into the equation yields sigma as 0.0083 in., which is the standard deviation of the left-hand side of the distribution only and will be referred to as sigma 1 or σ_1.

Using the same procedure for the right-hand side of the distribution (Table 4.10), the standard deviation for the right-hand side of the original distribution is determined and is equal to

$$\sigma_2 = 0.001\sqrt{\frac{384}{68}} = 0.00238 \text{ in.}$$

This is the standard deviation of the right-hand side of the distribution only and will be referred to as sigma 2 or σ_2.

As stated above, the *process capability* is equal to $3\sigma_1 + 3\sigma_2$ or $3 \times 0.0083 + 3 \times 0.00238 = 0.0096$ in. in this case. The expected spread would be: mode $\begin{matrix} -3\sigma_1 \\ +3\sigma_2 \end{matrix}$ which in this case is equal to: lower limit = 0.0025 inches and upper limit = 0.0071 inches.

4.7. APPLYING PROCESS CAPABILITIES TO THE DESIGN, MANUFACTURE, AND COST OF A PRODUCT

Now that the process capability has been calculated, what can be done with it? One important use for this information is to facilitate the design of the product. When the product is being designed, one of the most important considerations is the availability of machines and processes. Designing products to fit both existing equipment capabilities and highlighting the need for new equipment will not only help in quality planning but will mean reduced losses, smoother conversion to the new design, and faster scheduling. Another important use of the process capability results is in the acceptance of new or reconditioned equipment. Before accepting any new piece of equipment, it must be assured that the equipment is compatible with the job. This can only be done by performing a capability study. Capability studies can also be performed by the manufacturers of equipment to assure that the equipment meets its guaranteed performance.

In order to illustrate four other uses of the capability results, assume that there are three pieces of equipment used to generate the same type of quality characteristic. Capability studies are made on each of the three pieces of equipment and the results are tabulated below:

Machine A has a capability of 0.002 inches when the material being used is brass. When steel is used, the capability is 0.005 inches.
Machine B has a capability of 0.003 inches for brass and 0.006 for steel.
Machine C has a capability of 0.006 inches for brass and 0.008 inches for steel.

The difference between these machines is explained by the fact that each is a different make, size, age, and so on. In this particular case, it is discovered that the capability changes with the material used.

With this information, it is possible to schedule work to machines so that the tolerance specified on the drawing will actually dictate the machine to be used. It is also possible to establish control limits for equipment that has a narrow capability in comparison to the allowable tolerance band. This information can be used in setting up a machine for a production run and to determine the economic nominal around which to operate when the process capability is in excess of the tolerance band. Let's examine each of these uses individually.

4.8. SCHEDULING WORK ON MACHINES

Many product designs scheduled through a manufacturing situation may have different tolerances for the same type of characteristic to be generated. For example, if the turned diameter of a shaft is specified as 0.758 ±0.003 in., the total tolerance band is 0.006 in. Using the capability information given in the previous section, the only machines capable of holding this tolerance are A and B.

4.9. DETERMINING EQUIPMENT CONTROL LIMITS THAT HAVE A NARROW PROCESS CAPABILITY IN COMPARISON TO THE ALLOWABLE TOLERANCE BAND

In many types of operations the process capability of the equipment is narrow in comparison to the tolerance band of the parts it is producing. When such is the case, it may be worthwhile to take advantage of the entire allowable tolerance band. An example would be a process that has an inherent drift in the dimensional setting. Some examples of such types of operations or equipment are lathes, screw machines, jig borers, drill presses, and grinders.

The problem that arises in such cases is to set control limits that will allow one to take full advantage of the tolerance band and yet have assurance that defective parts are not being produced. A nomograph (Fig. 4.16) has been developed to assist in determining the proper control limits. It consists of four variables; and once two of these variables are chosen, the other two are fixed. The variables are:

1. The sample size used, n. The bottom of the nomograph.
2. P_1 percent, third scale from the top on the nomograph, is the minimum percent of parts out of tolerance that one could expect to find if the average of the sample fell directly on the control limit.
3. P_2 percent, the top scale of the nomograph, is the maximum percent of parts out of tolerance one could expect to find if the average of the sample fell directly on the control limit.
4. The distance the control limit is inside the specification limit and expressed as a percent of the process capability. Second scale from the top.

To illustrate, assume a shaft made of brass with a specification of 0.758 ± 0.003 inches in diameter. The tolerance band in this example is 0.006 inches. Aslo, it is assumed that for some reason this job is assigned to machine A, which has a capability of 0.002 inches. The

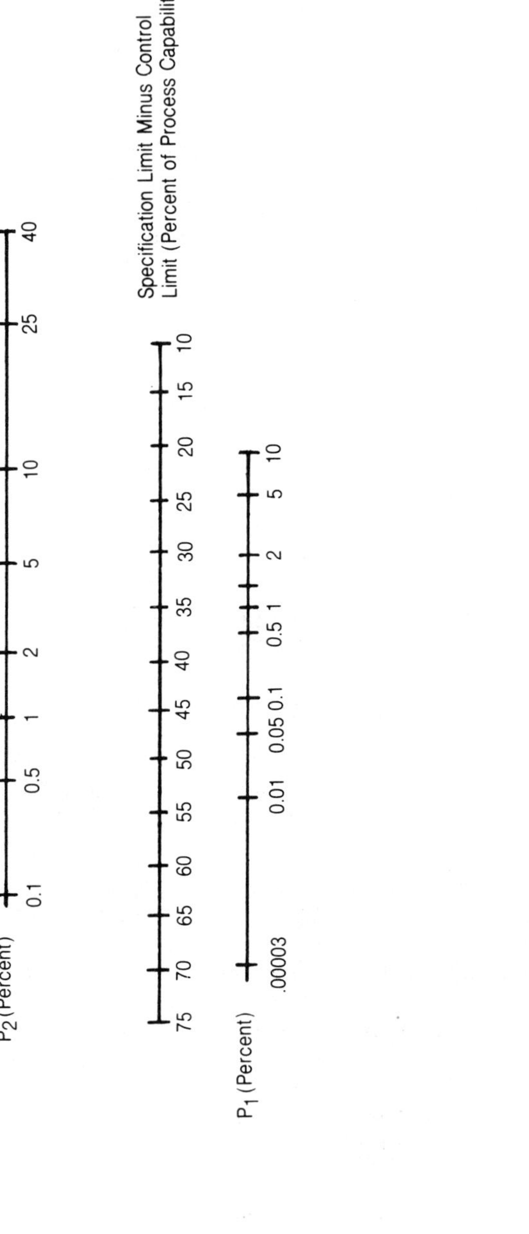

Figure 4.16 Control limit nomograph.

process capability in this case is small in comparison to the total tolerance band (33%). It was decided to use a sample of four. On the scale titled Sample size on the nomograph, locate the number 4. It is known that if the percent defective exceeds 2%, problems in succeeding operations will occur. The percent defective at any time should not exceed 2%, so this value is located on scale P_2 (percent) on the nomograph. A straight line is drawn between these two points. The value at which this drawn line crosses the Specification limit minus control limit (percent of process capability) column indicates that the distance between the control limit and the specification limit is to be 44% of the process capability. The process capability in this example is 0.002. The lower specification limit is 0.758 − 0.003 or 0.755 inches. The lower control limit, therefore, is 0.758 + (0.44) (0.002) or 0.7589. The upper specification limit is 0.758 + 0.003 or 0.761 inches. The upper control limit, therefore, is 0.758 − (.44) (0.002) or 0.7571 inches.

If a sample of four parts is chosen and the average of these four parts is between the limits of 0.7571 and 0.7589, then it can be said with confidence that acceptable parts are being produced. If the average should fall directly on one of the control limits, a maximum of 2% out of tolerance can be expected and a minimum of 5% out of tolerance.

4.10. SETTING UP A MACHINE FOR A PRODUCTION RUN

When setting up a machine for a production run, it is necessary to assure that the machine will produce acceptable parts. This is usually done by checking some of the first pieces produced and, on the basis of these checks, either approving or rejecting the setup. The questions to be answered are:

1. How many pieces must be measured?
2. Between what limits should these measurements fall?
3. What are the absolute rejection limits?

If the process capability and the acceptable quality limit (AQL) are known, these questions can be answered with the aid of the nomograph for set-up acceptance (Fig. 4.17). To illustrate, assume that we assign the production of the soft brass shaft to lathe B, which has a capability of 0.003 inches. In order to use the nomograph, the capability must be converted to a percent of the tolerance. This is accomplished by dividing the capability by the tolerance band and multiplying the answer by 100. Using the previous example, 0.003/0.006 times 100 equals 50%. Locate this value on the scale of the nomograph under process capability (% of tolerance). Choose an AQL of 2%. Now, locate this value on the scale of the nomograph for AQL. Draw a straight line through these two points and extend it through pivot line number 2. It has been decided to check two consecutively pro-

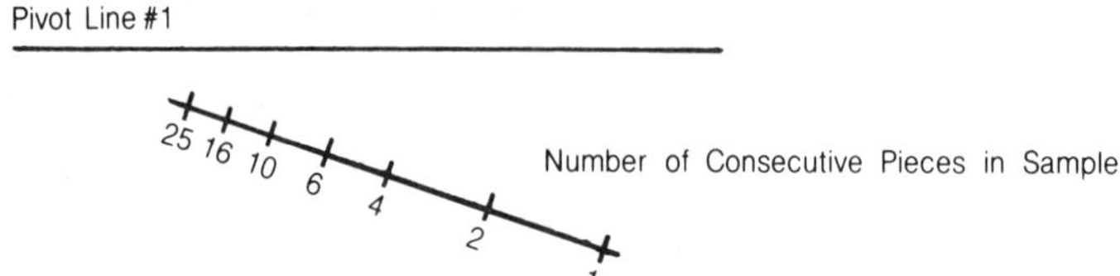

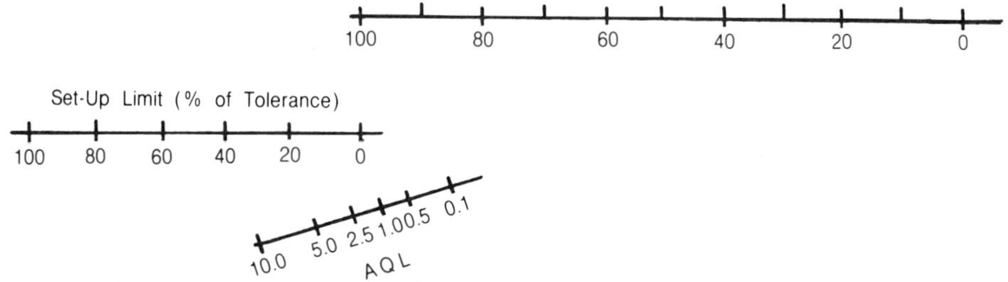

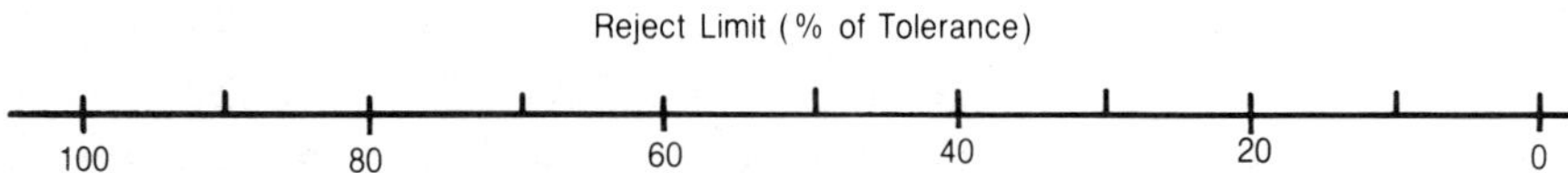

Figure 4.17 Nomograph for set-up acceptance.

duced parts. This value is located on the scale of the nomograph showing number of consecutive pieces in sample. Draw another straight line through this point and the process capability point on the scale, and extend this line through pivot line number 1.

Connecting the points at which the pivot lines are crossed, it is found that the set-up limit scale (% of tolerance) is crossed at the value of 54%. The tolerance in this example is ±0.003 inches. If the percent of tolerance is multiplied by the tolerance and this result is added to the nominal dimension, then the upper set-up limit is obtained. By subtracting this value from the nominal, the lower set-up limit is obtained. Since the sample's size is two, the *average* of the two measured samples must fall between the lower and upper set-up limits in order for the set up to be accepted. This does not mean that if the average of the two samples is beyond the set-up limits, the setup is automatically rejected. In order to determine the reject limits, extend the line connecting the points at which the pivot lines are crossed through the reject limit (% of tolerance) scale. If the average of the two samples falls at or beyond the value of the reject percent times the tolerance plus the nominal, the setup must be rejected. If it falls below the value obtained by subtracting the reject percent times the tolerance from the nominal, the setup must also be rejected. If the average falls between the accept value and the reject value, more samples are needed. In this example, the upper accept limit is equal to 0.54 X 0.003 + 0.758 or 0.7596. The lower accept limit is equal to 0.758 − 0.54 X 0.003 or 0.7564. The upper reject limit is 0.758 + 0.80 X 0.003 or 0.7604. The lower reject limit is 0.758 − 0.80 X 0.003 or 0.7556.

4.11. DETERMINING THE MOST ECONOMICAL NOMINAL DIMENSION FOR THE PROCESS TO OPERATE WHEN THE PROCESS CAPABILITY EXCEEDS THE TOLERANCE

Many dimensions are so set that an undersized part will result in scrap while an oversized part can be reworked into an acceptable part. The cost of scrap is often considerably more than the cost of rework. If the process capability should exceed the tolerance band and the process cannot be improved immediately, the result will be scrap and rework dollars down the drain. In a situation like this it may be desirable to shift the process average away from the nominal specified dimension. Naturally, this shift is always toward the least-cost situation. In other words, if the cost of scrapping a part is high in comparison to reworking the part, it would be advisable to make the average a little higher than the nominal so that fewer parts will have to be scrapped. This, of course, results in more parts being reworked. The problem is how far should the average be shifted away from the nominal to obtain the best-cost situation? The nomograph for determining minimum scrap and rework cost (Fig. 4.18) has been developed to assist in computing this shift. The nomograph consists of four variables:

1. Tolerance band
2. Process capability
3. The cost ratio of scrap to rework
4. The amount to shift the average from the nominal

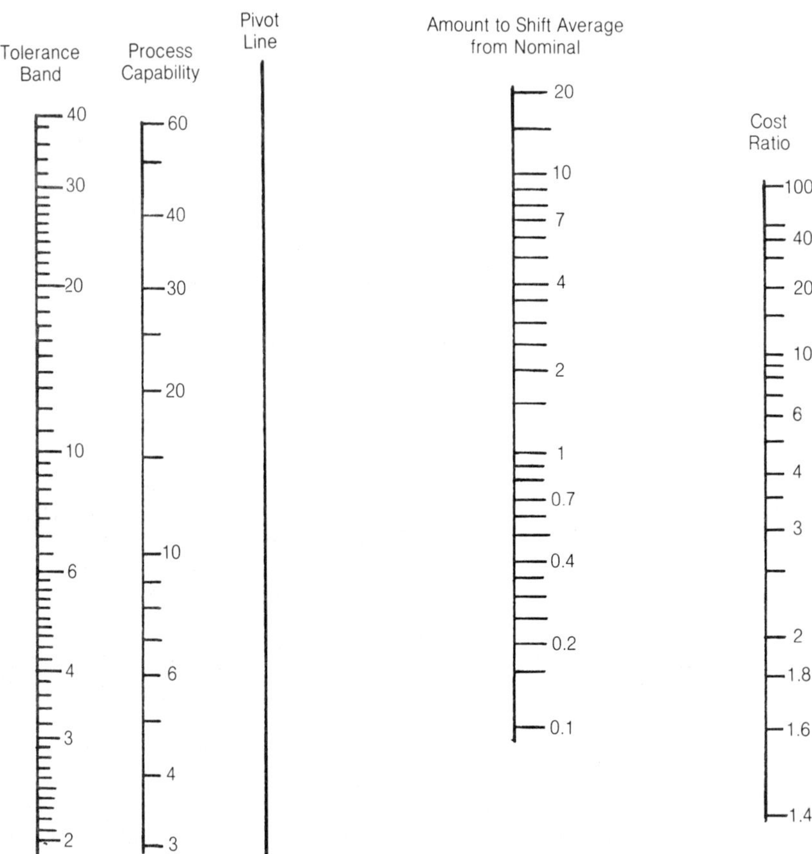

Figure 4.18 Nomograph for determining minimum scrap and rework cost.

To illustrate, consider once again the example of the turned shaft having the diameter specified as 0.758 ± 0.003. If the shaft is machined too small, it must be scrapped at a cost of $12. If the diameter is too large, it can be remachined at a cost of 75 cents. The process capability suddenly changes from 0.003 to 0.010 in., and spare parts needed for correction are temporarily unavailable. These shafts are needed to maintain production, so it is decided to continue the machining operation despite the scrap and rework costs. In order to calculate the "best" average for the process in terms of the least-cost situation, first locate the total tolerance band on the tolerance band scale of the nomograph. In this particular situation the total tolerance band is 0.006.

Now locate the process capability on the process capability scale of the nomograph. In this example the process capability is 0.010. The scales on the nomograph are in whole numbers. By multiplying the tolerance band and the process capability by 1000, these values can be converted into whole numbers. This procedure is sometimes referred to as *coding*. The coded values for the example are given as:

$1000 \times 0.006 = 6$ = tolerance band

$1000 \times 0.010 = 10$ = process capability

Now plot these points on the nomograph and connect them with a straight line, extending the straight line through the pivot line. The third point that must be located on the nomograph is the cost ratio. This cost ratio is calculated by dividing the scrap cost by the rework cost. In this example:

$$\frac{\$12}{\$.75} = 16$$

This cost ratio is now plotted on the cost ratio scale. A straight line is drawn between this point and the point at which the first drawn line crosses the pivot line. Notice that this second line crosses the amount to shift the average from nominal scale of the nomograph. This value is the amount in inches that the average must be shifted from the nominal dimension in order to realize the least-cost situation. In this example the value is 1.3. Since the tolerance band and the process capability is multiplied by 1000, this result of 1.3 must be divided by 1000 to obtain the actual amount of the shift desired.

$$\frac{1.3}{1000} = 0.0013 \text{ in.}$$

Because the shift must always be made toward the least-cost side, add 0.0013 to the nominal dimension of 0.758 to obtain the best average of 0.7593 for this process.

4.12. COMPUTER GRAPHICS APPLICATIONS

Manufacturing tolerances mainly depend on the manufacturing processes used to produce a feature. For instance, a hole in a plate can be produced by either using a drill or an end mill. The drilling operation is not as precise (amount of tolerance variation) compared to an end-milling operation, since the tooling and the construction differ between the two machines. Each manufacturing process has a tolerance variation associated with it and each particular machine used has its own tolerance variation characteristics within the manufacturing process. Therefore, designers must be aware of the capabilities, tolerance wise, of the various manufacturing processes and the machines that will be used in producing the various parts of machines and/or devices they are designing.

Most companies have a quality control or manufacturing department consisting of industrial engineers who are concerned with the most profitable methods of producing parts and/or products. One such industrial engineer, Dr. Demming, showed the industrial world how products can be optimumly manufactured by using statistical analysis. This philosophy has revolutionized today's industry, and computer graphics is now being used to determine the optimum dimensions and tolerances parts must have so they can be produced profitably. One must remember that great ideas and designs are worthless until they are produced profitably in our competitive industrial world. The following example illustrates the procedure the designer must follow in determining the optimum dimensions and tolerances of two mating parts. A shaft is to mate with a hole allowing a minimum clearance of +0.002 of an inch and a maximum clearance of +0.013 of an inch. A nominal 1-inch cold-rolled steel shaft will be used to satisfy the strength requirements of the design with a safety factor of 2.

The designer must choose what feature (hole or shaft diameter) will be used as the starting point (basic hole or shaft system) when dimensioning and tolerancing both features. Let's assume the basic shaft system will be used: since the shaft can be purchased as cold-rolled shafting with $^{+0.002}_{-0.000}$ tolerances or the shaft diameter is $1.000\ ^{+0.002}_{-0.000}$. Then the upper and lower limits of the hole would be 1.007/1.013 or $1.007\ ^{+0.006}_{-0.000}$ or $1.010\ \pm 0.003$ inches to satisfy the clearance specifications. A manufacturing process with a capability of producing parts within a ±0.003 tolerance variation must be used to produce this hole. The manufacturing department has a particular machine in the shop that will satisfy this requirement, but over a period of time it was observed that this particular machine actually produced parts according to the normal curve shown in Figure 4.19. The machine produces parts with a mean $\bar{x}$ of 1.011 inches and a standard deviation σ of 0.001 inches. Figure 4.20 shows the normal curve associated with the dimension and tolerances which were assigned by the designer to the hole. Superimposing these two normal curves (Fig. 4.21) indicates that a percentage of scrap (oversized holes) will be produced.

It is advantageous to have the normal curve related to the dimension and tolerances of the hole appearing on the drawing matched as closely as possible to the normal curve of the machine used in the shop to produce the hole or vice versa. Thus, the designer could change the hole's 1.010 ± 0.003 dimension and its tolerances to 1.011 ± 0.003 on the drawing and specify the shaft to be machined to a 1.002 ± 0.001 diameter. The shaft will cost more now due to the additional machining,

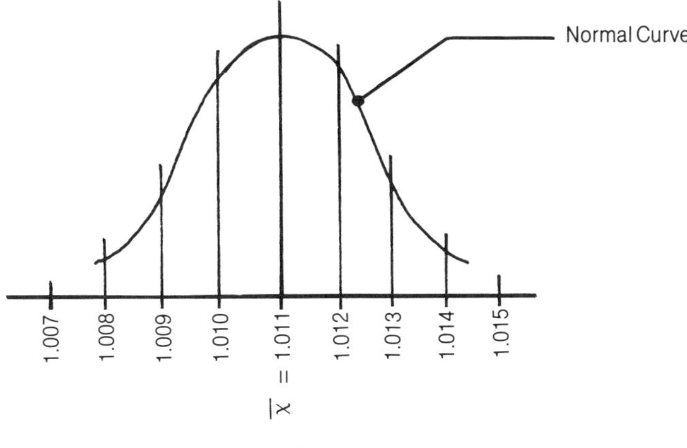

Figure 4.19 Machined holes normal curve.

but perhaps this cost would be less than the cost of producing scrap (oversized holes). For instance, a shaft with a minimum diameter of 1.001 inch must be purchased. Let's assume, the closest shaft diameter which can be purchased to this specification has a diameter of $1.005 \begin{smallmatrix} +0.005 \\ -0.000 \end{smallmatrix}$. Therefore, the maximum diameter of the shaft is 1.010 and minimum diameter is 1.005 inches. This means 0.007—0.009

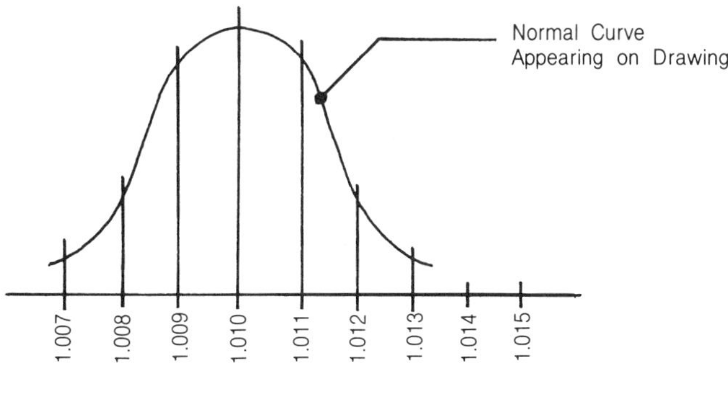

Figure 4.20 Normal curve associated with the dimension and tolerances assigned to the hole.

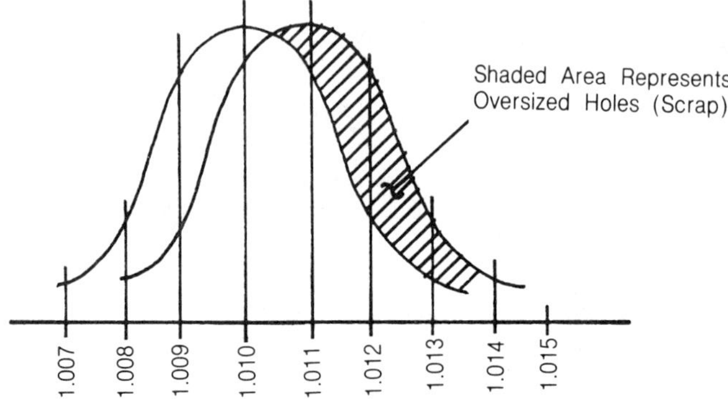

Figure 4.21 Superimposing the two normal curves indicates a percentage of scrap. Shaded area represents oversized holes (scrap).

inches of material can be removed from the 1.010-inch diameter and 0.002—0.004 inches from the 1.005 inch diameter. Each shaft must be measured to determine its size and the amount of material to be removed. This process is very expensive, and therefore, this alternative would not be used.

Another alternative to to change the mean of the machine, or, in other words, shift $\bar{x}$ = 1.011 of the machine toward $\bar{x}$ = 1.010 of the dimension. This particular machine has a tendency of producing holes 0.001 of an inch larger than the tool's diameter of 1.010 inch. This tendency could be due to several reasons, but the major cause is usually wear. To compensate for this wear, new tooling is required. For example, a tool which produces a 1.009 inch diameter hole could be used and the percentage of holes undersized are reamed to an acceptable size.

Determining the optimum dimensions and tolerances of mating features is very difficult and sometimes almost impossible. To simplify this task, the parts which are mass produced are measured at random and these sizes are documented before being assembled. The variation of size will indicate a trend. For instance, let's use the previous example of a shaft mating with a hole. If all the shafts and holes are measured before they are assembled and the majority of hole measurements are running in the direction of the maximum material condition (MMC) (smallest hole diameter) while the majority of the shaft measurements are running in the direction of the minimum or least condition (LMC) (smallest shaft diameter), then this statistical trend is satisfactory as long as both the hole and shaft are within their specified upper and lower limits of size. Now, if the measurements of the hole are running toward the maximum material condition (largest hole diameter) while the measurements of the shaft are running toward the minimum material condition (smallest shaft diameter), then this statistical trend is unsatisfactory. Therefore, the optimum dimensions and tolerances of mating parts need not be obtained as long as their statistical trends of size are in the correct direction. This same philosophy can be used when locating features on mating parts, too.

Most applications of computer graphics are in the creating and defining of three-dimensional objects in space. With the help of computer graphics, the designer is able to construct and manipulate the object's geometry (points, lines, and surfaces) easily while viewing it on a display screen or plotter. Since the object is explicitly defined as a mathematical model in space, important relationships which exist between the object's or objects' geometry can be easily acquired through the use of simple computer graphics commands. Therefore, computer graphics is not being used to optimize dimensions and tolerances, but instead, it is being used to understand the effect of dimensions and tolerances relative to the design.

4.13. WORK PROBLEMS

1. The bore of a 35 mm ball bearing may be between 1.3780 and 1.3775 inches. Dimension the shaft so that the fit ranges from an interference of 0.0008 to a clearance of 0.0001 inches.
2. Suppose that in problem 1 the limits of both the bearing bore and the shaft diameter represent a 3σ spread on each side of the mean. How many times in 1000 will the assembled fits fall between an interference of 0.0006 and an interference of 0.0001 inches?
3. A random sample of 50 bushings taken from production had outside diameters measured as follows:

0.6213	0.6219	0.6218	0.6218	0.6210
0.6211	0.6217	0.6220	0.6216	0.6212
0.6217	0.6216	0.6215	0.6215	0.6211
0.6220	0.6219	0.6215	0.6216	0.6212
0.6213	0.6217	0.6216	0.6221	0.6214
0.6213	0.6214	0.6212	0.6216	0.6215
0.6213	0.6214	0.6210	0.6217	0.6216
0.6213	0.6212	0.6209	0.6217	0.6216
0.6211	0.6216	0.6214	0.6218	0.6219
0.6208	0.6213	0.6215	0.6215	0.6214

a. Plot a distribution diagram.
b. Assume a normal distribution and compute the standard deviation σ.
c. Determine the limit dimensions for the bushing so that in normal production probably 99.73% of the parts will be acceptable.

5

Jigs, Fixtures, and Gages

5.1. ENSURING INTERCHANGEABILITY BY USING JIGS, FIXTURES, AND GAGING

The modern manufacturing of interchangeable parts is established on the fact that each part will be produced within a specified tolerance range. By using a jig or fixture, the part is presented to the tool in a fixed relationship, thus satisfying the prescribed tolerances. Gages and the gaging process are used to check the part so that it meets the tolerance specifications given to it for the purpose of interchangeability.

For a part to be statically stable while it is being machined, it must be in equilibrium in the x, y, and z directions. Therefore, the sum of the machining forces in the x, y, and z directions must be equal to zero. Also, the sum of the moments about the x, y, and z axes must be equal to zero if the part is to be in equilibrium. In otherwords, the free body of the part must be in equilibrium.

Since, a part may have three degrees of freedom of translation and three degrees of freedom of rotation, it is the purpose of a jig or fixture to restrict these six degrees of freedom of a part during the machining operation. These six degrees of freedom are shown in Figure 5.1.

The part would lose three degrees of freedom if placed and clamped on the locators numbered 1, 2, and 3 shown in Figure 5.2. If the part is then brought in contact with locators 4 and 5, two more degrees of freedom are restricted. The final degree of freedom is lost when the part is brought in contact with locator 6. The surfaces of the part resting against locators 1, 2, 3, 4, 5, and 6 are

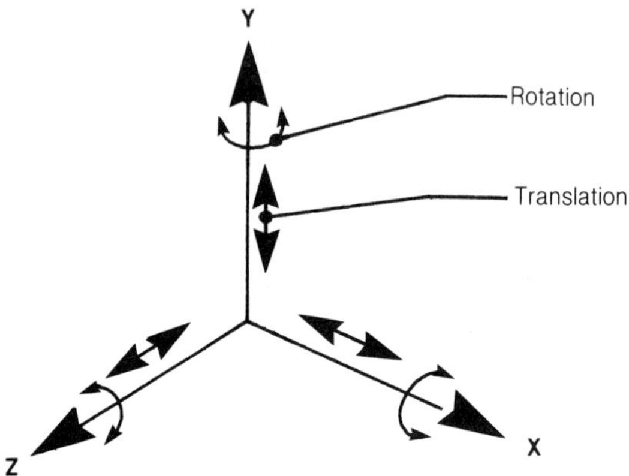

Figure 5.1 The six degrees of freedom.

referred to as the datum surfaces of the part (3-plane concept). Thus, the placement of locators on a jig or fixture ensures the reproducible location of parts each time their surfaces are brought in contact with these six locators.

To illustrate how a fixture separates form tolerances from position tolerances, the two mating parts shown in Figures 5.3 and 5.4 will be

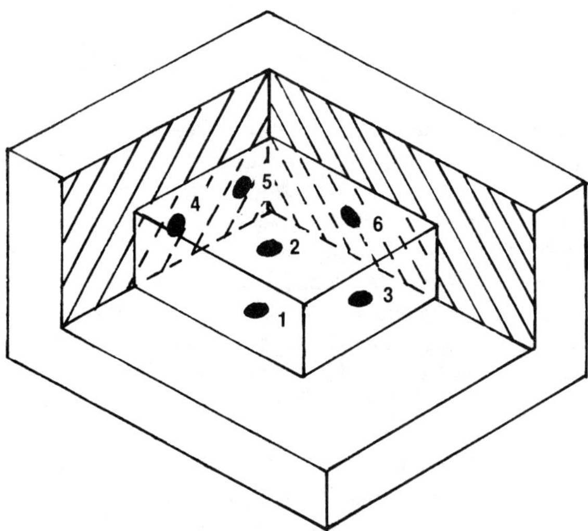

Figure 5.2 Restricting the six degrees of freedom.

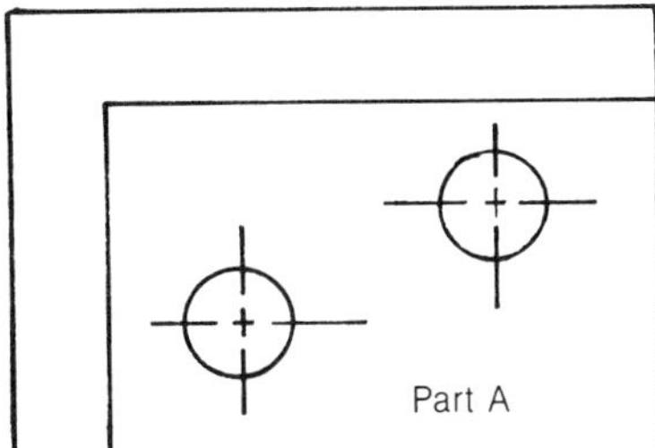

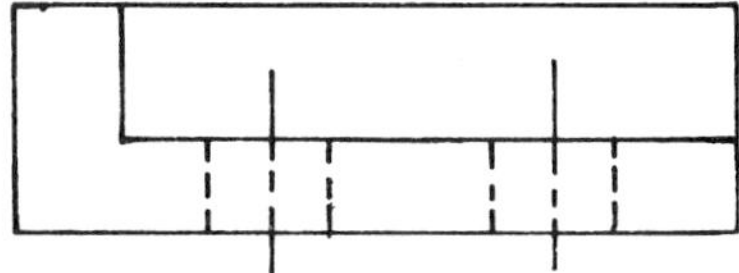

Figure 5.3 Part A.

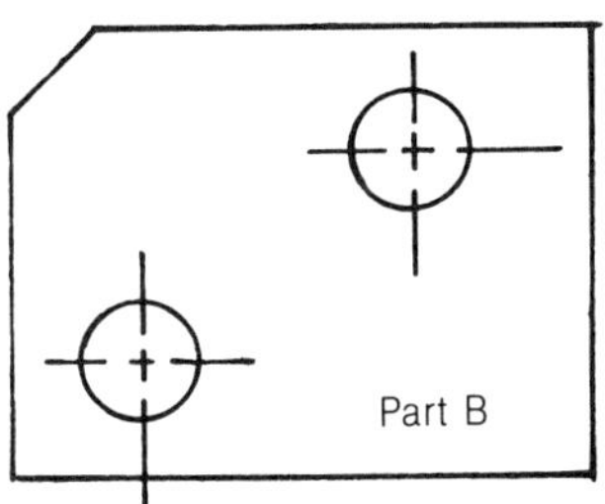

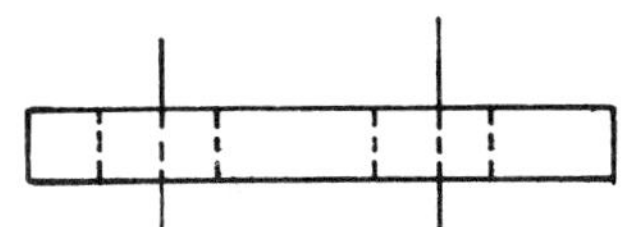

Figure 5.4 Part B.

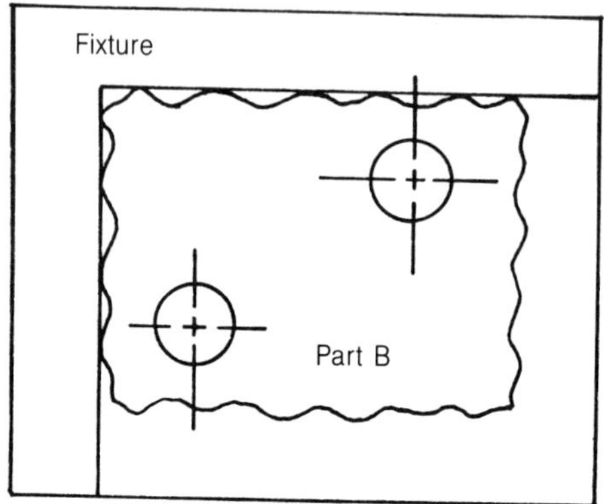

Figure 5.5 Fixture for holding part B.

used as an example. The two 0.500/0.560-inch diameter holes are drilled in part A to accommodate the two 0.495/0.490-inch diameter pins which are pressed-fitted in the holes of part B. Part A is located and held in a fixture which resembles part B. Likewise, a fixture resembling part A is used to locate and hold part B. Both fixtures with their respective parts are shown in Figures 5.5 and 5.6.

The surfaces of part A or part B do not influence of affect the location of their holes, since these holes are located from the fixture's perfectly flat and perpendicular datum surfaces designated as datums A, B, and C, instead of from the part's rough surfaces.

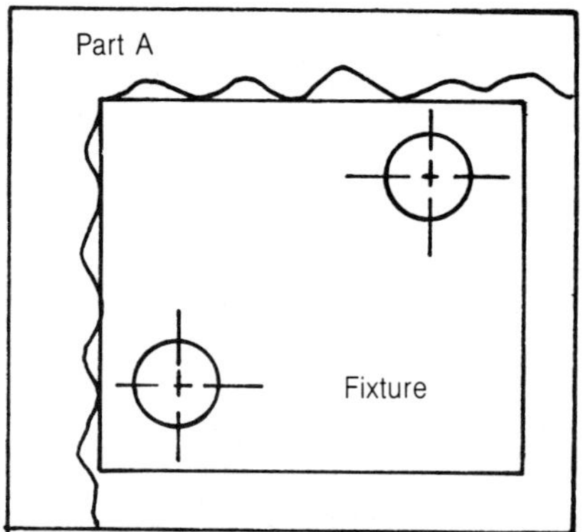

Figure 5.6 Fixture for holding part A.

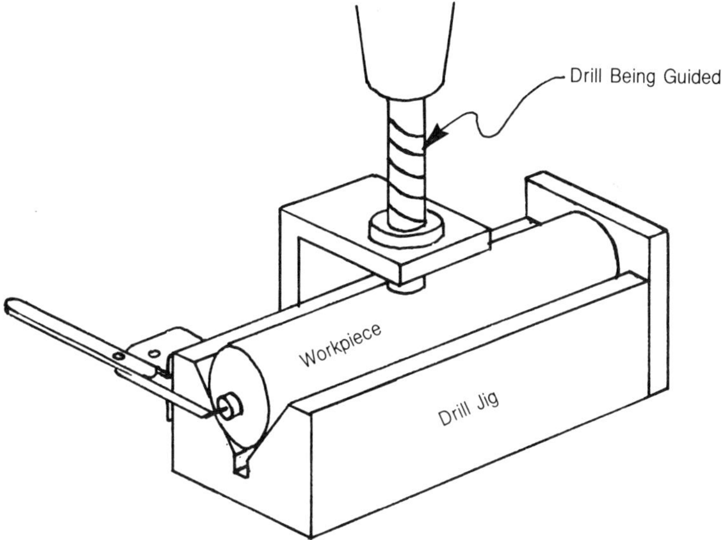

Figure 5.7 A jig.

5.2. JIGS AND FIXTURES

Jigs and fixtures are used primarily for holding parts during manufacturing. Jigs physically limit and control (guide) the path of a cutting tool during the machining process (Fig. 5.7). Fixtures do not guide the tool, but allow it to find its own path (Fig. 5.8). Both jigs and fixtures eliminate the six degrees of freedom of the part which is required during machining. There are other tools used for positioning parts relative to each other which are not used with the machining process and they are commonly called fixtures, too. They are known by the operation they perform such as assembly fixtures, welding fixtures, or soldering fixtures.

5.2.1. Analyzing the Operation

Before a jig or fixture can be designed for producing a part, the actual operation involved in producing the part must be first analyzed. An operation analysis of a part consists of a systematic procedure for finding all the facts that affect the production of that given part.

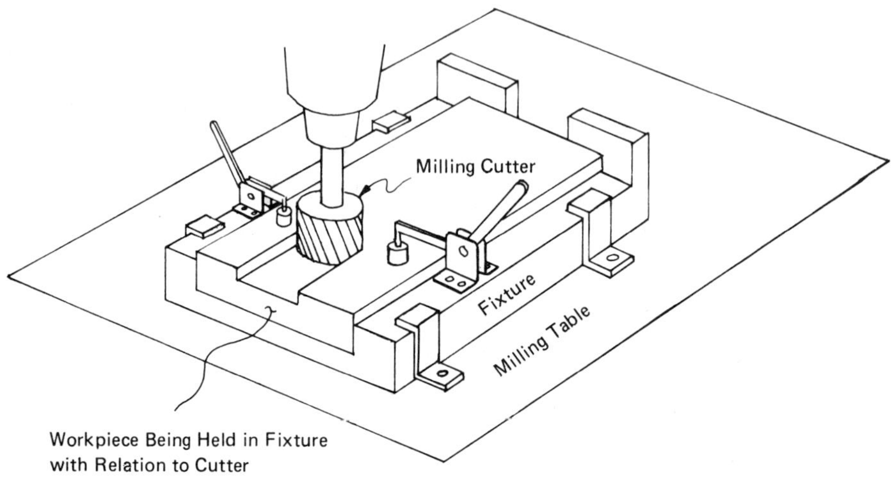

Figure 5.8 A fixture.

The procedure is based primarily on securing a complete overall understanding of every operation necessary to produce the part, from raw material to the finished product. It consists primarily of the following factors:

1. All operations performed on a part must be known
2. Purpose of each operation
3. Type and machineability of material used
4. Accuracy, tolerance, and inspection specifications associated with the part
5. Material handling methods used to transport the part
6. Machines and tools used to produce the part
7. Special tools that will be used such as jigs and fixtures, and their setups
8. Physical working area
9. Working conditions

The second stage of the operation analysis procedure involves the investigation of each single operation for the purpose of obtaining the most economical method. This investigation includes:

1. Total quantity of parts to be made
2. Production rate
3. Required operations to be performed on the part
4. Power and capacity requirements of the equipment that will be used to produce the part
5. Amount of features (holes, slots, etc.) to be made on the part

6. Amount of dimensions and tolerances used to ensure the proper relationships between features
7. Required work cycle time
8. Motion study involving the arrangement of work area and the motions used to place the part in a jig or fixture which would include load, clamp, position, machine, reposition, unclamp, unload, and clean

Besides positioning the part correctly in relation to the tool, jigs and fixtures are used to maintain the least amount of motion required by an operation or robot in the placement and removal of the part in a jig or fixture. Motion is divided into five elementary classes.

1. Finger
2. Finger and wrist
3. Finger, wrist, and forearm
4. Finger, wrist, forearm, and upper arm
5. Finger, wrist, forearm, upper arm, and body

Generally, motions involving the finger and wrist require less effort than a finger, wrist, forearm, and upper arm motion. Therefore, jigs and fixtures should be designed so that the least amount of motion is required to use them. Also, jigs and fixtures should be designed to eliminate unnecessary motions, balance the work of each hand, minimize the use of the eyes and eliminate the use of the hands for holding the part, and provide a comfortable working condition (eliminate unnecessary fatigue and strain).

The work to be performed on a part when using jigs and fixtures can be divided into four categories listed below, and are helpful in the design of jigs and fixtures.

1. Locating and positioning
2. Clamping
3. Chip control
4. Tool guides

5.2.2. Locating and Positioning

Locating and positioning a part in a jig or fixture refers to the establishment of a required relationship between the part and the jig or fixture. The actual relationship between the part and the tool, as established by the jig or fixture, is indirect. This fact is important since it ensures that parts produced from such jigs and fixtures are interchangeable.

For instance, a hole that is to be drilled in a shaft will be located exactly for each shaft which is to be drilled, because the path of the cutting tool (drill) is guided by bushings in the jig body as shown in

Figure 5.7. A part to be milled would be located in a fixture, which in turn is mounted to the milling table in direct relation between the jig or fixture, and the cutting tool remains constant while the work is indirectly positioned to the cutting tool by the special part incorporated into the jig or fixture.

Overall accuracy is thus dependent primarily on the accuracy with which the workpiece is consistently located within the jig or fixture. Imporper workpiece locating can easily result in excessive troubles and scrap.

Locating means and methods can best be determined during operation analysis from a study of the number, sequence, and tolerance of individual machining operations required, in addition to the individual characteristics of the particular machine tools available for the work.

In general, the first operation would produce surfaces, holes, or bores that can be utilized as locating points in all subsequent operations. As far as is practicable, these locating points should be on that portion of the work that is required to be the most accurate. If, however, the operation about to be performed is dimensioned relative to previously machined surfaces or holes other than those produced in the first operation, then it is advisable to locate from those points. It is occasionally advantageous to deliberately provide extra pads or surfaces that can be machined in the first operation and used as locating points, or to facilitate clamping, in subsequent operations.

To completely locate a part, it must be held against at least three points in a primary plane, at least two points in the secondary plane, and at least one point in the tertiary plane. This is referred to as the 3-plane concept where the three planes are mutually perpendicular to each other and the points in any one plane should be as far apart as possible. Each of these points eliminates one of the six degrees of freedom of the part in space, but in some instances it is not necessary to locate the part with respect to six degrees of freedom. Therefore, a fewer number of points can be used to locate the part, such as a part held in place by a chuck during a lathe-machining operation.

The proper location of a part in a jig or fixture can be affected by chips and dirt, locator wear, inadequate support, and locator design. A burr (minor burr) is produced at the start of a cut and at the end of a cut (major cut). Jigs and fixtures should be designed so that these burrs do not hinder the removal of the part. Suitable clearance grooves or slots should be provided to prevent sticking of the part upon removal.

Chips and dirt are byproducts of the machining process which cause locator wear and influence proper location. Chips and dirt are controlled by making the locators easy to clean, making them self-cleaning, or by protecting them. Therefore, locators should be made as small as possible within their wear specifications and mounted so that they are raised above the surrounding surfaces. Having the locator raised allows the chip to fall and be removed easily. The jig and fixture should be designed with openings and exits to provide for the removal of these chips.

Self-cleaning locators consist of sharp edges or grooves which scrape chips and dirt off the surfaces of the part as they slide across them. Providing reliefs around locating surfaces is another method of removing chips and dirt. The coolant used as the cooling and lubricating agent during machining can be used to flush chips and dirt away, or the chips and dirt can be pushed along or aside by using fixed wipers wiping against the fixture when it is moving on the machine table.

Several other methods of removing chips and dirt include using air suction if the chips are light and discontinuous or by gravity slides where the vibration of the machine induces chips to slide into a trough and a conveyor system removes them into containers.

A part should cover the entire locator when it is resting against it. This design helps protect the locator from chips and dirt. Enclosing locators such as slides and indexing pins and buttons helps to protect them from dirt and chips.

5.2.3. Locator Wear

Locating and positioning accuracy resulting from wear can be classified as rubbing, compression, erosive, or corrosive wear due to the particular cause. Wear due to rubbing occurs when surfaces slide over each other where a small amount of foreign material lies between the surfaces. Contact wear involves the actual atomic exchange of particles from the surface of the part to the locator or vice versa while wear due to abrasion occurs whenever the foreign particles are harder than either the part of locator, thus, removing a small amount of material from the part and/or locator resulting in wear.

If the compressive stresses in the locator exceed the proportional limit of the locator's material, peening, flaking, or chipping of the locator's surface will occur. This type of wear is referred to as compressive wear and it is caused by having a locator supporting a part which is too heavy. Erosive wear is caused by fluid movement such as coolant passing by the locator, and corrosive wear is caused by a chemical reaction occurring between the part and the locator such as an electrolytic chemical or common rusting.

Locator wear can be minimized by specifying high surface hardness, introducing a rolling instead of sliding contact, providing large contact areas, providing adequate lubrication of surfaces, and shielding or protecting wear surfaces as much as possible from dirt, and moisture.

5.2.4. Supporting the Part

Perhaps the major cause of inaccuracies in locating a part in a jig or fixture is the actual deflection of the jig or fixture during the machining operation. Forces induced into the part during machining are directly transmitted into the fixture, causing unwanted deflections of the part and dislodging the part from the surface of the locators. A free-body diagram of the fixture with stress and strain calculations must be performed to ensure that a jig or fixture is strong enough.

5.2.5. Methods of Location

A jig or fixture should be designed to be foolproof. A jig or fixture is foolproof when it is impossible to place the part in the jig or fixture in an improper position. The most common devices used for foolproofing jigs and fixtures are blocks and pins placed in such a manner so as to allow entry of correctly positioned parts and prevent incorrectly loaded parts from entering.

The part can be located in a jig or fixture by using sight, holes and pins, surfaces and so on. For example, the first-operation location of sand castings or welded assemblies in fixtures is done by sight, owing to the large dimensional variation encountered with sand castings and welded assemblies. Therefore, the part's irregularities must be balanced in such a manner so as to provide adequate material for machining purposes.

If holes are machined in the part, they can be used for locating purposes. It is recommended that two pins located as far apart as possible be used for accurate locating purposes. The use of two straight pins is not recommended, since a slight dimensional variation between holes may make it difficult to insert or remove the part. The most common means is to use one straight and one diamond pin. Pointing the diamond pin in the proper direction to hold the dimension makes it as satisfactory as a close-fitting round pin, in addition to facilitating the rapid insertion and removal of the part.

The most common method of locating a part in a jig or fixture is by using the surfaces (edges) of the part. Such an arrangement would provide for the surfaces of the part to touch the three mutually perpendicular sides of the fixture resulting in the proper location and containment of the part.

5.2.6. Clamping Methods

A clamp is used to exert a force on the part so that the part is forced against the locating surfaces of the jig or fixture to hold it in position against the forces caused by cutting or other manufacturing forces.

ASSEMBLY NUMBER											
Acorn Nut	Nut	Hand Knob	Cam	A	B	C	D	Thread E	F	Travel G	H
	ACN-220	ACN-220-HK	ACD-220	2	5/8	3/8	15/16	1/4-20	3/8	1/2	1 1/8
	ACN-225	ACN-225-HK	ACD-225	2 1/2						1	
ACN-430-NA	ACN-430	ACN-430-HK	ACD-430	3	1	1/2	1 3/4	3/8-16	9/16	1/2	1 15/16
ACN-435-NA	ACN-435	ACN-435-HK	ACD-435	3 1/2						1	
ACN-440-NA	ACN-440	ACN-440-HK	ACD-440	4						1 1/2	
ACN-535-NA	ACN-535	ACN-535-HK	ACD-535	3 1/2	1 1/4	5/8	2	1/2-13	9/16	11/16	2 1/4
ACN-540-NA	ACN-540	ACN-540-HK	ACD-540	4					3/4	1	
ACN-545-NA	ACN-545	ACN-545-HK	ACD-545	4 1/2						1 1/2	
ACN-555-NA	ACN-555	ACN-555-HK	ACD-555	5 1/2			2 1/2			2	2 3/4
ACN-560-NA	ACN-560	ACN-560-HK	ACD-560	6					1 1/4		
ACN-640-NA	ACN-640	ACN-640-HK	ACD-640	4	1 1/2	7/8	2 1/4	5/8-11	11/16	13/16	2 1/2
ACN-645-NA	ACN-645	ACN-645-HK	ACD-645	4 1/2					7/8	1 1/8	
ACN-650-NA	ACN-650	ACN-650-HK	ACD-650	5						1 5/8	
ACN-660-NA	ACN-660	ACN-660-HK	ACD-660	6			2 3/4			2 1/8	3
ACN-665-NA	ACN-665	ACN-665-HK	ACD-665	6 1/2					1 3/8		
ACN-750-NA	ACN-750	ACN-750-HK		5	1 3/4	1	2 1/2	3/4-10	7/8	1 3/8	2 3/4
ACN-760-NA	ACN-760	ACN-760-HK		6			3		1 1/8	1 5/8	3 1/4
ACN-765-NA	ACN-765	ACN-765-HK		6 1/2						2 1/8	
ACN-775-NA	ACN-775	ACN-775-HK		7 1/2			3 1/2			2 5/8	3 3/4
ACN-780-NA	ACN-780	ACN-780-HK		8					1 5/8		

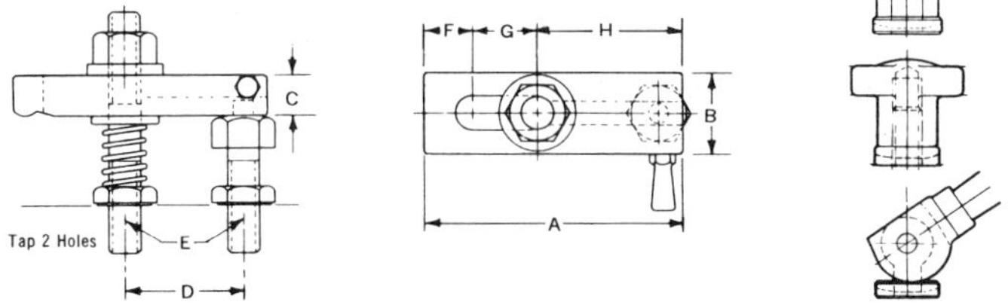

Figure 5.9 Standard clamp assemblies. (Courtesy of Engineering Products Inc., Farmington, Michigan.)

The clamp should be placed directly opposite the locating area on the jig or fixture, and preferably through heavy sections of the part. The force induced into the part from cutting should be directed into the locating area of the jig or fixture instead of the clamp, although some of this cutting force will be directed into the clamp during the cutting operation. For a clamp to be effective, it should exert a minimum force equal to the largest force imposed upon it during the cutting operation.

There are a variety of clamps available such as standard, cam, toggle, swinging, hook, etc. Figures 5.9 through 5.13 show various clamps from each category. The following criteria should be used when choosing a proper clamp.

1. Use only simple clamps. Complicated clamps lose their effectivness as they wear.

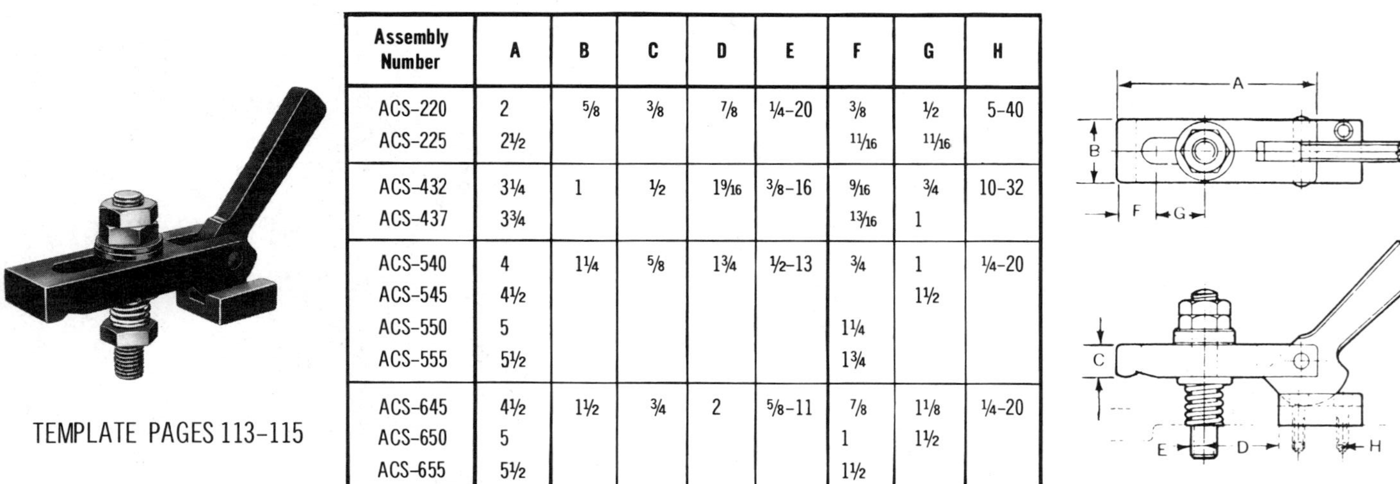

TEMPLATE PAGES 113–115

Assembly Number	A	B	C	D	E	F	G	H
ACS-220	2	5/8	3/8	7/8	1/4–20	3/8	1/2	5-40
ACS-225	2 1/2					11/16	11/16	
ACS-432	3 1/4	1	1/2	1 9/16	3/8–16	9/16	3/4	10-32
ACS-437	3 3/4					13/16	1	
ACS-540	4	1 1/4	5/8	1 3/4	1/2–13	3/4	1	1/4–20
ACS-545	4 1/2						1 1/2	
ACS-550	5					1 1/4		
ACS-555	5 1/2					1 3/4		
ACS-645	4 1/2	1 1/2	3/4	2	5/8–11	7/8	1 1/8	1/4–20
ACS-650	5					1	1 1/2	
ACS-655	5 1/2					1 1/2		

Figure 5.10 Single cam assemblies. (Courtesy of Engineering Products Inc., Farmington, Michigan.)

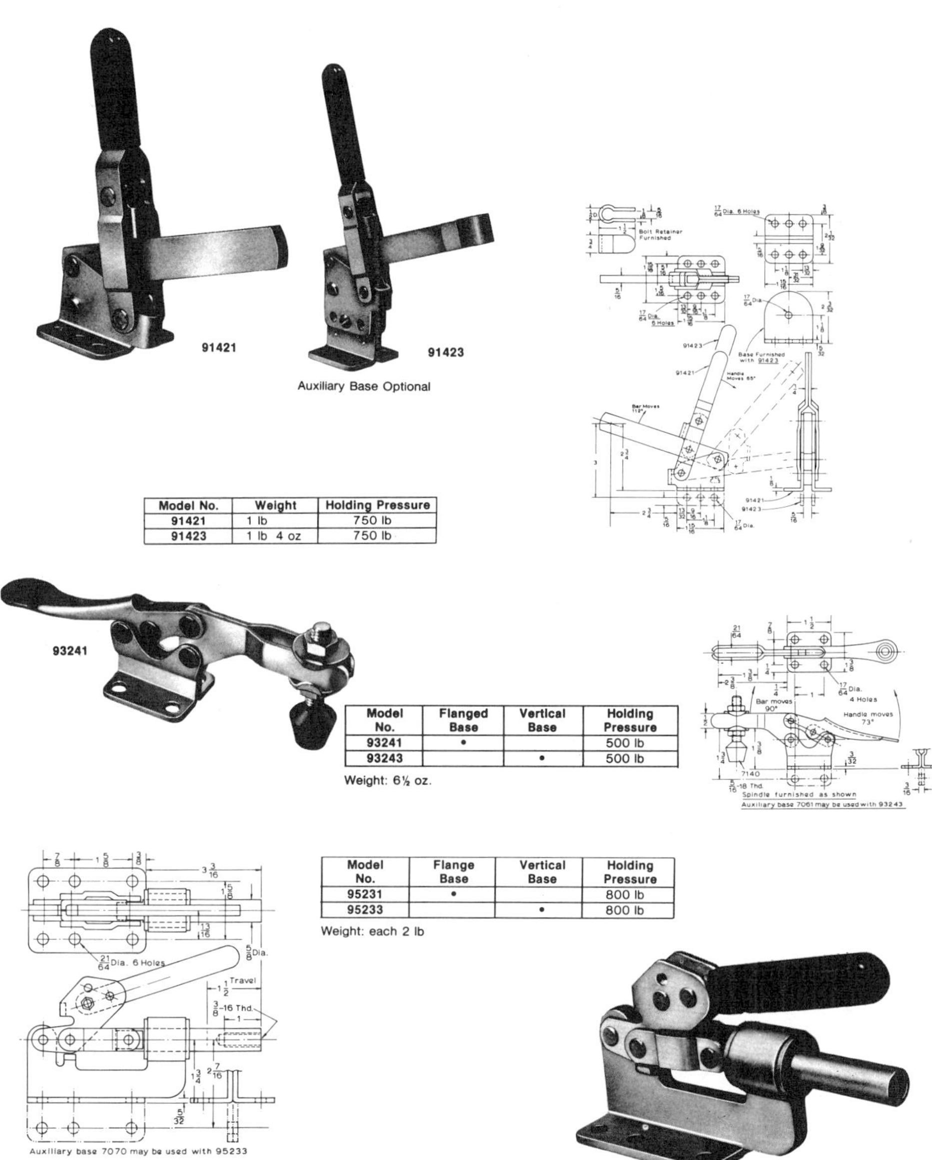

Model No.	Weight	Holding Pressure
91421	1 lb	750 lb
91423	1 lb 4 oz	750 lb

Model No.	Flanged Base	Vertical Base	Holding Pressure
93241	•		500 lb
93243		•	500 lb

Weight: 6½ oz.

Model No.	Flange Base	Vertical Base	Holding Pressure
95231	•		800 lb
95233		•	800 lb

Weight: each 2 lb

Figure 5.11 Toggle clamps. (Courtesy of Engineering Products Inc., Farmington, Michigan.)

Assembly Number	A	B	C	D	E	F	G Ream	H Screw	J Thd.	K Min.	K Max.
ASC-312	1½	—	—	—	—	13/16	9/16	—	5/16-18	1/4	7/16
ASC-502	2	—	—	—	—	1¼	.875	—	½-13	5/16	3/4
ASC-503	2¾	—	—	—	—	1¼	.875	—	½-13	5/16	3/4
ASC-623	3	—	—	—	—	1⅝	1.125	—	⅝-11	3/8	3/4
ASC-624	4	—	—	—	—	1⅝	1.125	—	⅝-11	3/8	3/4
ASC-312-B	1½	3/4	3/4	1	1/4	—	—	10-24	—	1/4	7/16
ASC-502-B	2	1 1/16	1⅛	1⅜	5/16	—	—	¼-20	—	5/16	3/4
ASC-503-B	2¾	1 1/16	1⅛	1⅜	5/16	—	—	¼-20	—	5/16	3/4
ASC-623-B	3	1 5/16	1⅜	1½	1/2	—	—	⅜-16	—	3/8	3/4
ASC-624-B	4	1 5/16	1⅜	1½	1/2	—	—	⅜-16	—	3/8	3/4

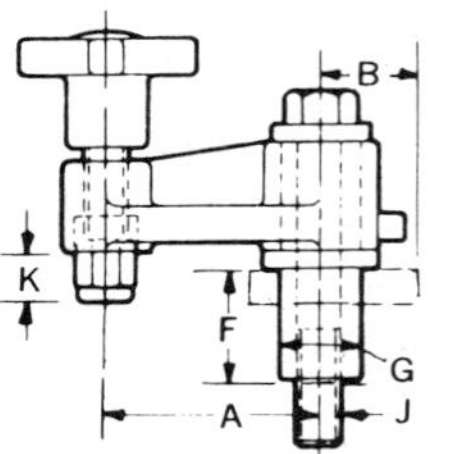

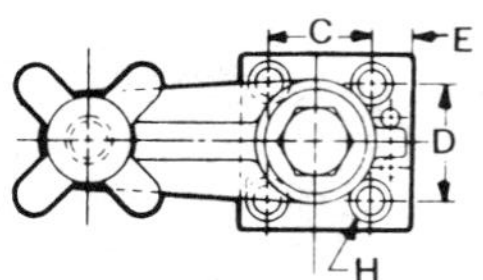

Figure 5.12 Swing clamp assemblies. (Courtesy of Engineering Products Inc., Farmington, Michigan.)

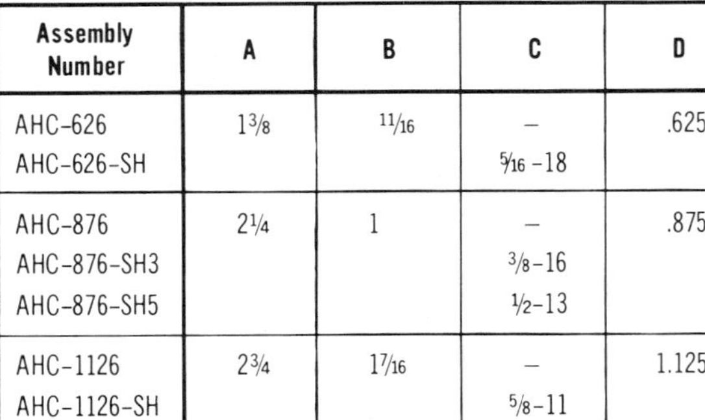

Assembly Number	A	B	C	D	E Thd.
AHC-626 AHC-626-SH	1 3/8	11/16	– 5/16 -18	.625	5/16 -18
AHC-876 AHC-876-SH3 AHC-876-SH5	2 1/4	1	– 3/8-16 1/2-13	.875	1/2-13 3/8-16 1/2-13
AHC-1126 AHC-1126-SH	2 3/4	1 7/16	– 5/8-11	1.125	5/8-11

Clamps are hardened.
Blocks are soft. Black oxide finish.

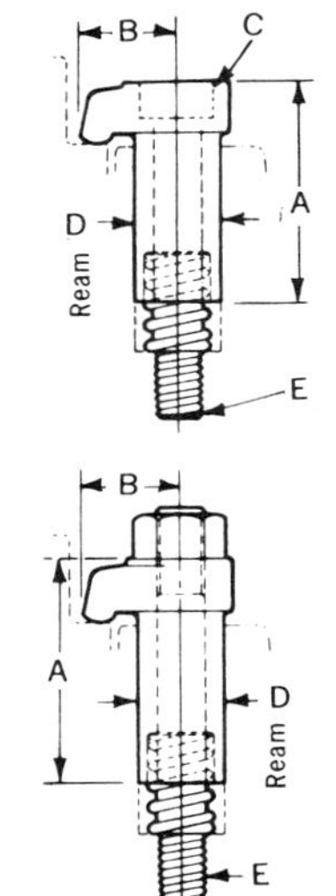

Assembly Number	A	B	C	D Screw
AHC-626-B	1 1/4	1 7/16	7/8	1/4
AHC-876-B	2 1/4	1 13/16	1 1/8	3/8
AHC-1126-B	2 3/4	2 1/4	1 1/2	1/2

Figure 5.13 Hook clamp assemblies. (Courtesy of Engineering Products Inc., Farmington, Michigan.)

2. Choose clamps suitable for the size and weight of the parts.
3. Rough parts with large dimensional variations need clamps having a longer travel in the clamping range, and which will dig into rough surfaces to hold them firmly.
4. The operation to which the clamps are applied determine the type of clamps which will be used.
5. Clamps should not make loading and unloading the part to and from the jig or fixture difficult.
6. Clamps which are able to move upon tightening, such as a plain straps, should not be used for production work. The part must be held rigid.
7. The number of loading and unloading cycles of parts in a jig or fixture determine if hydraulic clamping could be utilized.
8. Check on the safety features of the clamps for the operator.
9. The clamps should exert sufficient pressure on the part without too much effort on the part of the operator.
10. Always clamp over the jig's fixture's support areas.
11. If the operator is to use a wrench, all nuts should be the same size.

5.3. GAGE DESIGN

A gage is a device for determining the relative size or shape of a part. It is used either to control a part's dimensions within specified limits (tolerances) and/or to reject a part that is outside those specified limits. The term "gaging" refers to the process of investigating or determining if the part conforms dimensionally with the dimensions and tolerances on the drawing and if it satisfies the specifications called for on the drawing.

Generally, there are two methods used in gaging a part. A part can be gaged by using a fixed-limit gage to assure that the part is within the prescribed limits of size or it can be gaged with indicating gages to measure the extent of deviation of the part from a normal or basic size. Both fixed-limit and indicating gages are used to measure a part's limits, while indicating gages are used only to measure the part's deviation.

There are two types of fixed-limit gages. The first type is referred to as a GO-gage, which controls the minimum allowable looseness or maximum tightness in the fit between mating parts. In other words, they control the interchangeability between mating parts. The second type is referred to as a NO-GO gage or a minimum material limit gage. These gages control the maximum looseness or minimum interference between mating parts or, in other words, they control the tightness or looseness of the assembled parts. Using fixed-limit gages in measuring a part's limits provides a rapid method of determining whether or not the part is within its prescribed limits.

Indicating gages aid inspectors and lessen the time required for the gaging operation by indicating the measurements on a scale of some sort and incorporating some sort of magnifying device.

5.3.1. Gage Tolerances

Since a gage is also manufactured, it must have tolerances associated with it. Although the gage tolerances are small compared to the part's tolerances, they must be taken into account since they become a part of the part's tolerance.

A part is usually gaged twice, the first time at the production machine by a work or production gage, the second time in the inspection department by means of an inspection gage. Gage tolerances used for working gages are usually 10 percent of the part's total tolerance and up to 5 percent for inspection gages when no gage tolerance is specified.

Gagemakers' tolerances for plain cylindrical plugs and rings have been standardized into four classes, XX, X, Y, and Z according to nominal size and the degree of accuracy necessary in the gaging operation. The direction of tolerance is shown in Figure 5.14a. The cylindrical ring and plug gage symbols XX, X, Y, and Z are applied to specific gages. These symbols are listed below with their meanings.

Ring gage:

XX. Precision-lapped to very close tolerances and used principally for master gages and very close-toleranced parts.

X. Precision-lapped to very close tolerances and used principally as a master and reference gages. They are not used as working gages except when especially close tolerances are required.

Y. Precision-lapped and used in working gages and inspection gages of accurately ground parts.

Z. Used when extreme accuracy is not essential and when the working tolerances are fairly liberal. These gages have a ground-and-lapped finish.

Plug gage:

XX. Precision-lapped to very close tolerances and used principally for master gages, master disks, and setting gages for a close-toleranced part.

X. Precision-lapped to close tolerances. They are used for some types of master and inspection gages and for working gages when the tolerances are especially close and when a high grade of precision in the work is required.

Y. Lapped to slightly larger tolerances than the preceding classes and intended to be used as working gages.

Z. A commerical finish (i.e., ground-and-lapped). They are used

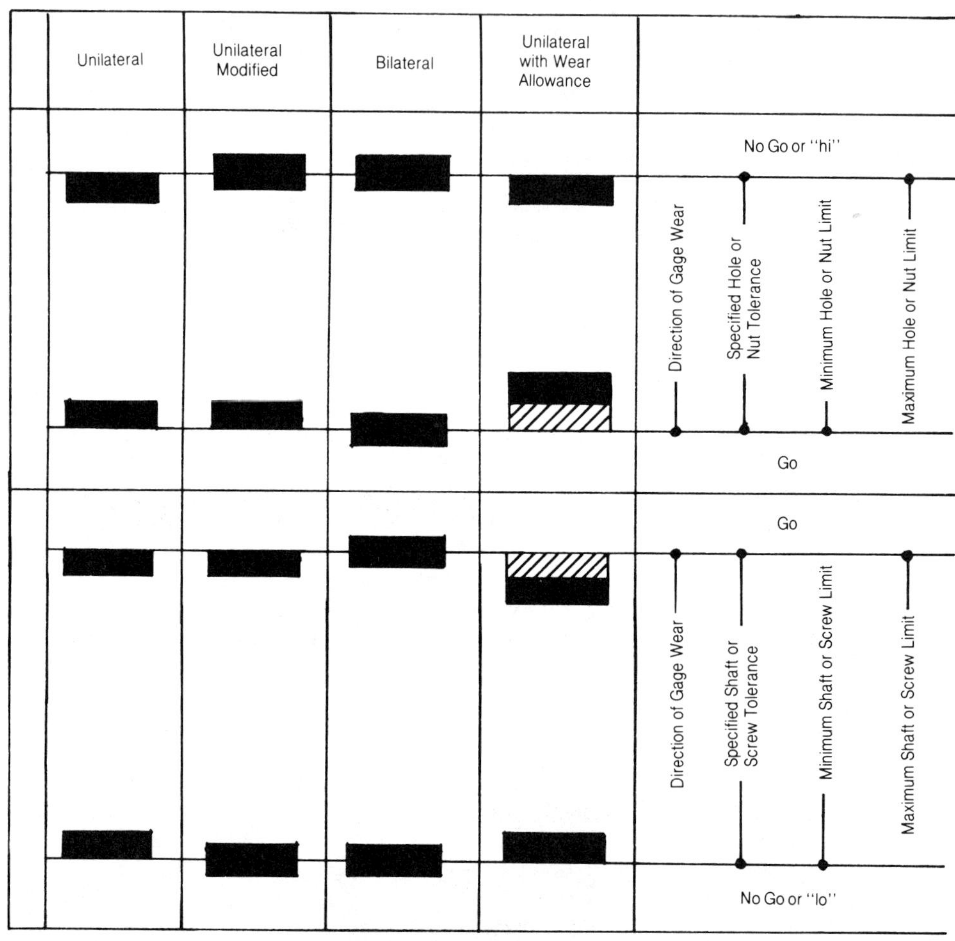

Figure 5.14a Methods of allocating gage tolerances.

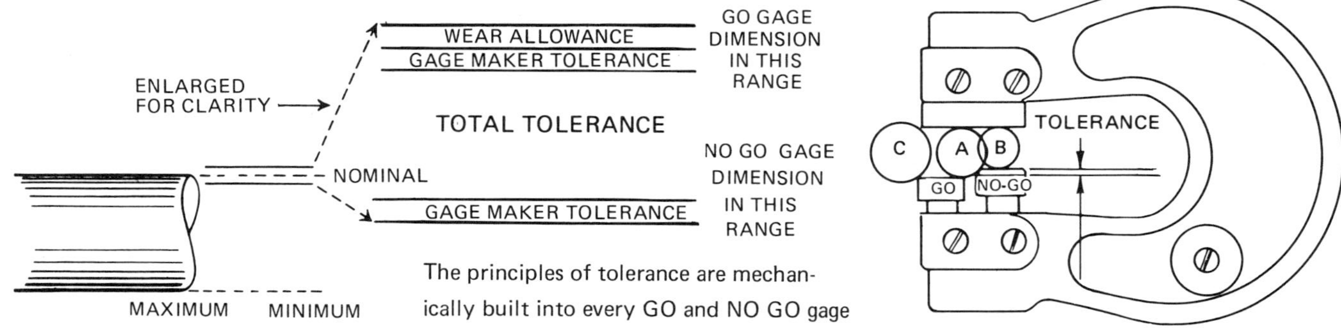

Figure 5.14b Tolerance and fixed gages.

as working gages when the tolerances are liberal and/or the quantities of pieces to be gaged are small.

The gagemaker's tolerances for plain cylindrical plugs and rings appear in Table 5.1, and how they are applied to fixed gages is shown in Figure 5.14b.

5.3.2. Gaging Systems

Bilateral or unilateral, or a combinations of the two tolerancing systems can be used with gages. The maximum or minimum work limits, or both, bisect the gage tolerance; in other words, gage tolerance is equally split and applied one-half plus and one-half minus. This is known as the bilateral tolerancing system. A unilateral gage tolerance is encompassed by the maximum and minimum work limits. The extreme size for all plain limit gages shall not exceed the extreme limits of the parts to be gaged. All variations in the gages, whatever their cause or purpose, shall bring these gages within these extreme limits. Thus, a gage which represents a minimum limit may be larger, but never smaller, than the minimum size specified for the part to be gaged; likewise, the gage which represents a maximum limit may be smaller, but never larger, than the maximum size specified for the part to be gaged.

Figure 5.15 illustrates how wear allowances are added to manufacturing tolerances on fixed-limit gages which are to be used to gage a female (hole) and male (shaft) features.

For example, a production GO-NO GO plug snap gage is to be designed to check a hole shaft diameter of 1.875 ± 0.0015. The tolerance (T_1) on the GO gage is obtained by taking 10% of the hole's specified tolerance range ($H_{max} - H_{min.} = 0.0030$) which yields 0.10 (0.0030) = 0.0003 inches. The wear allowance is 0.05 (0.0030) = 0.00015.

The GO plug is 1.8750 − 0.0015 = 1.8735 inches and the tolerance is applied in the plus direction or $1.8735^{+0.0003}_{-0.0000}$. The NO-GO snap gage is 1.8750 + 0.0015 = 1.8765 inches and the tolerance is applied in the minus direction or $1.8765^{+0.0000}_{-0.0003}$. The NO-GO snap gage is 1.8750 − 0.0015 = 1.8735 inches and the tolerance is applied in the plus direction or $1.8735^{+0.0003}_{-0.0000}$.

Now, add the allowance for wear to the plug-and-snap gages. The wear allowance is 5% (0.003) = 0.00015. The GO plug dimension would be 1.8735 + 0.00015 = 1.87365 and the tolerance applied would be $1.87365^{+0.0003}_{-0.0000}$. The NO-GO plug gage will wear in the direction of the tolerance range. Therefore, no allowances need be made for wear.

Table 5.1 Wear Allowances and Gagemakers' Tolerances for Plain Cylindrical Plugs and Rings

COMPONENT	GAGE			
TOTAL TOLERANCE	WEAR ALLOWANCE	TOLERANCE GO	TOLERANCE NOT GO	TOTAL TOLERANCE - INCH (.005 .010 .015 .020 .025)
.0005	.0000	.00005	.00005	
.001	.0001	.0001	.00005	
.002	.0001	.0001	.0001	
.003	.0001	.0002	.0001	
.004	.0002	.0002	.0002	
.005	.0003	.0002	.0002	
.006	.0004	.0002	.0002	
.007	.0004	.0003	.0002	
.008	.0005	.0003	.0002	
.009	.0005	.0004	.0002	
.010	.0005	.0005	.0003	
.012	.0006	.0006	.0003	
.014	.0006	.0008	.0004	
.015	.0006	.0009	.0005	
.016	.0006	.0010	.0005	
.018	.0006	.0010	.0006	
.020	.0006	.0010	.0007	
.022	.0006	.0010	.0008	
.024	.0006	.0010	.0009	
.025 AND UP	.0006	.0010	.0010	
Gagemakers Tolerance, Flush Pin and Adjustable Snap Gages				
.0005		.00005	.00005	
.001		.00005	.00005	
.002		.0001	.0001	
.003		.0001	.0001	
.004		.0002	.0002	
.010		.0003	.0003	
.015		.0004	.0004	
.020 AND UP		.0005	.0005	

NOTES:

This chart shows, both numerically and graphically, wear allowances and gagemakers tolerances for inspection gages.

For component tolerances not listed, the next smaller tolerance shall be used.

The first four columns show numerical values while the following portion of the table shows graphic values.

Column 1 shows the total tolerance on the product. Column 2 the wear allowance on the "Go" gage.

Column 3 the tolerance on the "Go" gage, and column 4 the tolerance on the "No-go" gage.

There is no wear allowance on the "No-go" gage because it wears toward the "Go" gage and in the direction of safety.

Wear allowances and tolerances used as standard practice by plants in United States ordnance production.

Limits of Hole = H_{max} / H_{min}
Specified Tolerance of Hole = $H_{max} - H_{min} = T$
Working Tolerance of Hole = $T + (T_1 + T_2 + W)$

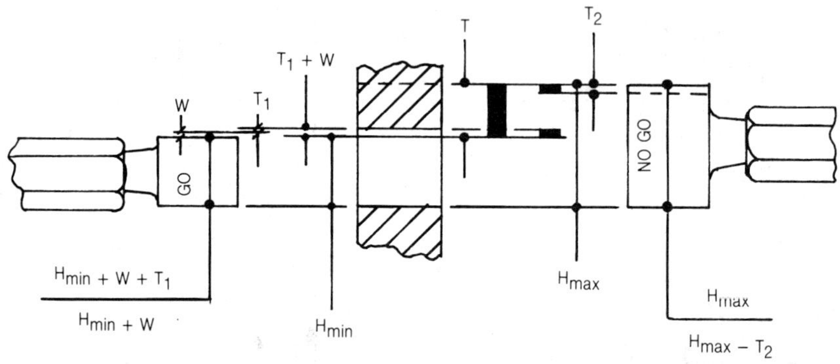

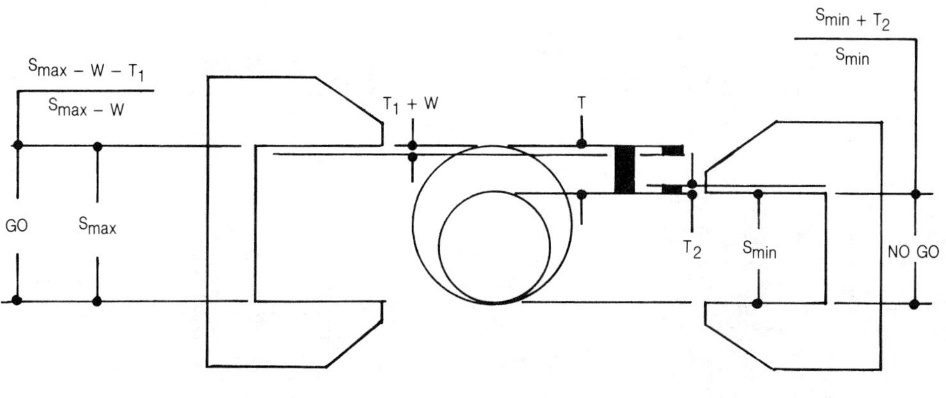

Limits of Shaft = S_{max} / S_{min}
Specified Tolerance of Shaft = $S_{max} - S_{min} = T$
Working Tolerance of Shaft = $T - (T_1 + T_2 + W)$

LEGEND

T_1 = Tolerance on Go Gage
H = Hole Diameter
W = Wear Allowance = $H_{min} - S_{max}$
T_2 = Tolerance on No Go Gage
S = Shaft Diameter

Figure 5.15 Wear allowances and manufacturing tolerance on fixed-limit gages.

The GO snap gage dimension is 1.87650 − 0.00015 = 1.87635 inches. The tolerance applied would be $1.87635\,{}^{+0.0000}_{-0.0003}$. The NO-GO snap gage will wear in the direction of the tolerance range. Therefore, no allowance need be made for wear. These dimensions are shown on the plug-and-snap gages shown in Figure 5.16a.

Again, consider mating parts such as the shaft and bore shown in Figure 5.16b (1). The nominal dimensions given are as follows: shaft diameter 1.000 inches, bore diameter 1.005 inches. This leaves a nominal clearance that cannot be maintained because the parts cannot be made dimensionally perfect. Consequently, a clearance range must be established which will afford satisfactory operation of the mating parts. For the sake of illustration assume that 0.005 inches is the maximum permissible clearance and 0.004 inches is the minimum permissible limit. The difference between these limits, or the *clearance tolerance,* is 0.001 inches.

Since the actual size of the parts determine the amount of clearance, the amount of clearance-tolerance should be allocated to the nominal part dimensions as a manufacturing tolerance so that the clearance limits are not exceeded. It is common practice to allocate one-half of the clearance-tolerance to each part to allow equal leeway in manufacturing. Thus, each part is given a tolerance of 0.001 inches divided by 2 or 0.0005 inches. These tolerances are now prescribed for each part so that when the permissible maximum error is cumulative (i.e., when the shaft is "under" and the bore is "over"), the clearance of 0.005 inches is not exceeded. Dimensions are now specified as shown in Figure 5.16b (2):

$$\text{Shaft diameter} = 1.0000\,{}^{+0.0000}_{-0.0005}$$

$$\text{Bore diameter} = 1.0040\,{}^{+0.0005}_{-0.0000}$$

Stated in terms of limits:

Shaft diameter = max 1.0000, min. 0.9995

Bore diameter = max 1.0045, min. 1.0040

If these tolerances are maintained, clearance between any two mating parts will always be between the desired limits 0.004 — 0.005 inches.

GO and NO-GO snap gages may be used to check the shaft diameters during production. Nominally, the GO gage should measure 1.0000 inches and the NO-GO gage 0.9995 inches. However, since gages also are a "made" item, tolerances must also be set for the gages. According to gagemakers' practice, the gage tolerance is required to be one-tenth of the total tolerance of the part to be gaged. In this example, each part has a tolerance of 0.0005 inches. The gage tolerance must therefore be 1/10 of 0.0005 inches, or 0.00005 inches (Fig. 5.16b(3)).

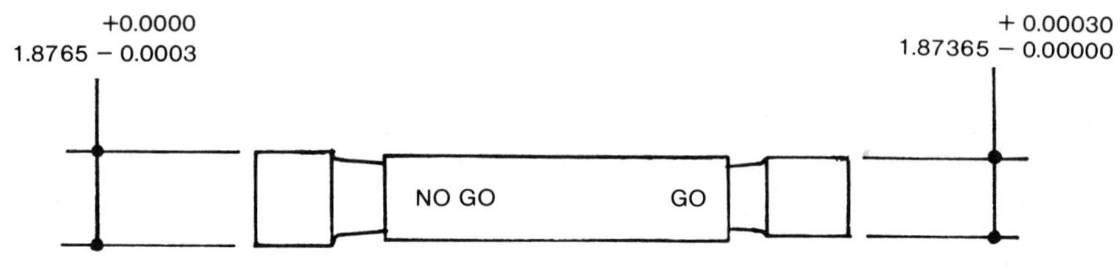

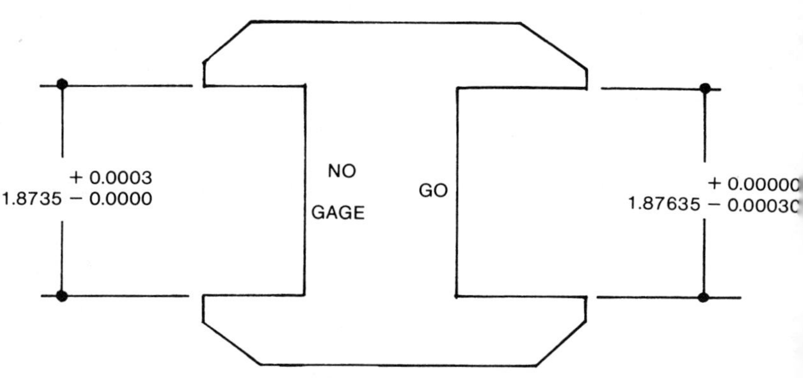

Figure 5.16a The dimensions and tolerances for the plug and snap gages.

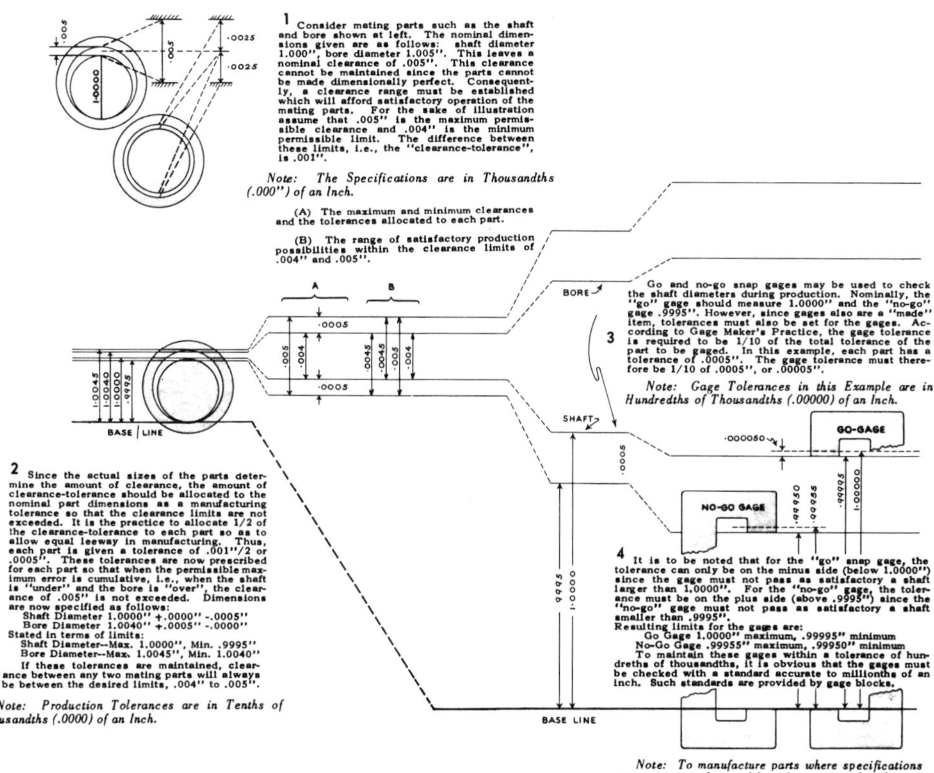

Figure 5.16b Visual clarification of clearance and tolerance.

It is noted that for the GO snap gage, the tolerance can only be on the minus side (below 1.0000 inches) since the gage must not pass as satisfactory a shaft larger than 1.0000 inches. For the NO-GO gage, the tolerance must be on the plus side (above 0.995 inches) since the NO-GO gage must not pass as satisfactory a shaft smaller than 0.9995 inches. Resulting limits for the gages are shown in Figure 5.16b(4) as:

GO gage = 1.0000 max, 0.9995 min.
NO-GO gage = 0.99955 max, 0.99950 min.

To maintain these gages within a tolerance of hundredths of thousandths, it is obvious that the gages must be checked with a standard accurate to millionths of an inch. Such standards are provided by gage blocks.

5.4. COMPUTER GRAPHICS APPLICATIONS

Jigs and fixtures can be easily designed and drawn using computer graphics, since the actual path of the cutting tool can be generated on the part, thus, simulating the actual machining process. Also, the various clamps, drill bushings, and so on used for holding the part and guiding the tool can be stored in the files of the computer and can be easily and quickly placed in any position on the display screen. There is even a program available which visually shows the cutting forces produced by the tool during the machining process and the effect these forces have on the part so that proper clamp selection and placement is made easier for the designer.

To illustrate how a drill jig would be designed using computer graphics, the orthographic drawing of the plug shown in Figure 5.17 will be used. A hole is to be drilled and threaded into the taper portion of the plug which means the drill must be guided during the drilling operation.

The first task is to draw the front, top, and left side view on the computer's display screen. The front view appears in the lower left quadrant (x,y), the top view appears in the upper left quadrant (x,z) and the left view appears in the lower right quadrant (y,z) portion of the display screen. A three-dimensional (3-D) model of the plug is automatically drawn in the upper right quadrant (x, y, z) of the display screen.

Figure 5.18 shows the plug and how it would be displayed on the computer's screen as the front, top, left, and 3-D views. Next, the front view only is placed on the display screen and the plug is rotated in the front view so the taper protion is horizontal to the x axis on the screen (Fig. 5.19). The plug is automatically rotated in the other three views by the computer.

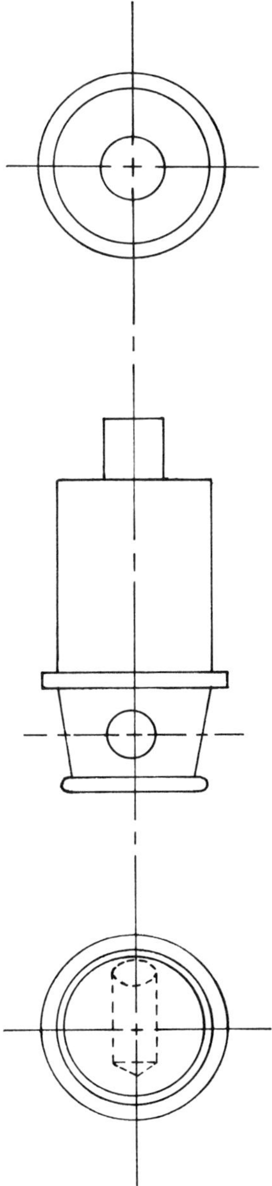

Figure 5.17 The detail drawing of the plug.

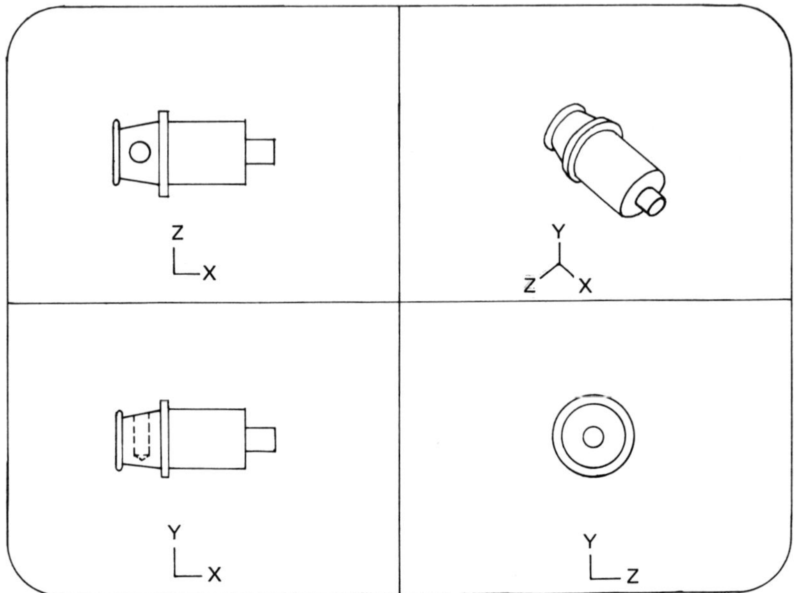

Figure 5.18 The plug displayed on the computer graphics screen.

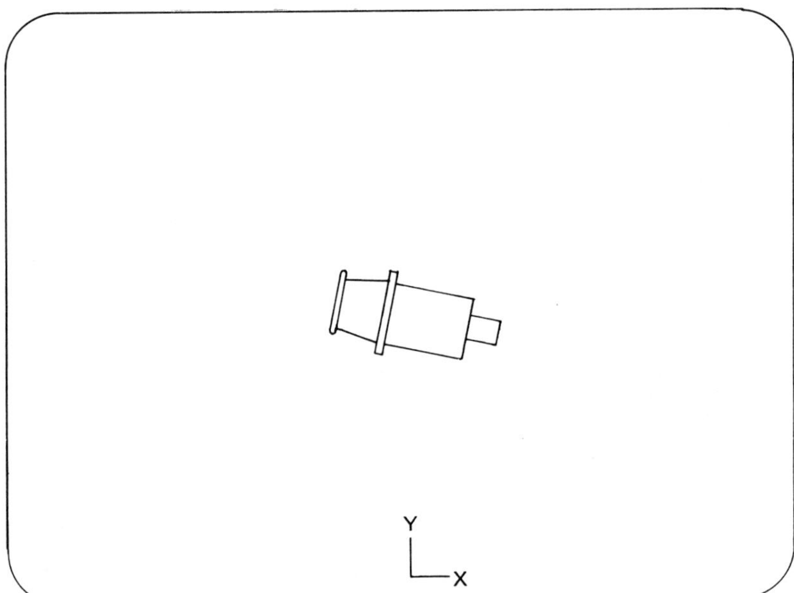

Figure 5.19 The plug rotated on the screen.

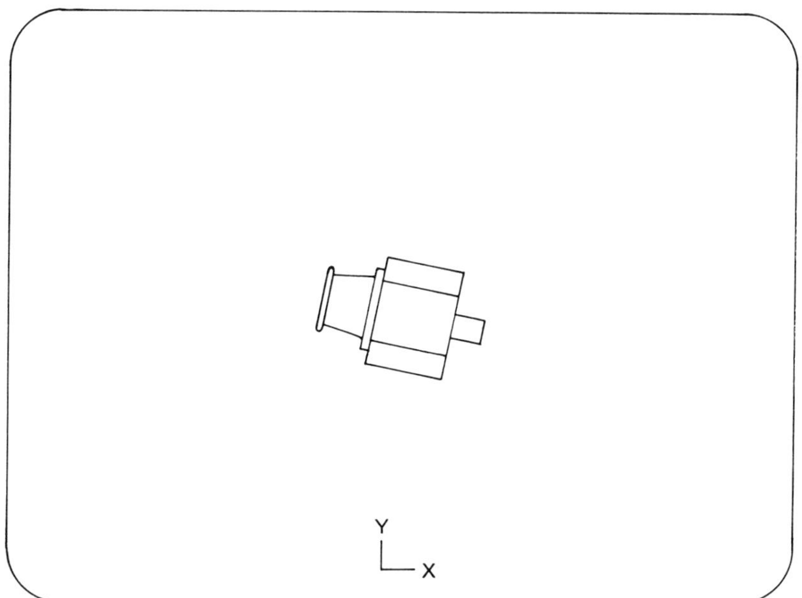

Figure 5.20 A bushing is placed around the plug on the screen.

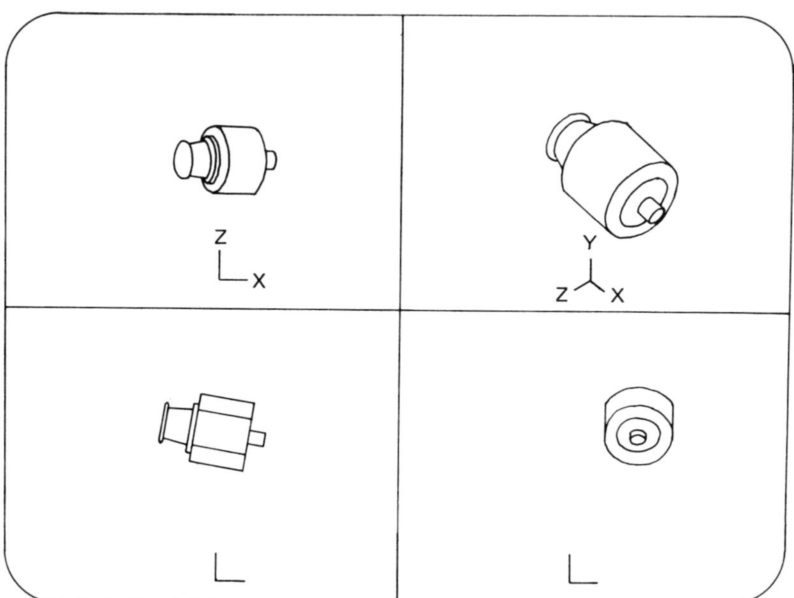

Figure 5.21 The bushing appearing in the four views on the screen.

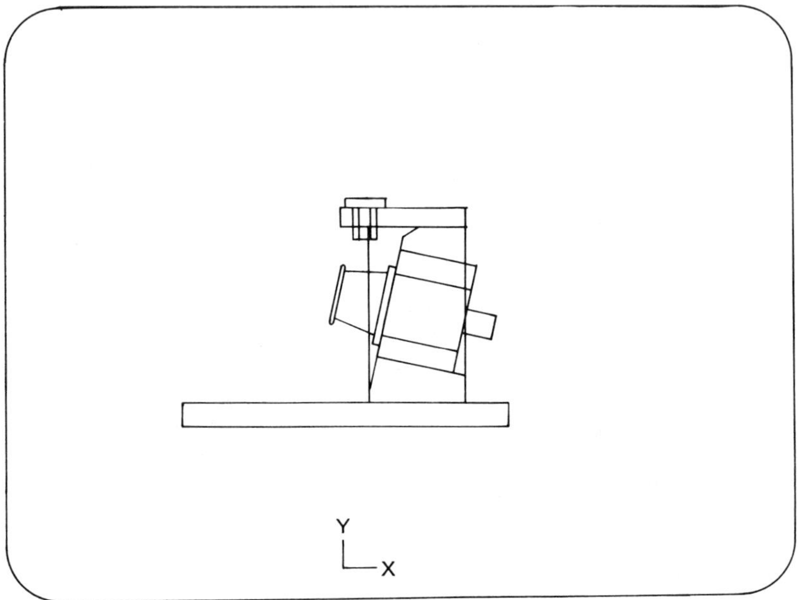

Figure 5.22 Drawing the drill jig in the front view on the computer graphics' display screen.

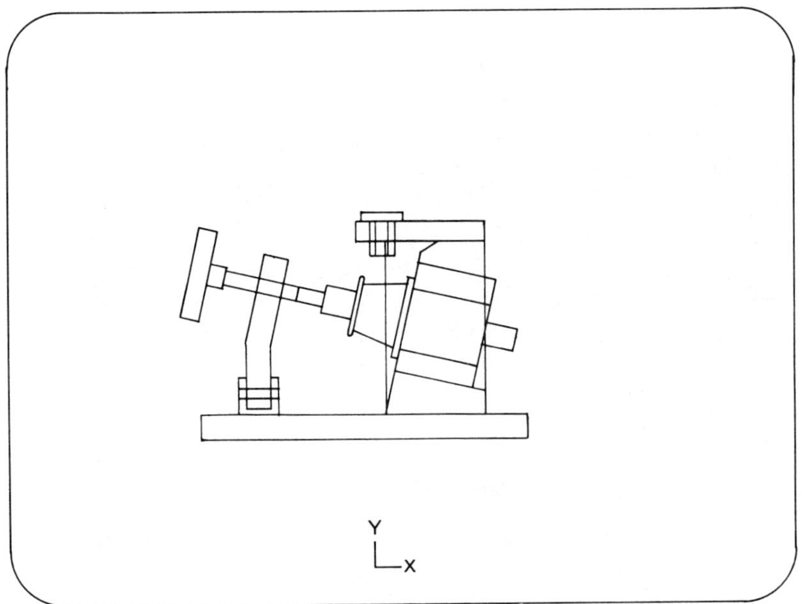

Figure 5.23 Continued drawing of the drill jig.

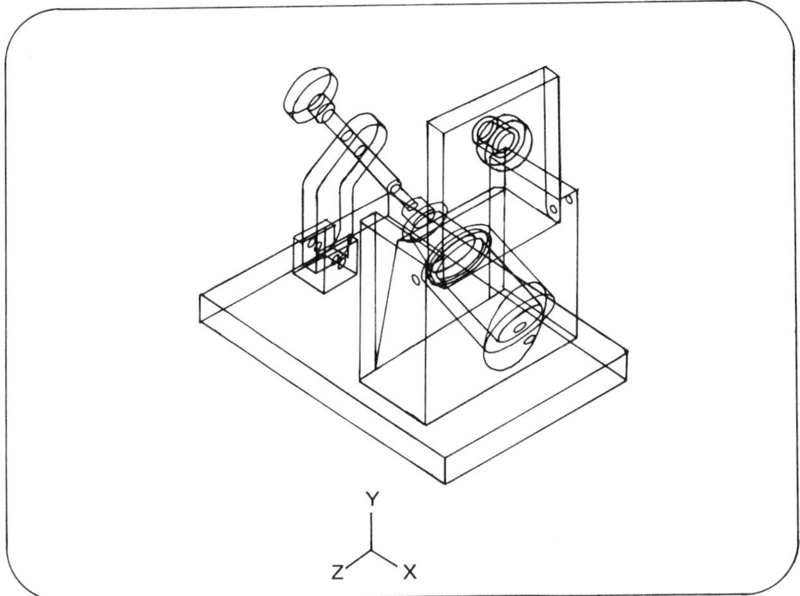

Figure 5.24 The final design of the drill jig appearing on the display screen.

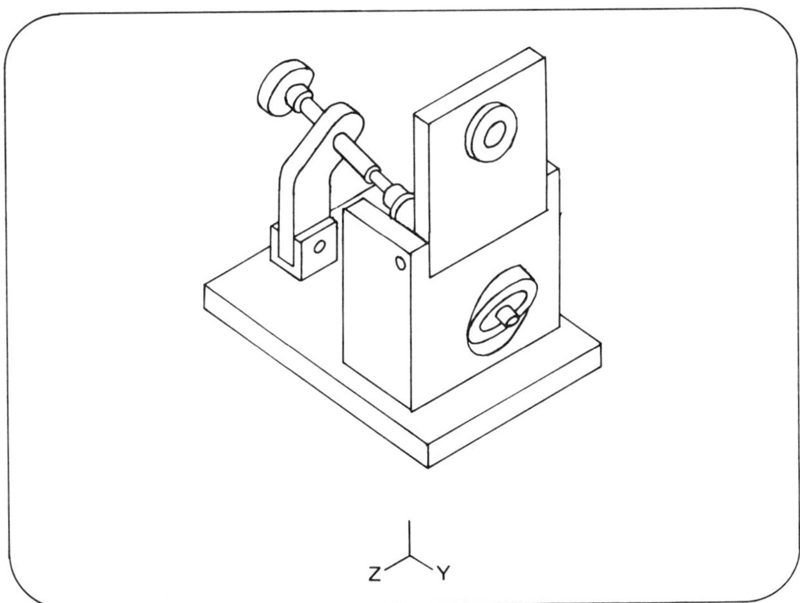

Figure 5.25 Hidden lines removed from the drill jig.

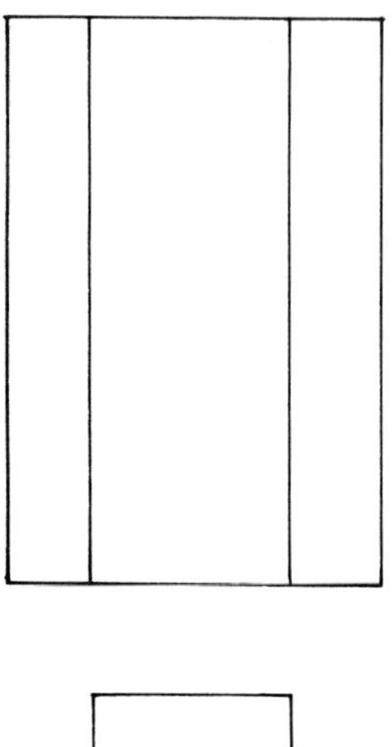

Figure 5.26 Steel block.

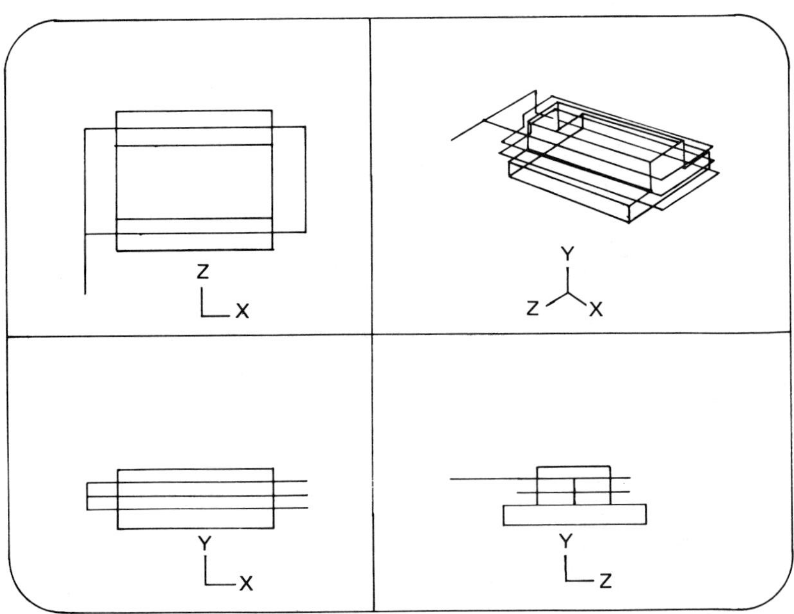

Figure 5.27 The cutter's path shown on the computer graphics' display screen.

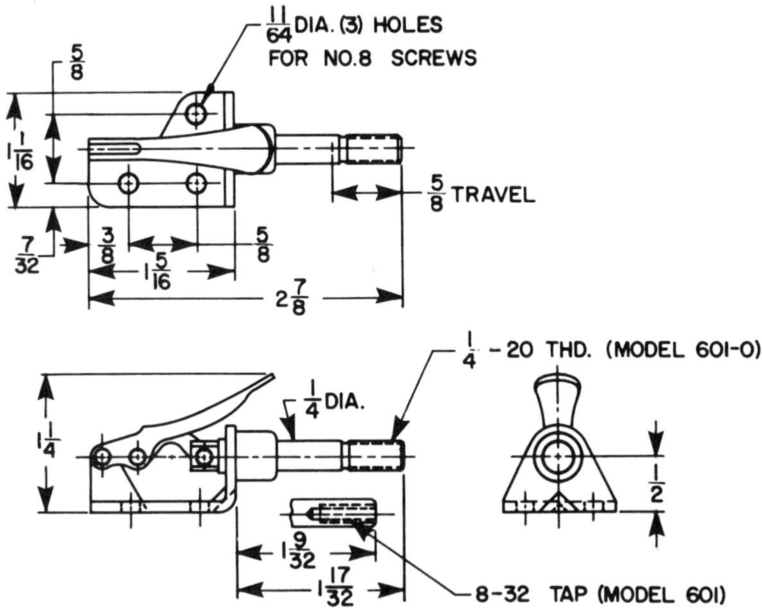

Figure 5.28 Detail drawing of a clamp.

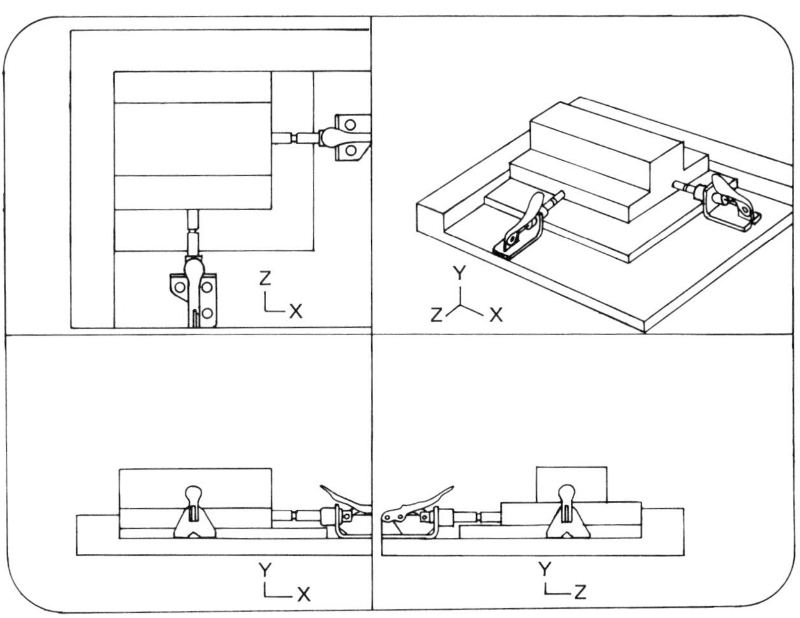

Figure 5.29 Placing the clamps on the computer graphics' display screen.

PLAIN WASHERS
TYPE W

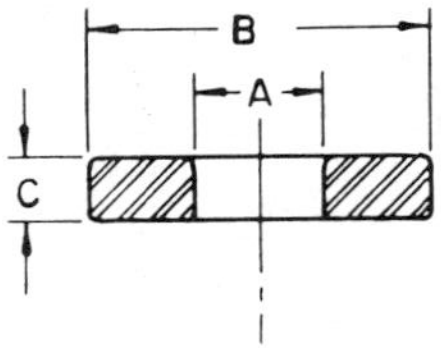

PLAIN WASHERS			
PART No.	A	B	C
W- 250	17/64	5/8	1/8
W- 312	11/32	3/4	1/8
W- 375	13/32	7/8	1/8
W- 376		1	3/16
W- 500	17/32	1 1/16	1/8
W- 501		1 1/4	3/16
W- 625	21/32	1 1/4	1/8
W- 626		1 1/2	1/4
W- 750	25/32	1 1/2	5/32
W- 751		1 5/8	1/4
W-1000	1 1/32	2	3/16

SPHERICAL WASHERS				
PART No.	A	B	C	R
WS- 250	9/32	5/8	3/16	3/4
WS- 312	11/32	7/8	1/4	1
WS- 375	13/32	1	1/4	1 1/2
WS- 500	17/32	1 3/16	5/16	1 3/4
WS- 625	21/32	1 3/8	5/16	1 3/4
WS- 750	25/32	1 5/8	3/8	2 1/4
WS- 875	15/16	1 7/8	3/8	2 1/2
WS-1000	1 1/16	2	1/2	2 3/4
WS-1125	1 3/16	2 1/4	1/2	3
WS-1250	1 15/16	2 5/8	1/2	3
WS-1500	1 9/16	2 7/8	1/2	3

Type W & WS Washers are Black Oxide and hardened. Bottom or top half of WS washers may be purchased separately.

SPHERICAL WASHERS
TYPE WS

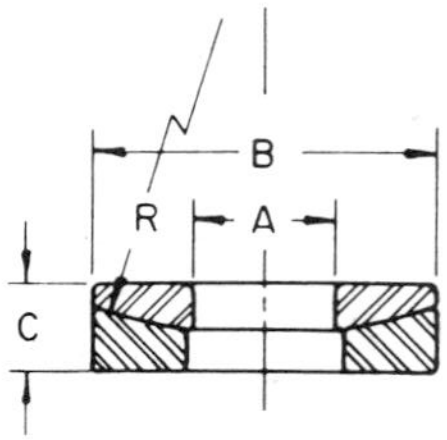

Figure 5.30 Flat and spherical washers. (Courtesy of Engineering Products Inc., Farmington, Michigan.)

PART No.	A	B	C
WC-250	9/32	3/4	1/4
WC-251		1	
WC-252		1 1/4	
WC-253		1 1/2	
WC-313	11/32	1 1/4	5/16
WC-375	13/32	1 1/4	3/8
WC-376		1 1/2	
WC-378		2	
WC-501	17/32	1 7/8	3/8
WC-625	21/32	2 1/8	3/8
WC-751	25/32	2 1/2	7/16

Washers are hardened and ground and have Black Oxide finish.

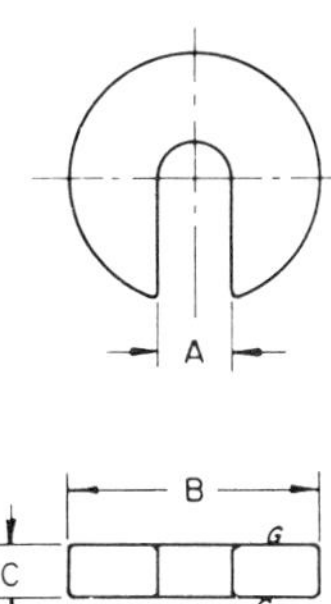

Figure 5.31 "C" washers. (Courtesy of Engineering Products Inc., Farmington, Michigan.)

A bushing is placed around the plug as part of the device for holding the plug in the drill jig (Fig. 5.20). Figure 5.21 shows how this bushing appears in the other three views. One must remember that the commands required to construct the bushing and the various parts of the drill jig are numerous, and therefore, will not be covered.

For the purpose of clarity, all of the hidden lines have been edited (removed) from the views. Figures 5.22 and 5.23 show the design of the drill jig being drawn in the front view with the final design appearing in Figure 5.24 as the 3-D view on the display screen. The 3-D view of the drill jig is a wire diagram with all lines appearing during the construction stage of drawing the drill jig on the computer. Figure 5.25 shows the same 3-D view after the lines which are hidden are removed.

The two sides of a steel block are to be removed by a milling operation producing the part shown in Figure 5.26. The actual path of the bottom of the milling cutter is shown on the computer's display screen in Figure 5.27. The fixture and its clamps for holding the block during the milling process may not interfere with the milling cutter path; otherwise, the cutter will hit the fixture and/or clamps resulting in damage to the milling cutter, fixture, and/or clamps or, worse yet, injury to the operator.

The various clamps used at a particular company from a particular vendor can be drawn and stored in the computer's files for immediate use. For instance, the clamp shown in Figure 5.28 will be used to provide the necessary force to hold the block in the fixture. This clamp is inserted into its proper position relative to the block and fixture by using several simple commands.

Another clamp is placed on the computer's screen using the same procedure (Fig. 5.29). By examining the computer's display screen the designer can now determine visually if either the fixture or clamps will interfere with the cutter's path.

PART No.		A	B	C	D	HEAD THICKNESS
Soft	Hardened					
SQ-190	SQ-190 H	10-32	1	½	½	3/16
SQ-250	SQ-250 H	¼-20	1¼	⅝	¾	¼
SQ-312 SQ-313	SQ-312 H SQ-313 H	5/16-18	1½ 2	¾	1	5/16
SQ-375 SQ-376	SQ-375 H SQ-376 H	⅜-16	1½ 2	1	1	⅜
SQ-500	SQ-500 H	½-13	1¾	1	1¼	½

All screws are Black Oxide finished.

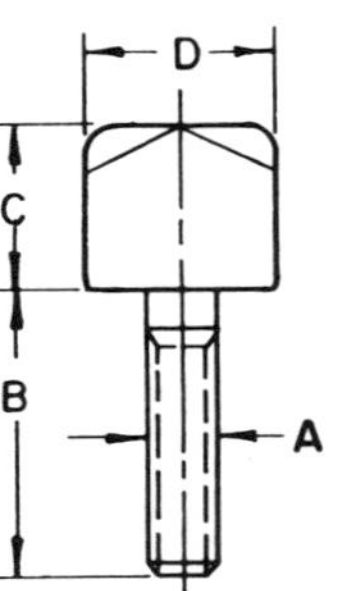

Figure 5.32 Quarter turn screws. (Courtesy of Engineering Products Inc., Farmington, Michigan.)

PART No.	A	B	C	D
BR-182	.188	3/8	3/8	1/4
BR-183				3/8
BR-185				1/2
BR-252	.2505	1/2	1/2	1/4
BR-253				3/8
BR-255				1/2
BR-312	.3135	1/2	1/2	1/4
BR-313				3/8
BR-314				1/2
BR-372	.3755	5/8	5/8	1/4
BR-373				3/8
BR-375				1/2
BR-376				5/8
BR-377				3/4

PART No.	A	B	C	D
BR-503	.5005	5/8	7/8	3/8
BR-504				7/16
BR-505				1/2
BR-506				5/8
BR-507				3/4
BR-623	.6255	3/4	1	3/8
BR-625				1/2
BR-626				5/8
BR-627				3/4
BR-753	.751	7/8	1 1/4	3/8
BR-755				1/2
BR-756				5/8
BR-757				3/4
BR-758				1

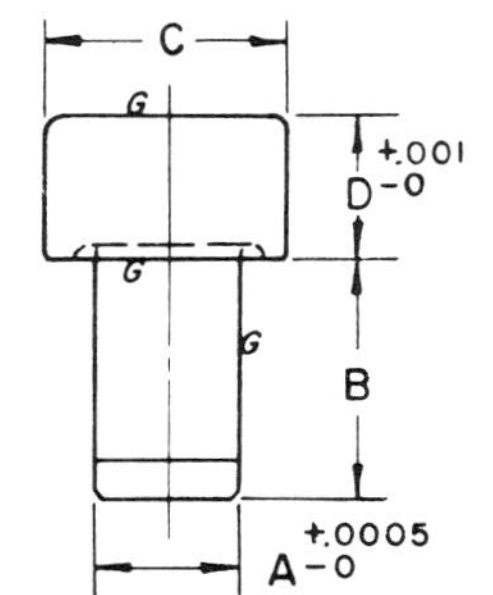

Hardened and ground. Lead diameter is ground on stem to aid in assembly. Black Oxide finish.

Figure 5.33 Rest buttons. (Courtesy of Engineering Products Inc., Farmington, Michigan.)

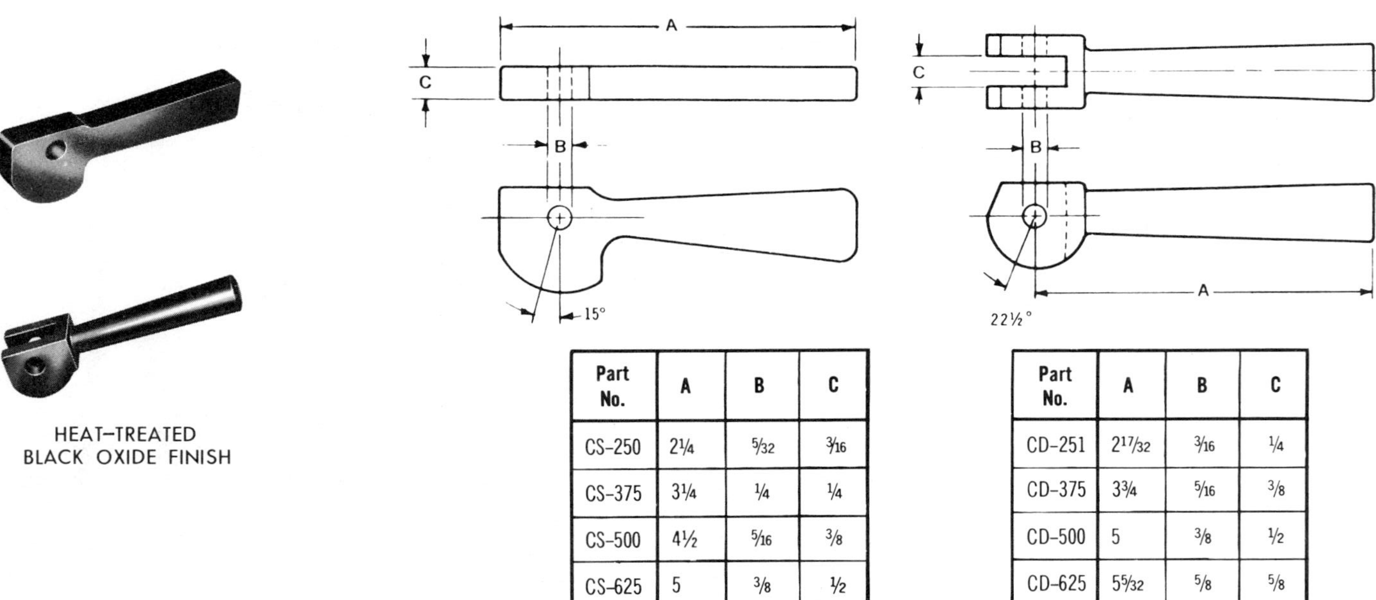

Part No.	A	B	C
CS-250	2¼	5/32	3/16
CS-375	3¼	¼	¼
CS-500	4½	5/16	3/8
CS-625	5	3/8	½

Part No.	A	B	C
CD-251	2 17/32	3/16	¼
CD-375	3¾	5/16	3/8
CD-500	5	3/8	½
CD-625	5 5/32	5/8	5/8

Figure 5.34 Cams. (Courtesy of Engineering Products Inc., Farmington, Michigan.)

HEAVY PRESSURE			LIGHT PRESSURE					+.001	
Standard	With Nylok	Press.	Standard	With Nylok	Press.	A	B	C	D
PS-6	PSN-6	3	PSL-6	PSLN-6	1	6-32	½	.045	1/16
PS-6A	PSN-6A	3	PSL-6A	PSLN-6A	1	6-32	⅜	.045	1/16
PS-8	PSN-8	5	PSL-8	PSLN-8	1½	8-32	⅝	.079	3/32
PS-10	PSN-10	7	PSL-10	PSLN-10	2	10-32	¾	.092	⅛
PS-25	PSN-25	8	PSL-25	PSLN-25	2½	¼-20	1	.118	3/16
PS-26	PSN-26	6	PSL-26	PSLN-26	1	¼-20	¾	.118	3/16
PS-27	PSN-27	8	PSL-27	PSLN-27	2½	¼-28	1	.118	3/16
PS-31	PSN-31	9	PSL-31	PSLN-31	3	5/16-18	1	.134	3/16
PS-37	PSN-37	10	PSL-37	PSLN-37	5	⅜-16	1⅛	.185	3/16
PS-50	PSN-50	12	PSL-50	PSLN-50	6	½-13	1¼	.247	¼
PS-62	PSN-62	18	PSL-62	PSLN-62	7	⅝-11	1½	.309	5/16
PS-75	PSN-75	22	PSL-75	PSLN-75	11	¾-10	1¾	.373	5/16
PS-100	PSN-100	42	PSL-100	PSLN-100	18	1-8	2¼	.498	½

Nylon pins available by adding suffix "N" to part number except in sizes 6, 6A, 75 and 100. Black Oxide finish.

When using "Nylok" it is recommended that the mating tapped hole be countersunk 90°-100° to 1/32 larger than the major diameter. Locking nylon feature to manufacturer's option.

Figure 5.35 Spring plungers. (Courtesy of Engineering Products Inc., Farmington, Michigan.)

CYANIDE HARDENED—BLACK OXIDE FINISH

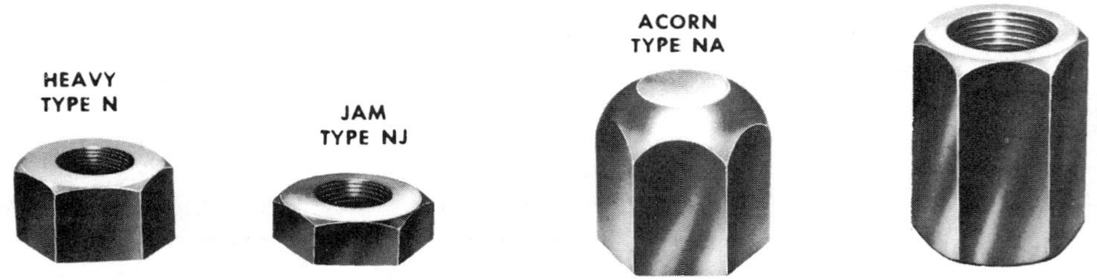

HEAVY TYPE N

JAM TYPE NJ

ACORN TYPE NA

EXTENSION TYPE NE

THREAD SIZE	ACROSS FLATS	HEAVY		JAM		ACORN			EXTENSION	
		PART No.	THICKNESS	PART No.	THICKNESS	PART No.	HEIGHT	THREAD DEPTH	PART No.	LENGTH
1/4-20	1/2	N- 250	15/64	NJ- 250	11/64	—	—	—	—	—
5/16-18	19/32	N- 312	19/64	NJ- 312	13/64	—	—	—	—	—
3/8-16	11/16	N- 375	23/64	NJ- 375	15/64	NA-375	3/4	9/16	NE- 375	1
1/2-13	7/8	N- 500	31/64	NJ- 500	19/64	NA-500	15/16	3/4	NE- 500	1 1/2
5/8-11	1 1/16	N- 625	39/64	NJ- 625	23/64	NA-625	1 3/16	15/16	NE- 625	1 3/4
3/4-10	1 1/4	N- 750	47/64	NJ- 750	27/64	NA-750	1 3/8	1 1/8	NE- 750	2
7/8- 9	1 7/16	N- 875	55/64	NJ- 875	31/64	—	—	—		
1″ - 8	1 5/8	N-1000	63/64	NJ-1000	35/64	—	—	—		

Figure 5.36 Hexagon nuts. (Courtesy of Engineering Products Inc., Farmington, Michigan.)

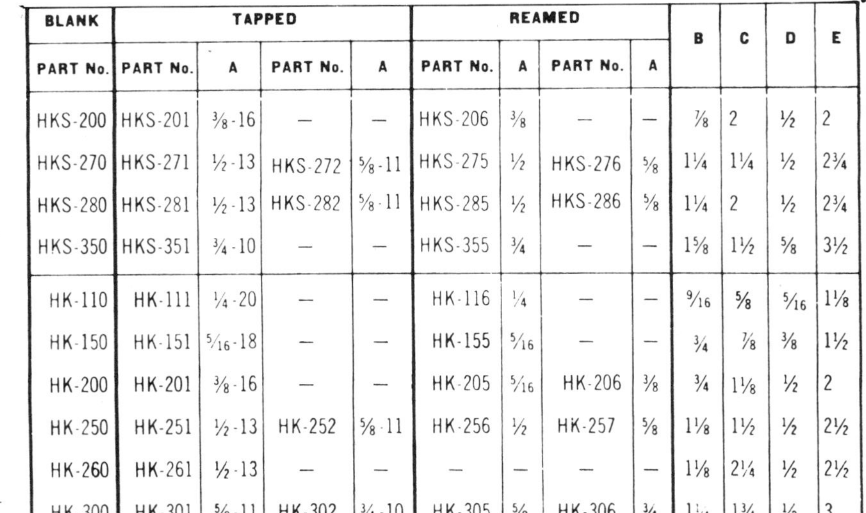

BLANK	TAPPED				REAMED				B	C	D	E
PART No.	PART No.	A	PART No.	A	PART No.	A	PART No.	A				
HKS-200	HKS-201	3/8-16	—	—	HKS-206	3/8	—	—	7/8	2	1/2	2
HKS-270	HKS-271	1/2-13	HKS-272	5/8-11	HKS-275	1/2	HKS-276	5/8	1 1/4	1 1/4	1/2	2 3/4
HKS-280	HKS-281	1/2-13	HKS-282	5/8-11	HKS-285	1/2	HKS-286	5/8	1 1/4	2	1/2	2 3/4
HKS-350	HKS-351	3/4-10	—	—	HKS-355	3/4	—	—	1 5/8	1 1/2	5/8	3 1/2
HK-110	HK-111	1/4-20	—	—	HK-116	1/4	—	—	9/16	5/8	5/16	1 1/8
HK-150	HK-151	5/16-18	—	—	HK-155	5/16	—	—	3/4	7/8	3/8	1 1/2
HK-200	HK-201	3/8-16	—	—	HK-205	5/16	HK-206	3/8	3/4	1 1/8	1/2	2
HK-250	HK-251	1/2-13	HK-252	5/8-11	HK-256	1/2	HK-257	5/8	1 1/8	1 1/2	1/2	2 1/2
HK-260	HK-261	1/2-13	—	—	—	—	—	—	1 1/8	2 1/4	1/2	2 1/2
HK-300	HK-301	5/8-11	HK-302	3/4-10	HK-305	5/8	HK-306	3/4	1 1/4	1 3/4	1/2	3

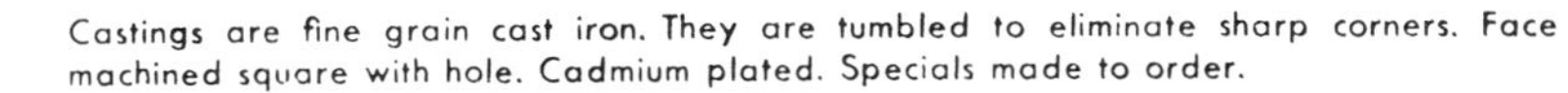

Castings are fine grain cast iron. They are tumbled to eliminate sharp corners. Face is machined square with hole. Cadmium plated. Specials made to order.

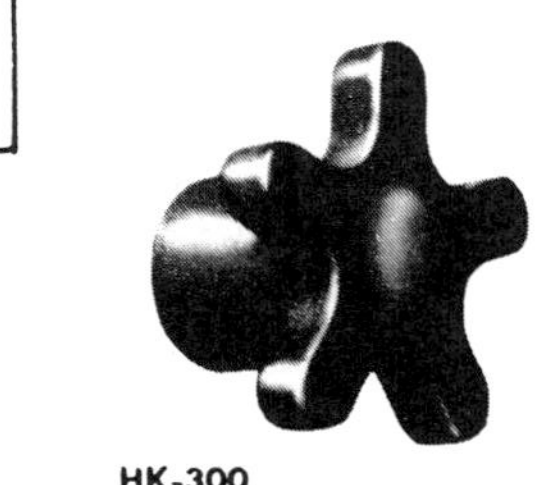

Figure 5.37 Cast iron knob handles. (Courtesy of Engineering Products Inc., Farmington, Michigan.)

A collection of tooling elements such as clamps, washers, and nuts can be drawn on a computer graphics system and stored in its files for future use on other tool design projects. For example, the plain and spherical washers shown in Figure 5.30 could all be placed into the computer's files as drawings for immediate placement on the display screen of the computer by using simple commands. Some other common items which are used extensively in tool design are shown in Figures 5.31 through 5.37. Again, each item would have a front, top, and side view of it stored in the files of the computer for immediate placement on the computer screen by the designer.

This view or drawing of each element can be moved about on the display screen easily and can be enlarged or reduced on the screen to match any scale during the design process.

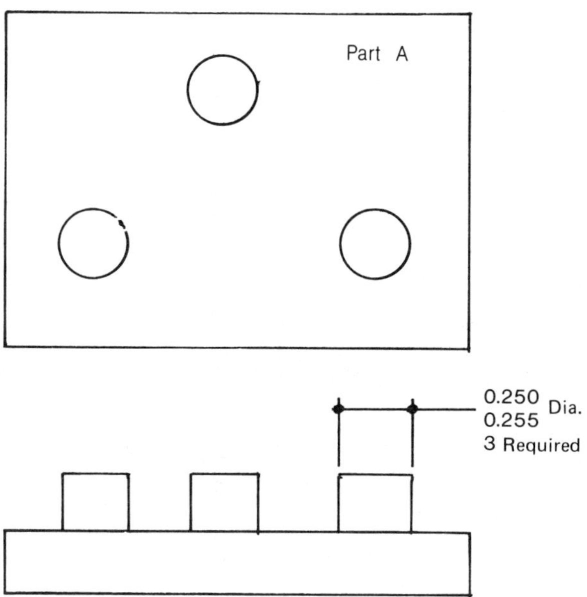

Figure 5.38 Part A.

A gage can be quickly drawn on a computer graphics system by using the mating part as the gage. For example, gages are required to check the pins in part A (Fig. 5.38) and the holes in part B (Fig. 5.39) which mate with each other. The gage for part A would resemble its mating part, which is B, except it would have the wear and gage allowances attached to the holes' dimensions.

The gage for part B would resemble its mating part, which is A, except it would have the wear and gage allowances attached to the pins' dimensions. Therefore, duplicate drawings of parts A and B can be made easily using computer graphics. These duplicate drawings will be used as gage drawings by renaming them with different file names and by changing their dimensions to gage dimensions.

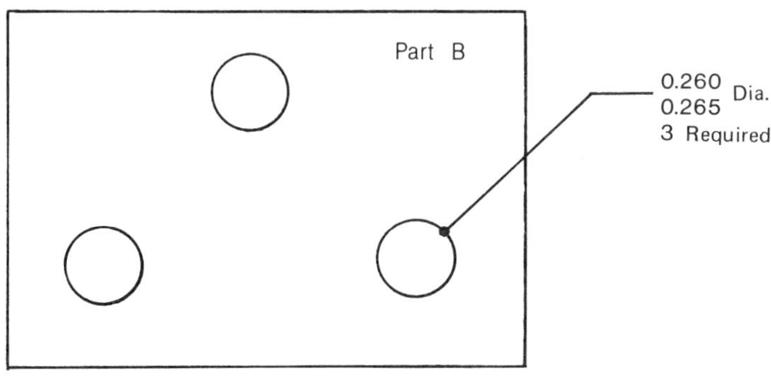

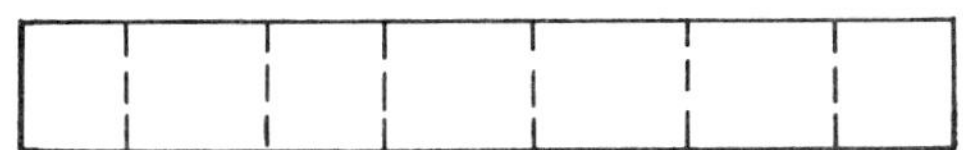

Figure 5.39 Part B.

5.5. WORK PROBLEMS

1. Assume a diameter of 1.625 ± 0.001. Apply the tolerances (a) to the GO and NO-GO plug gages and (b) to the snap gages. (c) Draw the plugs and snap gages and dimension them. (d) Calculate the wear allowances. (e) Draw the plugs and snap gages and dimension them according to part d.
2. Plug and snap gages are to be designed to check a dimension of 2.125 ± 0.005. (a) Calculate the plug gage dimensions and tolerances and insert the dimensions in a drawing of the gages. (b) Repeat part (a) for snap gages. Add 5% for wear.
3. Assembly and details of a drill jig to drill two holes in the bearing cap shown in Figure 5.40 are required. The two 0.360 diameter holes are to be drilled and reamed both to 0.3740 – 0.3755 diameter.

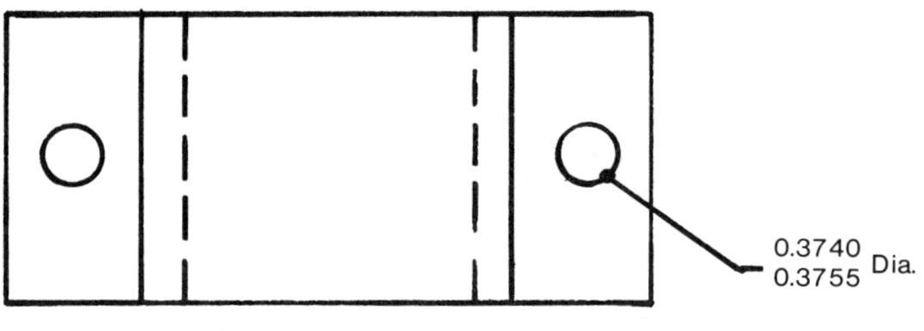

Figure 5.40 Bearing cap.

6

Metric Conversion

6.1. METRIC CONVERSION

The two most commonly used systems of measurement in the world are the English or *customary* (inch-pound) system and the *metric* system. Every industrialized nation on earth is on the metric system with the sole exception of the United States (Fig. 6.1).

The arguments for U.S. transition are impressive. Thanks to its mathematical logic and consistency, the metric system is not only easier to use than our current system, but it would also bring appreciable benefits to us economically and facilitate communication internationally, particularly in science and commerce.

All of us have an aversion to change — in moving from the known to the unknown. Once the metric system is understood, however, it is easier than many people think ("Basic Training Guide to the New Metrics and SI Units," ed. Robert C. Sellers & Assoc., National Tool, Die and Precision Machining Assn., Washington, D.C., 1972).

There are and have been variations to the metric system. The first practical system of metric units was based upon the centimeter as the unit of length, the gram as the unit of force, and the second as the unit of time, this was called the CGS system. At the turn of the century a more practical set of units was devised based upon the meter, the kilogram, and the second which became known as the MKS system. Around 1950 the ampere was added as the fourth basic unit so that mechanical units could be linked up with electromagnetic units. This changed the name to the MKSA system. More recently a system of units has been devised based upon the four MKSA units plus the kelvin

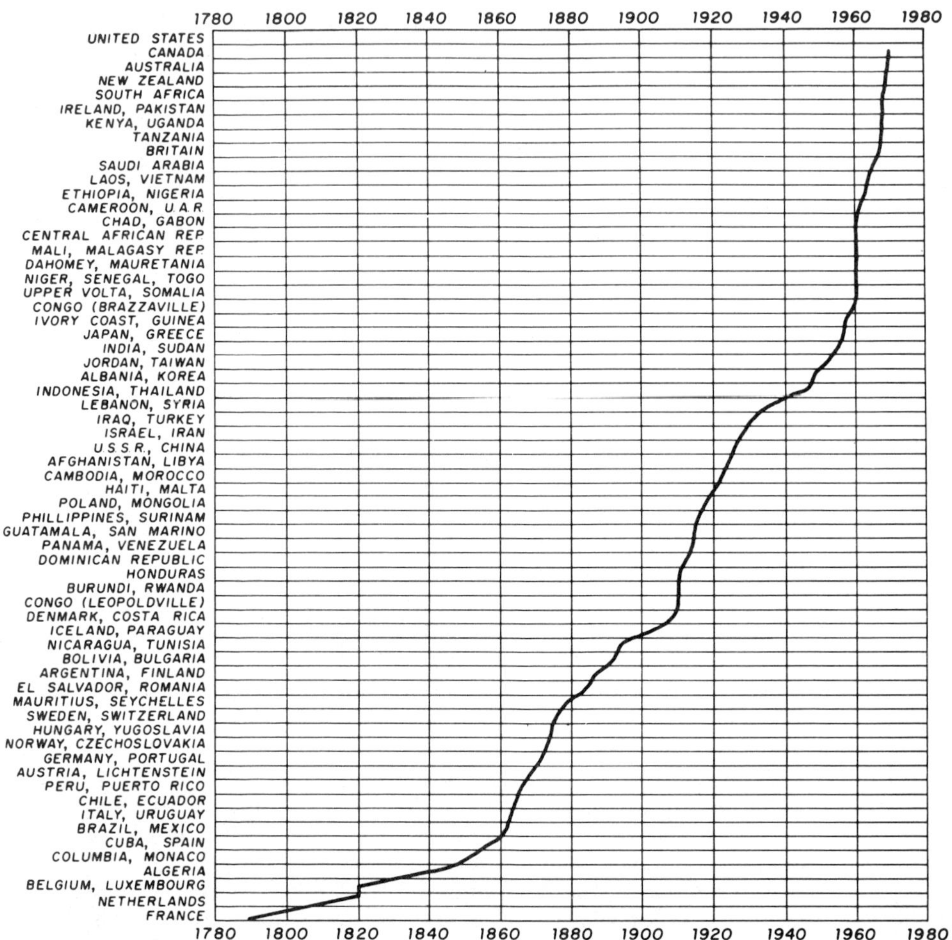

Figure 6.1 Metric usage in the world. (From R. C. Sellers, ed., *Basic Training Guide to the (L 6-1) New Metrics and SI Units*, National Tool, Die and Precision Machining Association, Washington, D.C., 1972.)

as the unit of temperature and the candela as the unit of luminous, and is commonly known as the metric-SI system.

The metric-SI system is a coherent system. A system of units are coherent if the product or quotient of any two unit quantities in the system is the unit of the resultant quantity.

Metrics is a family of measurement units based upon the decimal system. Any multiples or submultiples of any unit are related to the unit under consideration by powers of 10. For this reason, calculations in metrics can be made more rapidly, and the chances for error are appreciably reduced. For this reason alone, it is a simpler system to learn.

6.2. THE BASIC UNIT OF LENGTH: THE METER

The primary unit for measuring length is the meter. It is divided into smaller parts (tenths) quite unlike our foot which is divided into 12 inches. For purposes of comparison a meter equals roughly 3.3 feet.

Every multiple or submultiple in metrics has both a name and a numerical value which are used together. Each name has its own symbol which is directly related to its spelling. These are called prefixes. Most metric rules for everyday use will be divided into small increments using the millimeter (1/1000th of a meter). The ruler, tape, or scale have bold figures at every 10 mm mark. Figure 6.2 shows these millimeter increments on a metric rule.

Earlier it was mentioned that metrics worked from a base of 10. The power and base concept is used in math, for example, 4^2 meaning 4 is the base; 2 is the exponent and the expression "4 to the 2nd power" is used. When metrics uses 10 as a base and by moving to the left each place value increases by powers of 10. Thus, reading from left to right, the digit three places over is a multiple of 100. This is the heart of the decimal system. The powers of 10 are referred to as multiples and submultiples and in metrics have both names and symbols

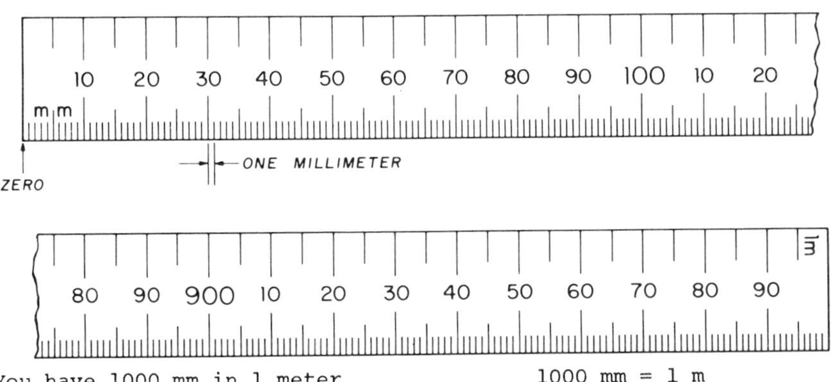

Figure 6.2 The meter.

Table 6.1 Prefixes and Symbols Used in Metrics

	Prefix	Symbol
1 000 000 000 000 = 10^{12}	Tera	T
1 000 000 000 = 10^{9}	Giga	G
1 000 000 = 10^{6}	Mega	M
1 000 = 10^{3}	Kilo	k
100 = 10^{2}	Hecto	h
10 = 10^{1}	Deca	da
0.1 = 10^{-1}	Deci	d
0.01 = 10^{-2}	Centi	c
0.001 = 10^{-3}	Milli	m
0.000 001 = 10^{-6}	Micro	μ
0.000 000 001 = 10^{-9}	Nano	n
0.000 000 000 001 = 10^{-12}	Pico	p

as shown in Table 6.1. The prefixes can be attached to the unit involved in discussing length resulting in Table 6.2.

On this basis, any measurement of length will utilize for almost all of the daily activities the meter and the millimeter. When long distances are to be expressed the kilometer is used. This is just over half the length of the customary mile, 0.6 to be exact. Therefore, for linear measure there are three metric terms: the meter, the millimeter, and the kilometer. The millimeter is used on mechanical drawings and during the machining operations in the shop as the standard metric unit of measurement.

Tables 6.3 – 6.6 show the various linear conversions from the customary inch system of units to the metric system of units.

Table 6.2 Prefixes Joined with the Suffix *-meter*

10^{-3}	meter	= a millimeter	(mm)	0.001
10^{-2}	meter	= a centimeter	(cm)	0.01
10^{-1}	meter	= a decimeter	(dm)	0.1
10^{0}	meter	= a meter	(m)	1
10^{1}	meters	= a dekameter	(da)	10
10^{2}	meters	= a hectometer	(ha)	100
10^{3}	meters	= a kilometer	(km)	1000

Table 6.3 Conversion Table: Inches to Decimals and Inches to Millimeters

in.	mm	in.	mm	in.	mm	in.	mm	in.	mm	in.	mm	in.	mm
0.001	0.0254	0.01	0.254	0.1	2.54	1	25.4	26	660.4	51	1295.4	76	1930.4
0.002	0.0508	0.02	0.508	0.2	5.08	2	50.8	27	685.8	52	1320.8	77	1955.8
0.003	0.0762	0.03	0.762	0.3	7.62	3	76.2	28	711.2	53	1346.2	78	1981.2
0.004	0.1016	0.04	1.016	0.4	10.16	4	101.6	29	736.6	54	1371.1	79	2006.8
0.005	0.1270	0.05	1.270	0.5	12.70	5	127.0	30	762.0	55	1397.0	80	2032.0
0.006	0.1524	0.06	1.524	0.6	15.24	6	152.4	31	787.4	56	1422.4	81	2057.4
0.007	0.1778	0.07	1.778	0.7	17.78	7	177.8	32	812.8	57	1447.8	82	2082.8
0.008	0.2032	0.08	2.032	0.8	20.32	8	203.2	33	838.2	58	1473.2	83	2108.2
0.009	0.2286	0.09	2.286	0.9	22.86	9	228.6	34	863.6	59	1498.6	84	2133.6
						10	254.0	35	889.0	60	1524.0	85	2159.0
						11	279.4	36	914.4	61	1549.4	86	2184.4
						12	304.8	37	939.8	62	1574.8	87	2209.8
						13	330.2	38	956.2	63	1600.2	88	2235.2
						14	355.6	39	990.6	64	1625.6	89	2260.6
						15	381.0	40	1016.0	65	1651.0	90	2286.0
						16	406.4	41	1041.4	66	1676.4	91	2311.4
						17	431.8	42	1066.8	67	1701.8	92	2336.8
						18	457.2	43	1092.2	68	1727.2	93	2362.2
						19	482.6	44	1117.6	69	1752.6	94	2387.6
						20	508.0	45	1143.0	70	1778.0	95	2413.0
						21	533.4	46	1168.4	71	1803.4	96	2438.4
						22	558.8	47	1193.8	72	1828.8	97	2463.8
						23	584.2	48	1219.2	73	1854.2	98	2489.2
						24	609.6	49	1244.6	74	1879.6	99	2514.6
						25	635.0	50	1270.0	75	1905.0	100	2540.0

Table 6.4 Conversion Table: Inches in Fractions to Millimeters

	in.	mm		in.	mm		in.	mm
1/64	0.015 625	0.396 875	25/64	0.390 625	9.921 875	49/64	0.765 525	19.446 875
1/32	0.031 250	0.793 750	13/32	0.406 250	10.318 750	25/32	0.781 250	19.843 750
3/64	0.046 875	1.190 625	27/64	0.421 875	10.715 625	51/64	0.796 875	20.240 655
1/16	0.062 500	1.587 500	7/16	0.437 500	11.112 500	13/16	0.812 500	20.637 500
5/64	0.078 125	1.984 375	29/64	0.453 125	11.509 375	53/64	0.828 125	21.034 372
3/32	0.093 750	2.381 250	15/32	0.468 750	11.906 250	27/32	0.843 750	21.431 250
7/64	0.109 375	2.778 125	31/64	0.484 375	12.303 125	55/64	0.859 375	21.828 125
1/8	0.125 000	3.175 000	1/2	0.500 000	12.700 000	7/8	0.875 000	22.225 000
9/64	0.140 625	3.571 875	33/64	0.515 625	13.096 875	57/64	0.890 625	22.621 875
5/32	0.156 250	3.968 750	17/32	0.531 250	13.493 750	29/32	0.906 250	23.018 750
11/64	0.171 875	4.365 625	35/64	0.546 875	13.890 625	59/64	0.921 875	23.415 625
3/16	0.187 500	4.762 500	9/16	0.562 500	14.287 500	15/16	0.937 500	23.812 500
13/64	0.203 125	5.159 375	37/64	0.578 125	14.684 375	61/64	0.953 125	24.209 375
7/32	0.218 750	5.556 250	19/32	0.593 750	15.081 250	31/32	0.968 750	24.606 250
15/64	0.234 375	5.953 125	39/64	0.609 375	15.478 125	63/64	0.984 375	25.003 125
1/4	0.250 000	6.350 000	5/8	0.625 000	15.875 000	1	1.000 000	25.400 000
17/64	0.265 625	6.746 875	41/64	0.640 625	16.271 875			
9/32	0.281 250	7.143 750	21/32	0.656 250	16.668 750			
19/64	0.296 875	7.540 625	43/64	0.671 875	17.065 625			
5/16	0.312 500	7.937 500	11/16	0.687 500	17.462 500			
21/64	0.328 125	8.334 375	45/64	0.703 125	17.859 375			
11/32	0.343 750	8.731 250	23/32	0.718 750	18.256 250			
23/64	0.359 375	9.128 125	47/64	0.734 375	18.653 125			
3/8	0.375 000	9.525 000	3/4	0.750 000	19.050 000			

Table 6.5 Conversion Table: Feet and Inches to Meters

Feet/Inches	0"	1"	2"	3"	4"	5"	6"	7"	8"	9"	10"	11"
	m	m	m	m	m	m	m	m	m	m	m	m
0		0.0254	0.0508	0.0762	0.1016	0.1270	0.1524	0.1778	0.2032	0.2286	0.2540	0.2794
1	0.3048	0.3302	0.3556	0.3810	0.4064	0.4318	0.4572	0.4826	0.5080	0.5334	0.5588	0.5842
2	0.6096	0.6350	0.6604	0.6858	0.7112	0.7366	0.7620	0.7874	0.8128	0.8382	0.8636	0.8890
3	0.9144	0.9398	0.9652	0.9906	1.0160	1.0414	1.0668	1.0922	1.1176	1.1430	1.1684	1.1938
4	1.2192	1.2446	1.2700	1.2954	1.3208	1.3462	1.3716	1.3970	1.4224	1.4478	1.4732	1.4986
5	1.5240	1.5494	1.5748	1.6002	1.6256	1.6510	1.6764	1.7018	1.7272	1.7526	1.7780	1.8034
6	1.8288	1.8542	1.8796	1.9050	1.9304	1.9558	1.9812	2.0066	2.0320	2.0574	2.0828	2.1082
7	2.1336	2.1590	2.1844	2.2098	2.2352	2.2606	2.2860	2.3114	2.3368	2.3622	2.3876	2.4130
8	2.4384	2.4638	2.4892	2.5146	2.5400	2.5654	2.5908	2.6162	2.6416	2.6670	2.6924	2.7178
9	2.7432	2.7686	2.7940	2.8194	2.8448	2.8702	2.8956	2.9210	2.9464	2.9718	2.9972	3.0226
10	3.0480	3.0734	3.0988	3.1242	3.1496	3.1750	3.2004	3.2258	3.2512	3.2766	3.3020	3.3274
11	3.3528	3.3782	3.4036	3.4290	3.4544	3.4798	3.5052	3.5306	3.5560	3.5814	3.6068	3.6322
12	3.6576	3.6830	3.7084	3.7338	3.7592	3.7846	3.8100	3.8354	3.8608	3.8862	3.9116	3.9370
13	3.9624	3.9878	4.0132	4.0386	4.0640	4.0894	4.1148	4.1402	4.1656	4.1910	4.2164	4.2418
14	4.2672	4.2926	4.3180	4.3434	4.3688	4.3942	4.4196	4.4450	4.4704	4.4958	4.5212	4.5466
15	4.5720	4.5974	4.6228	4.6482	4.6736	4.6990	4.7244	4.7498	4.7752	4.8006	4.8260	4.8514
16	4.8768	4.9022	4.9276	4.9530	4.9784	5.0038	5.0292	5.0546	5.0800	5.1054	5.1308	5.1562
17	5.1816	5.2070	5.2324	5.2578	5.2832	5.3086	5.3340	5.3594	5.3848	5.4102	5.4356	5.4610
18	5.4864	5.5118	5.5372	5.5626	5.5880	5.6134	5.6388	5.6642	5.6896	5.7150	5.7404	5.7658
19	5.7912	5.8166	5.8420	5.8674	5.8928	5.9182	5.9436	5.9690	5.9944	6.0198	6.0452	6.0706
20	6.0960	6.1214	6.1468	6.1722	6.1976	6.2230	6.2484	6.2738	6.2992	6.3246	6.3500	6.3754

Table 6.6 Conversion Table: Millimeters to Inches

mm	0	1	2	3	4	5	6	7	8	9
	in.	in.	in.	in.	in.	in.	in.	in.	in.	in.
0	-	0.0394	0.0787	0.1181	0.1575	0.1969	0.2362	0.2756	0.3150	0.3543
10	0.3937	0.4331	0.4724	0.5118	0.5512	0.5906	0.6299	0.6693	0.7087	0.7480
20	0.7874	0.8268	0.8661	0.9055	0.9449	0.9843	1.0236	1.0630	1.1024	1.1417
30	1.1811	1.2205	1.2598	1.2992	1.3386	1.3780	1.4173	1.4567	1.4961	1.5354
40	1.5748	1.6142	1.6535	1.6929	1.7323	1.7717	1.8110	1.8504	1.8898	1.9291
50	1.9685	2.0079	2.0472	2.0866	2.1260	2.1654	2.2047	2.2441	2.2835	2.3228
60	2.3622	2.4016	2.4409	2.4803	2.5197	2.5591	2.5984	2.6378	2.6772	2.7165
70	2.7559	2.7953	2.8346	2.8740	2.9134	2.9528	2.9921	3.0315	3.0709	3.1102
80	3.1496	3.1890	3.2283	3.2677	3.3071	3.3465	3.3858	3.4252	3.4646	3.5039
90	3.5433	3.5827	3.6220	3.6614	3.7008	3.7402	3.7795	3.8189	3.8583	3.8976
100	3.9370	3.9764	4.0157	4.0551	4.0945	4.1339	4.1732	4.2126	4.2520	4.2913
200	7.8740	7.9134	7.9528	7.9921	8.0315	8.0709	8.1102	8.1496	8.1890	8.2283
300	11.8110	11.8504	11.8898	11.9291	11.9685	12.0079	12.0472	12.0866	12.1260	12.1654
400	15.7480	15.7874	15.8268	15.8661	15.9055	15.9449	15.9843	16.0236	16.0630	16.1024
500	19.6850	19.7244	19.7638	19.8031	19.8425	19.8819	19.9213	19.9606	20.0000	20.0394
600	23.6220	23.6614	23.7008	23.7402	23.7795	23.8189	23.8583	23.8976	23.9370	23.9764
700	27.5591	27.5984	27.6378	27.6772	27.7165	27.7559	27.7953	27.8346	27.8740	27.9134
800	31.4961	31.5354	31.5748	31.6142	31.6535	31.6929	31.7323	31.7717	31.8110	31.8504
900	35.4331	35.4724	35.5118	35.5512	35.5906	35.6299	35.6693	35.7087	35.7480	35.7874
1000	39.3701									

6.3. THE METRIC SYSTEM OF FITS

The previous chapters dealt with the customary (inch-pound) system of specifying limits and fits between mating parts. It is necessary to use the limits and fits recommended by the International Standards Organization (ISO) when using the metric system. The standard metric terms and their definitions are listed below with Figure 6.3 illustrating these terms.

1. Limits of size. The maximum and minimum sizes permitted for a feature.
2. Basic size. The size by reference to which the limits of size are fixed. The basic size is the same for both members of a fit.
3. Upper deviation. The algebraic difference between the maximum limit of size and the corresponding basic size.
4. Lower deviation. The algebraic difference between the minimum limit of size and the corresponding basic size.
5. Zero line. In a graphical representation of limits and fits, the straight line to which the deviations are referred. The zero line is the line of zero deviation and represents the basic size.
6. Tolerance. The difference between the maximum limit of size and the minimum limit of size. It is an absolute value without sign.
7. Tolerance zone. In a graphical representation of tolerances, the zone comprised between the two lines representing the limits of tolerances and defined by its magnitude (tolerance) and by its position in relation to the zero line.
8. Fundamental deviation. That one of the two deviations, being the one nearest to the zero line, which is conventionally chosen to define the position of the tolerance zone in relation to the zero line.
9. Shaft basis system of fits. A system of fits in which the different clearances and interferences are obtained by associating various holes with a single shaft. In the ISO system, the basic shaft is the shaft the upper deviation of which is zero.
10. Hole basic system of fits. A system of fits in which the different clearances and interferences are obtained by associating various shafts with a single hole. In the ISO system, the basic hole is the hole the lower deviation of which is zero.
11. Tolerance symbols. Metric dimensions can be expressed on a mechanical drawing using three methods. The first method places the basic size as the primary dimension from which the tolerances are determined. This basic size is first part of the symbol followed by the fundamental position letter and the international tolerance (IT) number to give the tolerance zone (i.e., 30h7). The other two methods of showing this dimension would

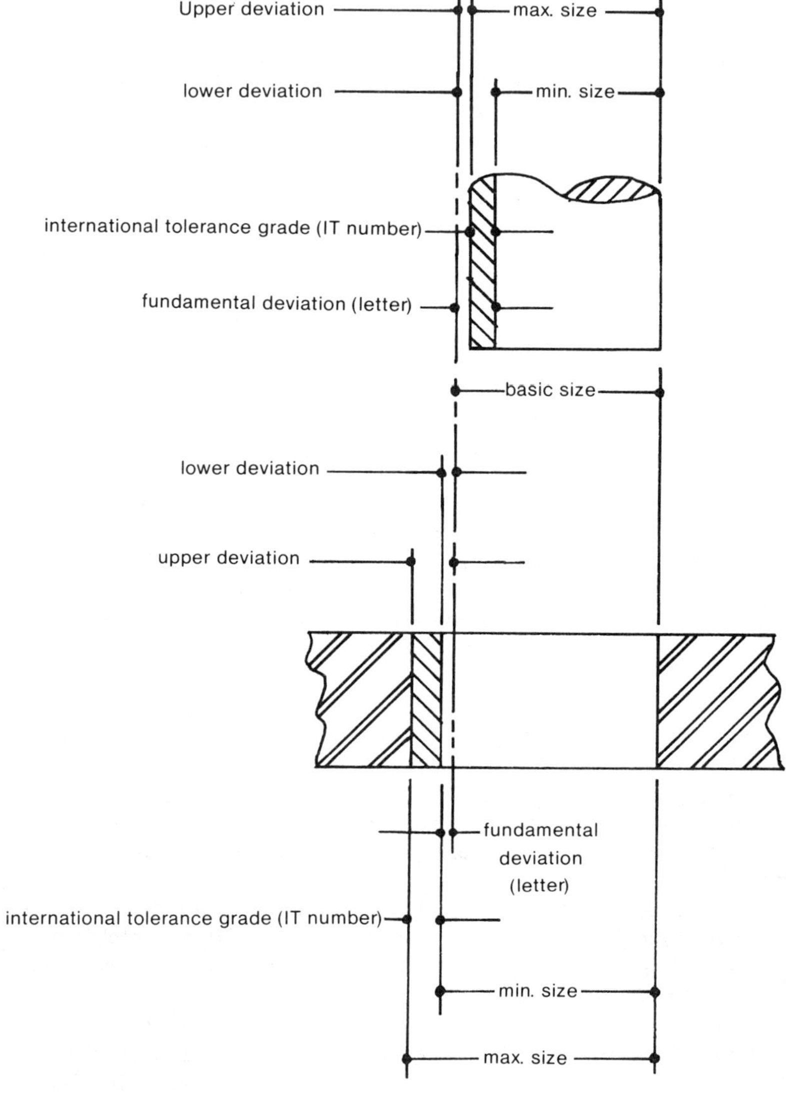

Figure 6.3 Terms used with metric fits.

be either 30h7 $\binom{30.000}{29.979}$ or $\frac{30.000}{29.979}$ (30h7).

Table 6.7 lists the basic sizes that are preferred when using the metric system. These preferred sizes are utilized as much as possible since metric raw materials usually follow these standards resulting in reduces costs.

Table 6.8 shows the preferred metric fits associated with the clearance, transition, and interference fits. The tolerance symbols used to designate these fits using either the hole basis system or the shaft basis system is also shown. When possible, these tables should be used for determining the mating parts' dimensions and their tolerances; otherwise, if sizes other than the preferred sizes are used, one must use different tables.

Figure 6.4 illustrates the three types of fits: a clearance fit, a transition fit, and an interference fit. These fits can be based either on the female feature (hole basis system) or the male feature (shaft basis system). The smallest (minimum) hole and the largest (maximum) shaft are the basic sizes or, in other words, the maximum material condition (MMC) of the hole and the shaft.

Figure 6.5 shows a series of fits for a hole basis system while Figure 6.6 shows a series of fits for a shaft basis system. In Figure 6.5, the sizes of the shafts are varied to give a variety of fits from a clearance fit of c11 to an interference fit of u6 and the lower deviation of the hole is zero. In Figure 6.6, the sizes of the holes are varied to give a variety of fits from a clearance fit of C11 to an interference fit of U7 and the lower deviation of the shaft is zero. The fits associated with these figures and Table 6.8 for various hole or shaft sizes are shown in Tables 6.9 – 6.12.

Suppose a free-running fit is required between a shaft and a hole with a basic diameter of 26 mm and the hole basis system is being used. The preferred size for the hole would be 25 mm selected from Table 6.7 and a preferred fit of H9/d9 is selected from Table 6.8. The upper and lower limits of the hole are determined by using Table 6.9 under column H9 and across from the basic size of 25 mm. The hole's upper and lower limits are 25.000 and 25.052 mm respectively. The upper and lower limits of the shaft are determined under column d9 and across from the basic size of 25 mm, yielding 24.935 and 24.883 mm, respectively. Figure 6.7 shows the three methods of noting these limits on the hole and shaft drawings.

The difference between the upper and lower limits is known as the international tolerance grade (IT). Table 6.13 shows the IT grades for various basic sizes. The previous example, has an IT grade of 14 since the difference between the hole's and shaft's upper limits is 0.052 mm. This value is found by finding the basic size of 25 mm and

Table 6.7 Preferred Metric Sizes

Basic Size, mm		Basic Size, mm		Basic Size, mm	
First Choice	Second Choice	First Choice	Second Choice	First Choice	Second Choice
1		10		100	
	1.1		11		110
1.2		12		120	
	1.4		14		140
1.6		16		160	
	1.8		18		180
2		20		200	
	2.2		22		220
2.5		25		250	
	2.8		28		280
3		30		300	
	3.5		35		350
4		40		400	
	4.5		45		450
5		50		500	
	5.5		55		550
6		60		600	
	7		70		700
8		80		800	
	9		90		900
				1000	

Table 6.8 Description of Preferred Metric sizes

	ISO SYMBOL		DESCRIPTION	
	Hole Basis	Shaft Basis		
Clearance Fits	H11/c11	C11/h11	*Loose running* fit for wide commercial tolerances or allowances on external members.	More Clearance
Clearance Fits	H9/d9	D9/h9	*Free running* fit not for use where accuracy is essential, but good for large temperature variations, high running speeds, or heavy journal pressures.	More Clearance
Clearance Fits	H8/f7	F8/h7	*Close running* fit for running on accurate machines and for accurate location at moderate speeds and journal pressures.	More Clearance
Clearance Fits	H7/g6	G7/h6	*Sliding* fit not intended to run freely, but to move and turn freely and locate accurately.	More Clearance
Clearance Fits	H7/h6	H7/h6	*Locational clearance* fit provides snug fit for locating stationary parts; but can be freely assembled and disassembled.	More Clearance
Transition Fits	H7/k6	K7/h6	*Locational transition* fit for accurate location, a compromise between clearance and interference.	More Interference
Transition Fits	H7/n6	N7/h6	*Locational transition* fit for more accurate location where greater interference is permissible.	More Interference
Interference Fits	H7/p6[1]	P7/h6	*Locational interference* fit for parts requiring rigidity and alignment with prime accuracy of location but without special bore pressure requirements.	More Interference
Interference Fits	H7/s6	S7/h6	*Medium drive* fit for ordinary steel parts or shrink fits on light sections, the tightest fit usable with cast iron.	More Interference
Interference Fits	H7/u6	U7/h6	*Force* fit suitable for parts which can be highly stressed or for shrink fits where the heavy pressing forces required are impractical.	More Interference

[1] Transition fit for basic sizes in range from 0 through 3 mm.

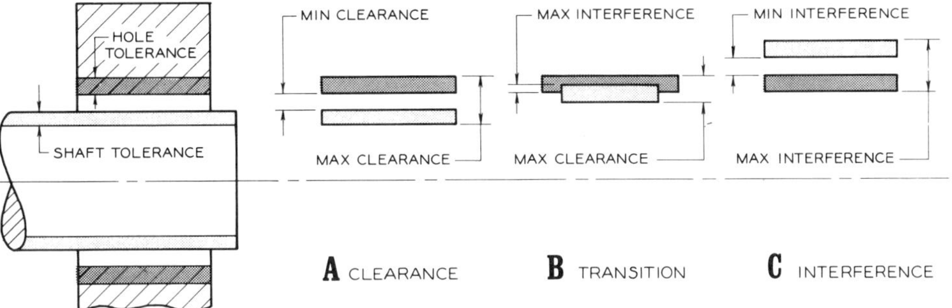

Figure 6.4 Clearance, transition, and interference fits.

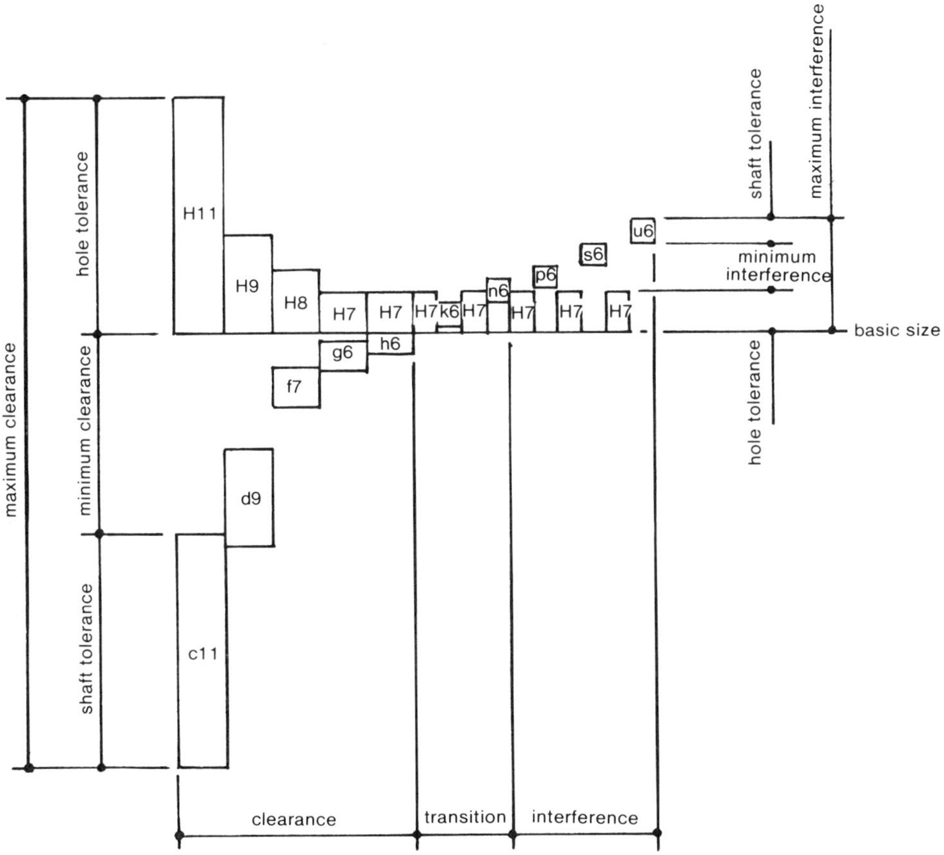

Figure 6.5 Tolerances associated with the hole basis system. (From J. H. Earle, *Drafting Technology*, Addison-Wesley, Reading, Massachusetts, 1982.)

Table 6.9 Preferred Hole Basis Clearance Fits: Cylindrical Fits (ANSI B4.2)

BASIC SIZE		LOOSE RUNNING			FREE RUNNING			CLOSE RUNNING			SLIDING			LOCATIONAL CLEARANCE		
		Hole H11	Shaft c11	Fit	Hole H9	Shaft d9	Fit	Hole H8	Shaft f7	Fit	Hole H7	Shaft g6	Fit	Hole H7	Shat h6	Fit
1	MAX	1.060	0.940	0.180	1.025	0.980	0.070	1.014	0.994	0.030	1.010	0.998	0.018	1.010	1.000	0.016
	MIN	1.000	0.880	0.060	1.000	0.955	0.020	1.000	0.984	0.006	1.000	0.992	0.002	1.000	0.994	0.000
1.2	MAX	1.260	1.140	0.180	1.225	1.180	0.070	1.214	1.194	0.030	1.210	1.198	0.018	1.210	1.200	0.016
	MIN	1.200	1.080	0.060	1.200	1.155	0.020	1.200	1.184	0.006	1.200	1.192	0.002	1.200	1.194	0.000
1.6	MAX	1.660	1.540	0.180	1.625	1.580	0.070	1.614	1.594	0.030	1.610	1.598	0.018	1.610	1.600	0.016
	MIN	1.600	1.480	0.060	1.600	1.555	0.020	1.600	1.584	0.006	1.600	1.592	0.002	1.600	1.594	0.000
2	MAX	2.060	1.940	0.180	2.025	1.980	0.070	2.014	1.994	0.030	2.010	1.998	0.018	2.010	2.000	0.016
	MIN	2.000	1.880	0.060	2.000	1.955	0.020	2.000	1.984	0.006	2.000	1.992	0.002	2.000	1.994	0.000
2.5	MAX	2.560	2.440	0.180	2.525	2.480	0.070	2.514	2.494	0.030	2.510	2.498	0.018	2.510	2.500	0.016
	MIN	2.500	2.380	0.060	2.500	2.455	0.020	2.500	2.484	0.006	2.500	2.492	0.002	2.500	2.494	0.000
3	MAX	3.060	2.940	0.180	3.025	2.980	0.070	3.014	2.994	0.030	3.010	2.998	0.018	3.010	3.000	0.016
	MIN	3.000	2.880	0.060	3.000	2.955	0.020	3.000	2.984	0.006	3.000	2.992	0.002	3.000	2.994	0.000
4	MAX	4.075	3.930	0.220	4.030	3.970	0.090	4.018	3.990	0.040	4.012	3.996	0.024	4.012	4.000	0.020
	MIN	4.000	3.855	0.070	4.000	3.940	0.030	4.000	3.978	0.010	4.000	3.988	0.004	4.000	3.992	0.000
5	MAX	5.075	4.930	0.220	5.030	4.970	0.090	5.018	4.990	0.040	5.012	4.996	0.024	5.012	5.000	0.020
	MIN	5.000	4.855	0.070	5.000	4.940	0.030	5.000	4.978	0.010	5.000	4.988	0.004	5.000	4.992	0.000
6	MAX	6.075	5.930	0.220	6.030	5.970	0.090	6.018	5.990	0.040	6.012	5.996	0.024	6.012	6.000	0.020
	MIN	6.000	5.855	0.070	6.000	5.940	0.030	6.000	5.978	0.010	6.000	5.988	0.004	6.000	5.992	0.000
8	MAX	8.090	7.920	0.260	8.036	7.960	0.112	8.022	7.987	0.050	8.015	7.995	0.029	8.015	8.000	0.024
	MIN	8.000	7.830	0.080	8.000	7.924	0.040	8.000	7.972	0.013	8.000	7.986	0.005	8.000	7.991	0.000
10	MAX	10.090	9.920	0.260	10.036	9.960	0.112	10.022	9.987	0.050	10.015	9.995	0.029	10.015	10.000	0.024
	MIN	10.000	9.830	0.080	10.000	9.924	0.040	10.000	9.972	0.013	10.000	9.986	0.005	10.000	9.991	0.000
12	MAX	12.110	11.905	0.315	12.043	11.950	0.136	12.027	11.984	0.061	12.018	11.994	0.035	12.018	12.000	0.029
	MIN	12.000	11.795	0.095	12.000	11.907	0.050	12.000	11.966	0.016	12.000	11.983	0.006	12.000	11.989	0.000
16	MAX	16.110	15.905	0.315	16.043	15.950	0.136	16.027	15.984	0.061	16.018	15.994	0.035	16.018	16.000	0.029
	MIN	16.000	15.795	0.095	16.000	15.907	0.050	16.000	15.966	0.016	16.000	15.983	0.006	16.000	15.989	0.000
20	MAX	20.130	19.890	0.370	20.052	19.935	0.169	20.033	19.980	0.074	20.021	19.993	0.041	20.021	20.000	0.034
	MIN	20.000	19.760	0.110	20.000	19.883	0.065	20.000	19.959	0.020	20.000	19.980	0.007	20.000	19.987	0.000
25	MAX	25.130	24.890	0.370	25.052	24.935	0.169	25.033	24.980	0.074	25.021	24.993	0.041	25.021	25.000	0.034
	MIN	25.000	24.760	0.110	25.000	24.883	0.065	25.000	24.959	0.020	25.000	24.980	0.007	25.000	24.987	0.000
30	MAX	30.130	29.890	0.370	30.052	29.935	0.169	30.033	29.980	0.074	30.021	29.993	0.041	30.021	30.000	0.034
	MIN	30.000	29.760	0.110	30.000	29.883	0.065	30.000	29.959	0.020	30.000	29.980	0.007	30.000	29.987	0.000

Table 6.9 (Continued)

BASIC SIZE		LOOSE RUNNING Hole H11	Shaft c11	Fit	FREE RUNNING Hole H9	Shaft d9	Fit	CLOSE RUNNING Hole H8	Shaft f7	Fit	SLIDING Hole H7	Shaft g6	Fit	LOCATIONAL CLEARANCE Hole H7	Shat h6	Fit
40	MAX	40.160	39.880	0.440	40.062	39.920	0.204	40.039	39.975	0.089	40.025	39.991	0.050	40.025	40.000	0.041
	MIN	40.000	39.720	0.120	40.000	39.858	0.080	40.000	39.950	0.025	40.000	39.975	0.009	40.000	39.984	0.000
50	MAX	50.160	49.870	0.450	50.062	49.920	0.204	50.039	49.975	0.089	50.025	49.991	0.050	50.025	50.000	0.041
	MIN	50.000	49.710	0.130	50.000	49.858	0.080	50.000	49.950	0.025	50.000	49.975	0.009	50.000	49.984	0.000
60	MAX	60.190	59.860	0.520	60.074	59.900	0.248	60.046	59.970	0.106	60.030	59.990	0.059	60.030	60.000	0.049
	MIN	60.000	59.670	0.140	60.000	59.826	0.100	60.000	59.940	0.030	60.000	59.971	0.010	60.000	59.981	0.000
80	MAX	80.190	79.850	0.530	80.074	79.900	0.248	80.046	79.970	0.106	80.030	79.990	0.059	80.030	80.000	0.049
	MIN	80.000	79.660	0.150	80.000	79.826	0.100	80.000	79.940	0.030	80.000	79.971	0.010	80.000	79.981	0.000
100	MAX	100.220	99.830	0.610	100.087	99.880	0.294	100.054	99.964	0.125	100.035	99.988	0.069	100.035	100.000	0.057
	MIN	100.000	99.610	0.170	100.000	99.793	0.120	100.000	99.929	0.036	100.000	99.966	0.012	100.000	99.978	0.000
120	MAX	120.220	119.820	0.620	120.087	119.880	0.294	120.054	119.964	0.125	120.035	119.988	0.069	120.035	120.000	0.057
	MIN	120.000	119.600	0.180	120.000	119.793	0.120	120.000	119.929	0.036	120.000	119.966	0.012	120.000	119.978	0.000
160	MAX	160.250	159.790	0.710	160.100	159.855	0.345	160.063	159.957	0.146	160.040	159.986	0.079	160.040	160.000	0.065
	MIN	160.000	159.540	0.210	160.000	159.755	0.145	160.000	159.917	0.043	160.000	159.961	0.014	160.000	159.975	0.000
200	MAX	200.290	199.760	0.820	200.115	199.830	0.400	200.072	199.950	0.168	200.046	199.985	0.090	200.046	200.000	0.075
	MIN	200.000	199.470	0.240	200.000	199.715	0.170	200.000	199.904	0.050	200.000	199.956	0.015	200.000	199.971	0.000
250	MAX	250.290	249.720	0.860	250.115	249.830	0.400	250.072	249.950	0.168	250.046	249.985	0.090	250.046	250.000	0.075
	MIN	250.000	249.430	0.280	250.000	249.715	0.170	250.000	249.904	0.050	250.000	249.956	0.015	250.000	249.971	0.000
300	MAX	300.320	299.670	0.970	300.130	299.810	0.450	300.081	299.944	0.189	300.052	299.983	0.101	300.052	300.000	0.084
	MIN	300.000	299.350	0.330	300.000	299.680	0.190	300.000	299.892	0.056	300.000	299.951	0.017	300.000	299.968	0.000
400	MAX	400.360	399.600	1.120	400.140	399.790	0.490	400.089	399.938	0.208	400.057	399.982	0.111	400.057	400.000	0.093
	MIN	400.000	399.240	0.400	400.000	399.650	0.210	400.000	399.881	0.062	400.000	399.946	0.018	400.000	399.964	0.000
500	MAX	500.400	499.520	1.280	500.155	499.770	0.540	500.097	499.932	0.228	500.063	499.980	0.123	500.063	500.000	0.103
	MIN	500.000	499.120	0.480	500.000	499.615	0.230	500.000	499.869	0.068	500.000	499.940	0.020	500.000	499.960	0.000

Source: J. H. Earle, *Drafting Technology*, Addison-Wesley, Reading, Massachusetts, 1982.

Table 6.10 Preferred Hole Basis Transition and Interference Fits: Cylindrical Fits (ANSI B4.2)

BASIC SIZE		LOCATIONAL TRANSN.			LOCATIONAL TRANSN.			LOCATIONAL INTERF.			MEDIUM DRIVE			FORCE		
		Hole H7	Shaft k6	Fit	Hole H7	Shaft n6	Fit	Hole H7	Shaft p6	Fit	Hole H7	Shaft s6	Fit	Hole H7	Shaft u6	Fit
1	MAX	1.010	1.006	0.010	1.010	1.010	0.006	1.010	1.012	0.004	1.010	1.020	-0.004	1.010	1.024	-0.008
	MIN	1.000	1.000	-0.006	1.000	1.004	-0.010	1.000	1.006	-0.012	1.000	1.014	-0.020	1.000	1.018	-0.024
1.2	MAX	1.210	1.206	0.010	1.210	1.210	0.006	1.210	1.212	0.004	1.210	1.220	-0.004	1.210	1.224	-0.008
	MIN	1.200	1.200	-0.006	1.200	1.204	-0.010	1.200	1.206	-0.012	1.200	1.214	-0.020	1.200	1.218	-0.024
1.6	MAX	1.610	1.606	0.010	1.610	1.610	0.006	1.610	1.612	0.004	1.610	1.620	-0.004	1.610	1.624	-0.008
	MIN	1.600	1.600	-0.006	1.600	1.604	-0.010	1.600	1.606	-0.012	1.600	1.614	-0.020	1.600	1.618	-0.024
2	MAX	2.010	2.006	0.010	2.010	2.010	0.006	2.010	2.012	0.004	2.010	2.020	-0.004	2.010	2.024	-0.008
	MIN	2.000	2.000	-0.006	2.000	2.004	-0.010	2.000	2.006	-0.012	2.000	2.014	-0.020	2.000	2.018	-0.024
2.5	MAX	2.510	2.506	0.010	2.510	2.510	0.006	2.510	2.512	0.004	2.510	2.520	-0.004	2.510	2.524	-0.008
	MIN	2.500	2.500	-0.006	2.500	2.504	-0.010	2.500	2.506	-0.012	2.500	2.514	-0.020	2.500	2.518	-0.024
3	MAX	3.010	3.006	0.010	3.010	3.010	0.006	3.010	3.012	0.004	3.010	3.020	-0.004	3.010	3.024	-0.008
	MIN	3.000	3.000	-0.006	3.000	3.004	-0.010	3.000	3.006	-0.012	3.000	3.014	-0.020	3.000	3.018	-0.024
4	MAX	4.012	4.009	0.011	4.012	4.016	0.004	4.012	4.020	0.000	4.012	4.027	-0.007	4.012	4.031	-0.011
	MIN	4.000	4.001	-0.009	4.000	4.008	-0.016	4.000	4.012	-0.020	4.000	4.019	-0.027	4.000	4.023	-0.031
5	MAX	5.012	5.009	0.011	5.012	5.016	0.004	5.012	5.020	0.000	5.012	5.027	-0.007	5.012	5.031	-0.011
	MIN	5.000	5.001	-0.009	5.000	5.008	-0.016	5.000	5.012	-0.020	5.000	5.019	-0.027	5.000	5.023	-0.031
6	MAX	6.012	6.009	0.011	6.012	6.016	0.004	6.012	6.020	0.000	6.012	6.027	-0.007	6.012	6.031	-0.011
	MIN	6.000	6.001	-0.009	6.000	6.008	-0.016	6.000	6.012	-0.020	6.000	6.019	-0.027	6.000	6.023	-0.031
8	MAX	8.015	8.010	0.014	8.015	8.019	0.005	8.015	8.024	0.000	8.015	8.032	-0.008	8.015	8.037	-0.013
	MIN	8.000	8.001	-0.010	8.000	8.010	-0.019	8.000	8.015	-0.024	8.000	8.023	-0.032	8.000	8.028	-0.037
10	MAX	10.015	10.010	0.014	10.015	10.019	0.005	10.015	10.024	0.000	10.015	10.032	-0.008	10.015	10.037	-0.013
	MIN	10.000	10.001	-0.010	10.000	10.010	-0.019	10.000	10.015	-0.024	10.000	10.023	-0.032	10.000	10.028	-0.037
12	MAX	12.018	12.012	0.017	12.018	12.023	0.006	12.018	12.029	0.000	12.018	12.039	-0.010	12.018	12.044	-0.015
	MIN	12.000	12.001	-0.012	12.000	12.012	-0.023	12.000	12.018	-0.029	12.000	12.028	-0.039	12.000	12.033	-0.044
16	MAX	16.018	16.012	0.017	16.018	16.023	0.006	16.018	16.029	0.000	16.018	16.039	-0.010	16.018	16.044	-0.015
	MIN	16.000	16.001	-0.012	16.000	16.012	-0.023	16.000	16.018	-0.029	16.000	16.028	-0.039	16.000	16.033	-0.044
20	MAX	20.021	20.015	0.019	20.021	20.028	0.006	20.021	20.035	-0.001	20.021	20.048	-0.014	20.021	20.054	-0.020
	MIN	20.000	20.002	-0.015	20.000	20.015	-0.028	20.000	20.022	-0.035	20.000	20.035	-0.048	20.000	20.041	-0.054
25	MAX	25.021	25.015	0.019	25.021	25.028	0.006	25.021	25.035	-0.001	25.021	25.048	-0.014	25.021	25.061	-0.027
	MIN	25.000	25.002	-0.015	25.000	25.015	-0.028	25.000	25.022	-0.035	25.000	25.035	-0.048	25.000	25.048	-0.061
30	MAX	30.021	30.015	0.019	30.021	30.028	0.006	30.021	30.035	-0.001	30.021	30.048	-0.014	30.021	30.061	-0.027
	MIN	30.000	30.002	-0.015	30.000	30.015	-0.028	30.000	30.022	-0.035	30.000	30.035	-0.048	30.000	30.048	-0.061

Table 6.10 (Continued)

BASIC SIZE		LOCATIONAL TRANSN. Hole H7	Shaft k6	Fit	LOCATIONAL TRANSN. Hole H7	Shaft n6	Fit	LOCATIONAL INTERF. Hole H7	Shaft p6	Fit	MEDIUM DRIVE Hole H7	Shaft s6	Fit	FORCE Hole H7	Shaft u6	Fit
40	MAX	40.025	40.018	0.023	40.025	40.033	0.008	40.025	40.042	-0.001	40.025	40.059	-0.018	40.025	40.076	-0.035
	MIN	40.000	40.002	-0.018	40.000	40.017	-0.033	40.000	40.026	-0.042	40.000	40.043	-0.059	40.000	40.060	-0.076
50	MAX	50.025	50.018	0.023	50.025	50.033	0.008	50.025	50.042	-0.001	50.025	50.059	-0.018	50.025	50.086	-0.045
	MIN	50.000	50.002	-0.018	50.000	50.017	-0.033	50.000	50.026	-0.042	50.000	50.043	-0.059	50.000	50.070	-0.086
60	MAX	60.030	60.021	0.028	60.030	60.039	0.010	60.030	60.051	-0.002	60.030	60.072	-0.023	60.030	60.106	-0.057
	MIN	60.000	60.002	-0.021	60.000	60.020	-0.039	60.000	60.032	-0.051	60.000	60.053	-0.072	60.000	60.087	-0.106
80	MAX	80.030	80.021	0.028	80.030	80.039	0.010	80.030	80.051	-0.002	80.030	80.078	-0.029	80.030	80.121	-0.072
	MIN	80.000	80.002	-0.021	80.000	80.020	-0.039	80.000	80.032	-0.051	80.000	80.059	-0.078	80.000	80.102	-0.121
100	MAX	100.035	100.025	0.032	100.035	100.045	0.012	100.035	100.059	-0.002	100.035	100.093	-0.036	100.035	100.146	-0.089
	MIN	100.000	100.003	-0.025	100.000	100.023	-0.045	100.000	100.037	-0.059	100.000	100.071	-0.093	100.000	100.124	-0.146
120	MAX	120.035	120.025	0.032	120.035	120.045	0.012	120.035	120.059	-0.002	120.035	120.101	-0.044	120.035	120.166	-0.109
	MIN	120.000	120.003	-0.025	120.000	120.023	-0.045	120.000	120.037	-0.059	120.000	120.079	-0.101	120.000	120.144	-0.166
160	MAX	160.040	160.028	0.037	160.040	160.052	0.013	160.040	160.068	-0.003	160.040	160.125	-0.060	160.040	160.215	-0.150
	MIN	160.000	160.003	-0.028	160.000	160.027	-0.052	160.000	160.043	-0.068	160.000	160.100	-0.125	160.000	160.190	-0.215
200	MAX	200.046	200.033	0.042	200.046	200.060	0.015	200.046	200.079	-0.004	200.046	200.151	-0.076	200.046	200.265	-0.190
	MIN	200.000	200.004	-0.033	200.000	200.031	-0.060	200.000	200.050	-0.079	200.000	200.122	-0.151	200.000	200.236	-0.265
250	MAX	250.046	250.033	0.042	250.046	250.060	0.015	250.046	250.079	-0.004	250.046	250.169	-0.094	250.046	250.313	-0.238
	MIN	250.000	250.004	-0.033	250.000	250.031	-0.060	250.000	250.050	-0.079	250.000	250.140	-0.169	250.000	250.284	-0.313
300	MAX	300.052	300.036	0.048	300.052	300.066	0.018	300.052	300.088	-0.004	300.052	300.202	-0.118	300.052	300.382	-0.298
	MIN	300.000	300.004	-0.036	300.000	300.034	-0.066	300.000	300.056	-0.088	300.000	300.170	-0.202	300.000	300.350	-0.382
400	MAX	400.057	400.040	0.053	400.057	400.073	0.020	400.057	400.098	-0.005	400.057	400.244	-0.151	400.057	400.471	-0.378
	MIN	400.000	400.004	-0.040	400.000	400.037	-0.073	400.000	400.062	-0.098	400.000	400.208	-0.244	400.000	400.435	-0.471
500	MAX	500.063	500.045	0.058	500.063	500.080	0.023	500.063	500.108	-0.005	500.063	500.292	-0.189	500.063	500.580	-0.477
	MIN	500.000	500.005	-0.045	500.000	500.040	-0.080	500.000	500.068	-0.108	500.000	500.252	-0.292	500.000	500.540	-0.580

Source: J. H. Earle, *Drafting Technology*, Addison-Wesley, Reading, Massachusetts, 1982.

Table 6.11 Preferred Shaft Basis Clearance Fits: Cylindrical Fits (ANSI B4.2)

BASIC SIZE		LOOSE RUNNING Hole C11	Shaft h11	Fit	FREE RUNNING Hole D9	Shaft h9	Fit	CLOSE RUNNING Hole F8	Shaft h7	Fit	SLIDING Hole G7	Shaft h6	Fit	LOCATIONAL CLEARANCE Hole H7	Shaft h6	Fit
1	MAX	1.120	1.000	0.180	1.045	1.000	0.070	1.020	1.000	0.030	1.012	1.000	0.018	1.010	1.000	0.016
	MIN	1.060	0.940	0.060	1.020	0.975	0.020	1.006	0.990	0.006	1.002	0.994	0.002	1.000	0.994	0.000
1.2	MAX	1.320	1.200	0.180	1.245	1.200	0.070	1.220	1.200	0.030	1.212	1.200	0.018	1.210	1.200	0.016
	MIN	1.260	1.140	0.060	1.220	1.175	0.020	1.206	1.190	0.006	1.202	1.194	0.002	1.200	1.194	0.000
1.6	MAX	1.720	1.600	0.180	1.645	1.600	0.070	1.620	1.600	0.030	1.612	1.600	0.018	1.610	1.600	0.016
	MIN	1.660	1.540	0.060	1.620	1.575	0.020	1.606	1.590	0.006	1.602	1.594	0.002	1.600	1.594	0.000
2	MAX	2.120	2.000	0.180	2.045	2.000	0.070	2.020	2.000	0.030	2.012	2.000	0.018	2.010	2.000	0.016
	MIN	2.060	1.940	0.060	2.020	1.975	0.020	2.006	1.990	0.006	2.002	1.994	0.002	2.000	1.994	0.000
2.5	MAX	2.620	2.500	0.180	2.545	2.500	0.070	2.520	2.500	0.030	2.512	2.500	0.018	2.510	2.500	0.016
	MIN	2.560	2.440	0.060	2.520	2.475	0.020	2.506	2.490	0.006	2.502	2.494	0.002	2.500	2.494	0.000
3	MAX	3.120	3.000	0.180	3.045	3.000	0.070	3.020	3.000	0.030	3.012	3.000	0.018	3.010	3.000	0.016
	MIN	3.060	2.940	0.060	3.020	2.975	0.020	3.006	2.990	0.006	3.002	2.994	0.002	3.000	2.994	0.000
4	MAX	4.145	4.000	0.220	4.060	4.000	0.090	4.028	4.000	0.040	4.016	4.000	0.024	4.012	4.000	0.020
	MIN	4.070	3.925	0.070	4.030	3.970	0.030	4.010	3.988	0.010	4.004	3.992	0.004	4.000	3.992	0.000
5	MAX	5.145	5.000	0.220	5.060	5.000	0.090	5.028	5.000	0.040	5.016	5.000	0.024	5.012	5.000	0.020
	MIN	5.070	4.925	0.070	5.030	4.970	0.030	5.010	4.988	0.010	5.004	4.992	0.004	5.000	4.992	0.000
6	MAX	6.145	6.000	0.220	6.060	6.000	0.090	6.028	6.000	0.040	6.016	6.000	0.024	6.012	6.000	0.020
	MIN	6.070	5.925	0.070	6.030	5.970	0.030	6.010	5.988	0.010	6.004	5.992	0.004	6.000	5.992	0.000
8	MAX	8.170	8.000	0.260	8.076	8.000	0.112	8.035	8.000	0.050	8.020	8.000	0.029	8.015	8.000	0.024
	MIN	8.080	7.910	0.080	8.040	7.964	0.040	8.013	7.985	0.013	8.005	7.991	0.005	8.000	7.991	0.000
10	MAX	10.170	10.000	0.260	10.076	10.000	0.112	10.035	10.000	0.050	10.020	10.000	0.029	10.015	10.000	0.024
	MIN	10.080	9.910	0.080	10.040	9.964	0.040	10.013	9.985	0.013	10.005	9.991	0.005	10.000	9.991	0.000
12	MAX	12.205	12.000	0.315	12.093	12.000	0.136	12.043	12.000	0.061	12.024	12.000	0.035	12.018	12.000	0.029
	MIN	12.095	11.890	0.095	12.050	11.957	0.050	12.016	11.982	0.016	12.006	11.989	0.006	12.000	11.989	0.000
16	MAX	16.205	16.000	0.315	16.093	16.000	0.136	16.043	16.000	0.061	16.024	16.000	0.035	16.018	16.000	0.029
	MIN	16.095	15.890	0.095	16.050	15.957	0.050	16.016	15.982	0.016	16.006	15.989	0.006	16.000	15.989	0.000
20	MAX	20.240	20.000	0.370	20.117	20.000	0.169	20.053	20.000	0.074	20.028	20.000	0.041	20.021	20.000	0.034
	MIN	20.110	19.870	0.110	20.065	19.948	0.065	20.020	19.979	0.020	20.007	19.987	0.007	20.000	19.987	0.000
25	MAX	25.240	25.000	0.370	25.117	25.000	0.169	25.053	25.000	0.074	25.028	25.000	0.041	25.021	25.000	0.034
	MIN	25.110	24.870	0.110	25.065	24.948	0.065	25.020	24.979	0.020	25.007	24.987	0.007	25.000	24.987	0.000
30	MAX	30.240	30.000	0.370	30.117	30.000	0.169	30.053	30.000	0.074	30.028	30.000	0.041	30.021	30.000	0.034
	MIN	30.110	29.870	0.110	30.065	29.948	0.065	30.020	29.979	0.020	30.007	29.987	0.007	30.000	29.987	0.000

Table 6.11 (Continued)

BASIC SIZE		LOOSE RUNNING Hole C11	Shaft h11	Fit	FREE RUNNING Hole D9	Shaft h9	Fit	CLOSE RUNNING Hole F8	Shaft h7	Fit	SLIDING Hole G7	Shaft h6	Fit	LOCATIONAL CLEARANCE Hole H7	Shaft h6	Fit
40	MAX	40.280	40.000	0.440	40.142	40.000	0.204	40.064	40.000	0.089	40.034	40.000	0.050	40.025	40.000	0.041
	MIN	40.120	39.840	0.120	40.080	39.938	0.080	40.025	39.975	0.025	40.009	39.984	0.009	40.000	39.984	0.000
50	MAX	50.290	50.000	0.450	50.142	50.000	0.204	50.064	50.000	0.089	50.034	50.000	0.050	50.025	50.000	0.041
	MIN	50.130	49.840	0.130	50.080	49.938	0.080	50.025	49.975	0.025	50.009	49.984	0.009	50.000	49.984	0.000
60	MAX	60.330	60.000	0.520	60.174	60.000	0.248	60.076	60.000	0.106	60.040	60.000	0.059	60.030	60.000	0.049
	MIN	60.140	59.810	0.140	60.100	59.926	0.100	60.030	59.970	0.030	60.010	59.981	0.010	60.000	59.981	0.000
80	MAX	80.340	80.000	0.530	80.174	80.000	0.248	80.076	80.000	0.106	80.040	80.000	0.059	80.030	80.000	0.049
	MIN	80.150	79.810	0.150	80.100	79.926	0.100	80.030	79.970	0.030	80.010	79.981	0.010	80.000	79.981	0.000
100	MAX	100.390	100.000	0.610	100.207	100.000	0.294	100.090	100.000	0.125	100.047	100.000	0.069	100.035	100.000	0.057
	MIN	100.170	99.780	0.170	100.120	99.913	0.120	100.036	99.965	0.036	100.012	99.978	0.012	100.000	99.978	0.000
120	MAX	120.400	120.000	0.620	120.207	120.000	0.294	120.090	120.000	0.125	120.047	120.000	0.069	120.035	120.000	0.057
	MIN	120.180	119.780	0.180	120.120	119.913	0.120	120.036	119.965	0.036	120.012	119.978	0.012	120.000	119.978	0.000
160	MAX	160.460	160.000	0.710	160.245	160.000	0.345	160.106	160.000	0.146	160.054	160.000	0.079	160.040	160.000	0.065
	MIN	160.210	159.750	0.210	160.145	159.900	0.145	160.043	159.960	0.043	160.014	159.975	0.014	160.000	159.975	0.000
200	MAX	200.530	200.000	0.820	200.285	200.000	0.400	200.122	200.000	0.168	200.061	200.000	0.090	200.046	200.000	0.075
	MIN	200.240	199.710	0.240	200.170	199.885	0.170	200.050	199.954	0.050	200.015	199.971	0.015	200.000	199.971	0.000
250	MAX	250.570	250.000	0.860	250.285	250.000	0.400	250.122	250.000	0.168	250.061	250.000	0.090	250.046	250.000	0.075
	MIN	250.280	249.710	0.280	250.170	249.885	0.170	250.050	249.954	0.050	250.015	249.971	0.015	250.000	249.971	0.000
300	MAX	300.650	300.000	0.970	300.320	300.000	0.450	300.137	300.000	0.189	300.069	300.000	0.101	300.052	300.000	0.084
	MIN	300.330	299.680	0.330	300.190	299.870	0.190	300.056	299.948	0.056	300.017	299.968	0.017	300.000	299.968	0.000
400	MAX	400.760	400.000	1.120	400.350	400.000	0.490	400.151	400.000	0.208	400.075	400.000	0.111	400.057	400.000	0.093
	MIN	400.400	399.640	0.400	400.210	399.860	0.210	400.062	399.943	0.062	400.018	399.964	0.018	400.000	399.964	0.000
500	MAX	500.880	500.000	1.280	500.385	500.000	0.540	500.165	500.000	0.228	500.083	500.000	0.123	500.063	500.000	0.103
	MIN	500.480	499.600	0.480	500.230	499.845	0.230	500.068	499.937	0.068	500.020	499.960	0.020	500.000	499.960	0.000

Source: J. H. Earle, *Drafting Technology*, Addison-Wesley, Reading, Massachusetts, 1982.

Table 6.12 Preferred Shaft Basis Transition and Interference Fits: Cylindrical Fits (ANSI B4.2)

BASIC SIZE		LOCATIONAL TRANSN. Hole K7	Shaft h6	Fit	LOCATIONAL TRANSN. Hole N7	Shaft h6	Fit	LOCATIONAL INTERF. Hole P7	Shaft h6	Fit	MEDIUM DRIVE Hole S7	Shaft h6	Fit	FORCE Hole U7	Shaft h6	Fit
1	MAX	1.000	1.000	0.006	0.996	1.000	0.002	0.994	1.000	0.000	0.986	1.000	-0.008	0.982	1.000	-0.012
	MIN	0.990	0.994	-0.010	0.986	0.994	-0.014	0.984	0.994	-0.016	0.976	0.994	-0.024	0.972	0.994	-0.028
1.2	MAX	1.200	1.200	0.006	1.196	1.200	0.002	1.194	1.200	0.000	1.186	1.200	-0.008	1.182	1.200	-0.012
	MIN	1.190	1.194	-0.010	1.186	1.194	-0.014	1.184	1.194	-0.016	1.176	1.194	-0.024	1.172	1.194	-0.028
1.6	MAX	1.600	1.600	0.006	1.596	1.600	0.002	1.594	1.600	0.000	1.586	1.600	-0.008	1.582	1.600	-0.012
	MIN	1.590	1.594	-0.010	1.586	1.594	-0.014	1.584	1.594	-0.016	1.576	1.594	-0.024	1.572	1.594	-0.028
2	MAX	2.000	2.000	0.006	1.996	2.000	0.002	1.994	2.000	0.000	1.986	2.000	-0.008	1.982	2.000	-0.012
	MIN	1.990	1.994	-0.010	1.986	1.994	-0.014	1.984	1.994	-0.016	1.976	1.994	-0.024	1.972	1.994	-0.028
2.5	MAX	2.500	2.500	0.006	2.496	2.500	0.002	2.494	2.500	0.000	2.486	2.500	-0.008	2.482	2.500	-0.012
	MIN	2.490	2.494	-0.010	2.486	2.494	-0.014	2.484	2.494	-0.016	2.476	2.494	-0.024	2.472	2.494	-0.028
3	MAX	3.000	3.000	0.006	2.996	3.000	0.002	2.994	3.000	0.000	2.986	3.000	-0.008	2.982	3.000	-0.012
	MIN	2.990	2.994	-0.010	2.986	2.994	-0.014	2.984	2.994	-0.016	2.976	2.994	-0.024	2.972	2.994	-0.028
4	MAX	4.003	4.000	0.011	3.996	4.000	0.004	3.992	4.000	0.000	3.985	4.000	-0.007	3.981	4.000	-0.011
	MIN	3.991	3.992	-0.009	3.984	3.992	-0.016	3.980	3.992	-0.020	3.973	3.992	-0.027	3.969	3.992	-0.031
5	MAX	5.003	5.000	0.011	4.996	5.000	0.004	4.992	5.000	0.000	4.985	5.000	-0.007	4.981	5.000	-0.011
	MIN	4.991	4.992	-0.009	4.984	4.992	-0.016	4.980	4.992	-0.020	4.973	4.992	-0.027	4.969	4.992	-0.031
6	MAX	6.003	6.000	0.011	5.996	6.000	0.004	5.992	6.000	0.000	5.985	6.000	-0.007	5.981	6.000	-0.011
	MIN	5.991	5.992	-0.009	5.984	5.992	-0.016	5.980	5.992	-0.020	5.973	5.992	-0.027	5.969	5.992	-0.031
8	MAX	8.005	8.000	0.014	7.996	8.000	0.005	7.991	8.000	0.000	7.983	8.000	-0.008	7.978	8.000	-0.013
	MIN	7.990	7.991	-0.010	7.981	7.991	-0.019	7.976	7.991	-0.024	7.968	7.991	-0.032	7.963	7.991	-0.037
10	MAX	10.005	10.000	0.014	9.996	10.000	0.005	9.991	10.000	0.000	9.983	10.000	-0.008	9.978	10.000	-0.013
	MIN	9.990	9.991	-0.010	9.981	9.991	-0.019	9.976	9.991	-0.024	9.968	9.991	-0.032	9.963	9.991	-0.037
12	MAX	12.006	12.000	0.017	11.995	12.000	0.006	11.989	12.000	0.000	11.979	12.000	-0.010	11.974	12.000	-0.015
	MIN	11.988	11.989	-0.012	11.977	11.989	-0.023	11.971	11.989	-0.029	11.961	11.989	-0.039	11.956	11.989	-0.044
16	MAX	16.006	16.000	0.017	15.995	16.000	0.006	15.989	16.000	0.000	15.979	16.000	-0.010	15.974	16.000	-0.015
	MIN	15.988	15.989	-0.012	15.977	15.989	-0.023	15.971	15.989	-0.029	15.961	15.989	-0.039	15.956	15.989	-0.044
20	MAX	20.006	20.000	0.019	19.993	20.000	0.006	19.986	20.000	-0.001	19.973	20.000	-0.014	19.967	20.000	-0.020
	MIN	19.985	19.987	-0.015	19.972	19.987	-0.028	19.965	19.987	-0.035	19.952	19.987	-0.048	19.946	19.987	-0.054
25	MAX	25.006	25.000	0.019	24.993	25.000	0.006	24.986	25.000	-0.001	24.973	25.000	-0.014	24.960	25.000	-0.027
	MIN	24.985	24.987	-0.015	24.972	24.987	-0.028	24.965	24.987	-0.035	24.952	24.987	-0.048	24.939	24.987	-0.061
30	MAX	30.006	30.000	0.019	29.993	30.000	0.006	29.986	30.000	-0.001	29.973	30.000	-0.014	29.960	30.000	-0.027
	MIN	29.985	29.987	-0.015	29.972	29.987	-0.028	29.965	29.987	-0.035	29.952	29.987	-0.048	29.939	29.987	-0.061

Table 6.12 (Continued)

BASIC SIZE		LOCATIONAL TRANSN. Hole K7	Shaft h6	Fit	LOCATIONAL TRANSN. Hole N7	Shaft h6	Fit	LOCATIONAL INTERF. Hole P7	Shaft h6	Fit	MEDIUM DRIVE Hole S7	Shaft h6	Fit	FORCE Hole U7	Shaft h6	Fit
40	MAX	40.007	40.000	0.023	39.992	40.000	0.008	39.983	40.000	-0.001	39.966	40.000	-0.018	39.949	40.000	-0.035
	MIN	39.982	39.984	-0.018	39.967	39.984	-0.033	39.958	39.984	-0.042	39.941	39.984	-0.059	39.924	39.984	-0.076
50	MAX	50.007	50.000	0.023	49.992	50.000	0.008	49.983	50.000	-0.001	49.966	50.000	-0.018	49.939	50.000	-0.045
	MIN	49.982	49.984	-0.018	49.967	49.984	-0.033	49.958	49.984	-0.042	49.941	49.984	-0.059	49.914	49.984	-0.086
60	MAX	60.009	60.000	0.028	59.991	60.000	0.010	59.979	60.000	-0.002	59.958	60.000	-0.023	59.924	60.000	-0.087
	MIN	59.979	59.981	-0.021	59.961	59.981	-0.039	59.949	59.981	-0.051	59.928	59.981	-0.072	59.894	59.981	-0.106
80	MAX	80.009	80.000	0.028	79.991	80.000	0.010	79.979	80.000	-0.002	79.952	80.000	-0.029	79.909	80.000	-0.072
	MIN	79.979	79.981	-0.021	79.961	79.981	-0.039	79.949	79.981	-0.051	79.922	79.981	-0.078	79.879	79.981	-0.121
100	MAX	100.010	100.000	0.032	99.990	100.000	0.012	99.976	100.000	-0.002	99.942	100.000	-0.036	99.889	100.000	-0.089
	MIN	99.975	99.978	-0.025	99.955	99.978	-0.045	99.941	99.978	-0.059	99.907	99.978	-0.093	99.854	99.978	-0.146
120	MAX	120.010	120.000	0.032	119.990	120.000	0.012	119.976	120.000	-0.002	119.934	120.000	-0.044	119.869	120.000	-0.109
	MIN	119.975	119.978	-0.025	119.955	119.978	-0.045	119.941	119.978	-0.059	119.899	119.978	-0.101	119.834	119.978	-0.166
160	MAX	160.012	160.000	0.037	159.988	160.000	0.013	159.972	160.000	-0.003	159.915	160.000	-0.060	159.825	160.000	-0.150
	MIN	159.972	159.975	-0.028	159.948	159.975	-0.052	159.932	159.975	-0.068	159.875	159.975	-0.125	159.785	159.975	-0.215
200	MAX	200.013	200.000	0.042	199.986	200.000	0.015	199.967	200.000	-0.004	199.895	200.000	-0.076	199.781	200.000	-0.190
	MIN	199.967	199.971	-0.033	199.940	199.971	-0.060	199.921	199.971	-0.079	199.849	199.971	-0.151	199.735	199.971	-0.265
250	MAX	250.013	250.000	0.042	249.986	250.000	0.015	249.967	250.000	-0.004	249.877	250.000	-0.094	249.733	250.000	-0.238
	MIN	249.967	249.971	-0.033	249.940	249.971	-0.060	249.921	249.971	-0.079	249.831	249.971	-0.169	249.687	249.971	-0.313
300	MAX	300.016	300.000	0.048	299.986	300.000	0.018	299.964	300.000	-0.004	299.850	300.000	-0.118	299.670	300.000	-0.298
	MIN	299.964	299.968	-0.036	299.934	299.968	-0.066	299.912	299.968	-0.088	299.798	299.968	-0.202	299.618	299.968	-0.382
400	MAX	400.017	400.000	0.053	399.984	400.000	0.020	399.959	400.000	-0.005	399.813	400.000	-0.151	399.586	400.000	-0.378
	MIN	399.960	399.964	-0.040	399.927	399.964	-0.073	399.902	399.964	-0.098	399.756	399.964	-0.244	399.529	399.964	-0.471
500	MAX	500.018	500.000	0.058	499.983	500.000	0.023	499.955	500.000	-0.005	499.771	500.000	-0.189	499.483	500.000	-0.477
	MIN	499.955	499.960	-0.045	499.920	499.960	-0.080	499.892	499.960	-0.108	499.708	499.960	-0.292	499.420	499.960	-0.580

Source: J. H. Earle, *Drafting Technology*, Addison-Wesley, Reading, Massachusetts, 1982.

Table 6.13 The International Tolerance Grades

Basic sizes		Tolerance grades[3]																	
Over	Up to and including	IT01	IT0	IT1	IT2	IT3	IT4	IT5	IT6	IT7	IT8	IT9	IT10	IT11	IT12	IT13	IT14	IT15	IT16
0	3	0.0003	0.0005	0.0008	0.0012	0.002	0.003	0.004	0.006	0.010	0.014	0.025	0.040	0.060	0.100	0.140	0.250	0.400	0.600
3	6	0.0004	0.0006	0.001	0.0015	0.0025	0.004	0.005	0.008	0.012	0.018	0.030	0.048	0.075	0.120	0.180	0.300	0.480	0.750
6	10	0.0004	0.0006	0.001	0.0015	0.0025	0.004	0.006	0.009	0.015	0.022	0.036	0.058	0.090	0.150	0.220	0.360	0.580	0.900
10	18	0.0005	0.0008	0.0012	0.002	0.003	0.005	0.008	0.011	0.018	0.027	0.043	0.070	0.110	0.180	0.270	0.430	0.700	1.100
18	30	0.0006	0.001	0.0015	0.0025	0.004	0.006	0.009	0.013	0.021	0.033	0.052	0.084	0.130	0.210	0.330	0.520	0.840	1.300
30	50	0.0006	0.001	0.0015	0.0025	0.004	0.007	0.011	0.016	0.025	0.039	0.062	0.100	0.160	0.250	0.390	0.620	1.000	1.600
50	80	0.0008	0.0012	0.002	0.003	0.005	0.008	0.013	0.019	0.030	0.046	0.074	0.120	0.190	0.300	0.460	0.740	1.200	1.900
80	120	0.001	0.0015	0.0025	0.004	0.006	0.010	0.015	0.022	0.035	0.054	0.087	0.140	0.220	0.350	0.540	0.870	1.400	2.200
120	180	0.0012	0.002	0.0035	0.005	0.008	0.012	0.018	0.025	0.040	0.063	0.100	0.160	0.250	0.400	0.630	1.000	1.600	2.500
180	250	0.002	0.003	0.0045	0.007	0.010	0.014	0.020	0.029	0.046	0.072	0.115	0.185	0.290	0.460	0.720	1.150	1.850	2.900
250	315	0.0025	0.004	0.006	0.008	0.012	0.016	0.023	0.032	0.052	0.081	0.130	0.210	0.320	0.520	0.810	1.300	2.100	3.200
315	400	0.003	0.005	0.007	0.009	0.013	0.018	0.025	0.036	0.057	0.089	0.140	0.230	0.360	0.570	0.890	1.400	2.300	3.600
400	500	0.004	0.006	0.008	0.010	0.015	0.020	0.027	0.040	0.063	0.097	0.155	0.250	0.400	0.630	0.970	1.550	2.500	4.000
500	630	0.0045	0.006	0.009	0.011	0.016	0.022	0.030	0.044	0.070	0.110	0.175	0.280	0.440	0.700	1.100	1.750	2.800	4.400
630	800	0.005	0.007	0.010	0.013	0.018	0.025	0.035	0.050	0.080	0.125	0.200	0.320	0.500	0.800	1.250	2.000	3.200	5.000
800	1000	0.0055	0.008	0.011	0.015	0.021	0.029	0.040	0.056	0.090	0.140	0.230	0.360	0.560	0.900	1.400	2.300	3.600	5.600
1000	1250	0.0065	0.009	0.013	0.018	0.024	0.034	0.046	0.066	0.105	0.165	0.260	0.420	0.660	1.050	1.650	2.600	4.200	6.600
1250	1600	0.008	0.011	0.015	0.021	0.029	0.040	0.054	0.078	0.125	0.195	0.310	0.500	0.780	1.250	1.950	3.100	5.000	7.800
1600	2000	0.009	0.013	0.018	0.025	0.035	0.048	0.065	0.092	0.150	0.230	0.370	0.600	0.920	1.500	2.300	3.700	6.000	9.200
2000	2500	0.011	0.015	0.022	0.030	0.041	0.057	0.077	0.110	0.175	0.280	0.440	0.700	1.100	1.750	2.800	4.400	7.000	11.000
2500	3150	0.013	0.018	0.026	0.036	0.050	0.069	0.093	0.135	0.210	0.330	0.540	0.860	1.350	2.100	3.300	5.400	8.600	13.500

[3] IT Values for tolerance grades larger than IT16 can be calculated by using the following formulas:
IT17 = IT12 × 10; IT18 = IT13 × 10; etc.

Source: J. H. Earle, *Drafting Technology*, Addison-Wesley, Reading, Massachusetts, 1982.

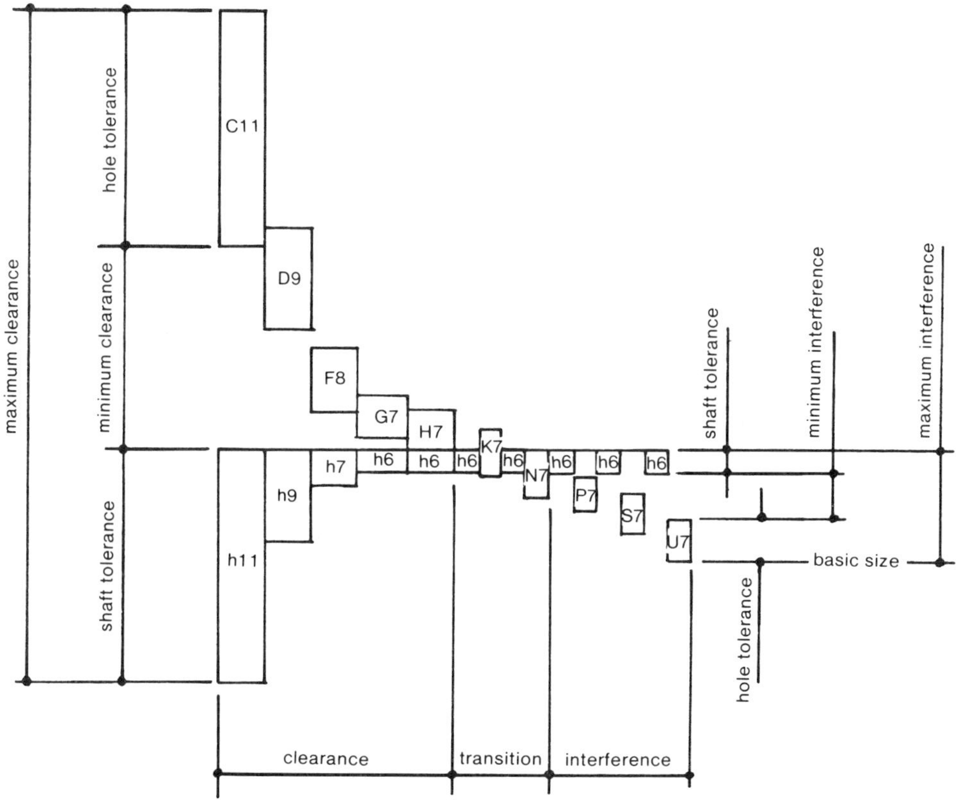

Figure 6.6 Tolerances associated with the shaft basis system. (From J. H. Earle, *Drafting Technology*, Addison-Wesley, Reading, Massachusetts, 1982.)

moving across until 0.052 is located in the IT9 column. Using Figure 6.8 indicates that a boring or reaming operation would be used to machine the hole while a turning operation would be used to machine the shaft.

Now let's use a shaft basis system with a basic diameter of 106 mm which must have a force fit. The preferred size of 100 mm is selected from Table 6.7 with a preferred fit of U7/h6 selected from Table 6.8. The dimensions on the shaft would be 100.000/99.978 mm and the dimensions on the hole would be 99.889/99.854 mm. These values are obtained from Table 6.12 by locating 100 mm in the basic size column and moving along the row horizontally until the force fit U7/h6 column is reached. The values of 100.000/99.978 mm for the shaft and 99.889/ 99.854 mm for the hole appear under their respective U7 and h6

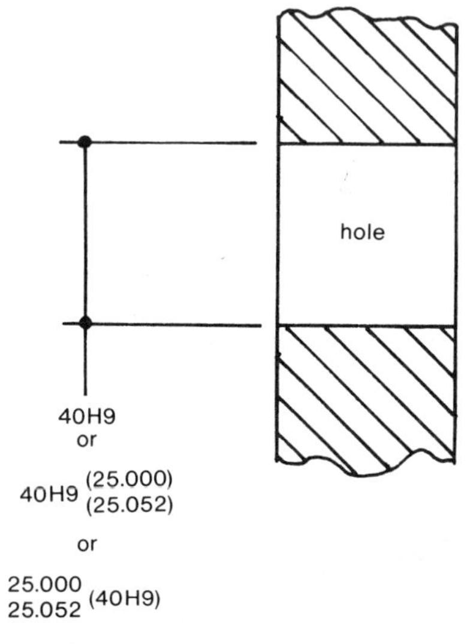

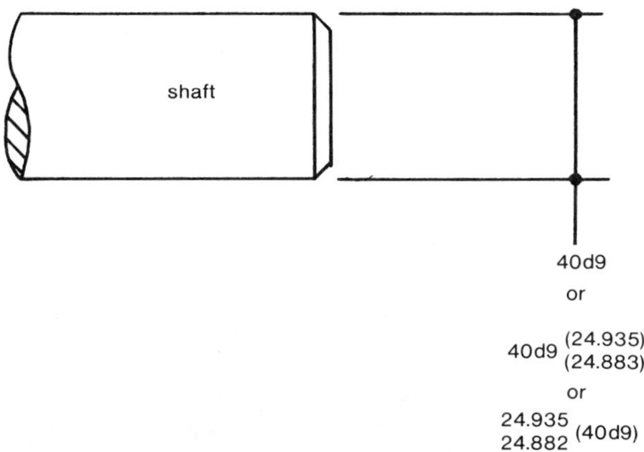

Figure 6.7 Methods of dimensioning and tolerancing mating features using metrics.

columns in this row. Figure 6.9 shows these dimensions placed on the shaft and hole for this force fit. Using Table 6.13 it is found that a grade of IT6 can be used for producing the shaft (100.000 − 99.978 = 0.022 mm) while a grade of IT7 (99.889 − 99.854 = 0.035 mm) can be used to produce the hole. Using Figure 6.8 will determine what machining process could be used to obtain this tolerance range.

There are many features on a part that do not mate with other parts, and therefore, do not have a fit associated with them. But all dimen-

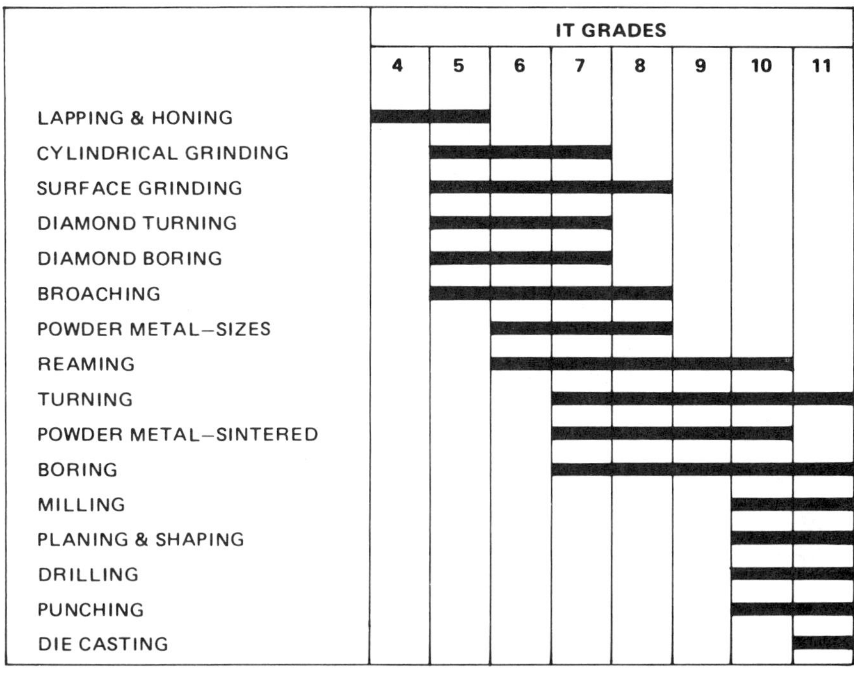

Figure 6.8 Manufacturing processes related to the international tolerance grades. (From J. H. Earle, *Drafting Technology*, Addison-Wesley, Reading, Massachusetts, 1982.)

sions on a drawing are understood to have tolerances, and the amount of tolerance must be noted. Figure 6.10 shows the international tolerance grades and their applications in industry. Since mass-produced items may use grades IT12 through IT16, general tolerances using these grades may be expressed as a note such as:

Unless otherwise specified all untoleranced

dimensions are $\pm \frac{\text{IT } 15}{2}$

This means a tolerance of ±1.400 mm can be used for a dimension between 500 and 630 mm. This value is obtained from Table 6.13 by finding the column of IT15 and locating the row with the basic size range of 500 to 630 mm. This row intersects the IT15 column yielding 2.800 mm as the tolerance range that can be used as the general tolerance. Dividing this value by 2 yields ±1.400 mm as its bilateral general tolerance associated with the dimensions which lies between 500 and 630 mm.

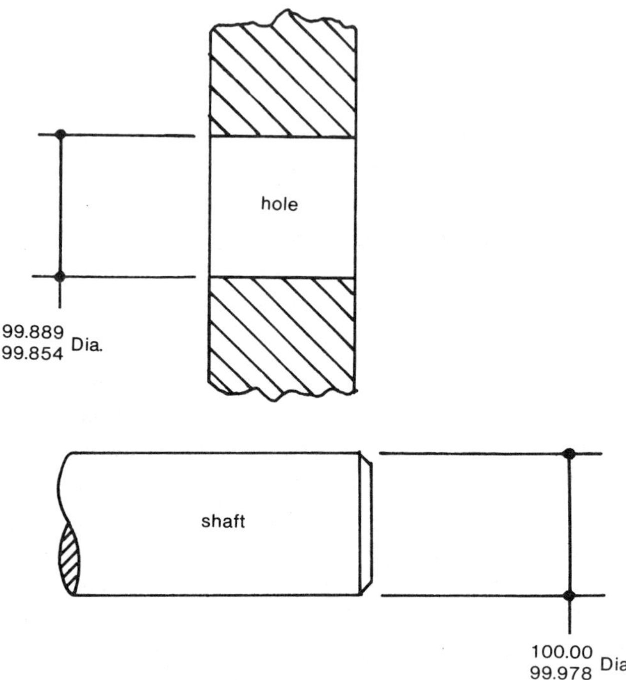

Figure 6.9 Upper and lower limits of a mating hole and shaft using metrics.

Another method of specifying general tolerances pertaining to dimensions appearing on a drawing is by using Table 6.14. This table classifies general linear tolerances according to a fine, medium, and coarse series for dimensions of graduated sizes. For example, a fine tolerance can be specified by the following note:

General tolerance specified in ANSI B4.3

fine series apply

or by placing the information in a table as shown in Figure 6.11 and placing this table on the drawing to provide the tolerances for various ranges of sizes of dimensions on the drawing.

If the dimensions have one or no decimal place, then the general tolerances can be expressed in a table as shown in Figure 6.12 or by a note:

Unless otherwise specified all

untoleranced dimensions are ± 0.06 mm

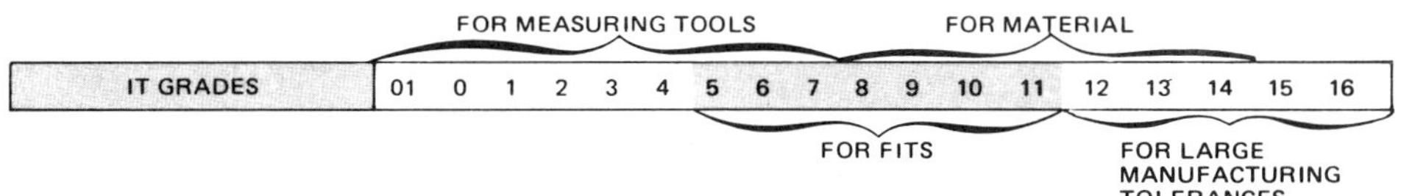

Figure 6.10 The usage of the various international grades. (From J. H. Earle, *Drafting Technology*, Addison-Wesley, Reading, Massachusetts, 1982.)

Dimensions in mm

General Tolerance
Unless Otherwise Specified The Following
Tolerances are Applicable

Linear	Over to	0.5 6	6 30	30 120	120 315	315 1000	1000 2000
Tol	±	0.1	0.2	0.3	0.5	0.8	1.2

Figure 6.11 Specifying general tolerances by using grades. (From J. H. Earle, *Drafting Technology*, Addison-Wesley, Reading, Massachusetts, 1982.)

Dimensions in mm

General Tolerance
Unless Otherwise Specified The Following
Tolerances Are Applicable

Linear	Over To	— 120	120 315	315 1000	1000 —
Tol	One Decimal ±	0.3	0.5	0.8	1.2
	No Decimals ±	0.8	1.2	2	3

Figure 6.12 Specifying general tolerances by using series. (From J. H. Earle, *Drafting Technology*, Addison-Wesley, Reading, Massachusetts, 1982.)

Table 6.14 Permissible Metric Variations Using Fine, Medium, and Coarse Series

		Variations in mm						
Basic dimensions mm		0.5 to 3	over 3 to 6	over 6 to 30	over 30 to 120	over 120 to 315	over 315 to 1,000	over 1,000 to 2,000
Permissible variations	Fine series	±0.05	±0.05	±0.1	±0.15	±0.2	±0.3	±0.5
	Medium series	±0.1	±0.1	±0.2	±0.3	±0.5	±0.8	±1.2
	Coarse series		±0.2	±0.5	±0.8	±1.2	±2	±3

Source: J. H. Earle, *Drafting Technology*, Addison-Wesley, Reading, Massachusetts, 1982.

and should be used only where the dimensions on a drawing have slight differences of magnitude.

6.4. SOFT CONVERSION VERSUS HARD CONVERSION OF DIMENSIONS AND TOLERANCES ON MECHANICAL DRAWINGS

Presently, many companies in the United States are involved in converting from the inch to the metric system pertaining to their technical drawings and manufacturing processes. This conversion can be either a soft type of conversion or a hard type of conversion. A soft inch-to-millimeter conversion consists of changing the dimensions and tolerances on a technical drawing directly to millimeters by using the conversion 1 inch = 25.4 millimeters. Figure 6.13 shows a typical soft type of conversion of the part's dimensions and tolerances from inches to millimeters. A hard inch-to-millimeter conversion consists of redoing the entire system of dimensions and tolerances appearing on the technical drawing as if the product was designed from the beginning using metrics. In other words, the dimensions and tolerances must not only be in millimeters, but must satisfy all the metric standards, such as nominal metric sizes, metric tolerances, metric gaging, and so on.

For example, a metric radial bearing would have a bore size of 25 mm as its acceptable nominal size instead of a bore of 25.4 mm. Therefore, metric machine elements will never mate correctly with parts that have been softly converted from the inch system to the metric system since they are essentially still in the inch system.

To correctly convert mechanical drawings from the inch system to the metric system, one must first determine the tolerance zones present with the inch system. In other words, a tolerance stackup between mating parts must be performed as the first step in this conversion.

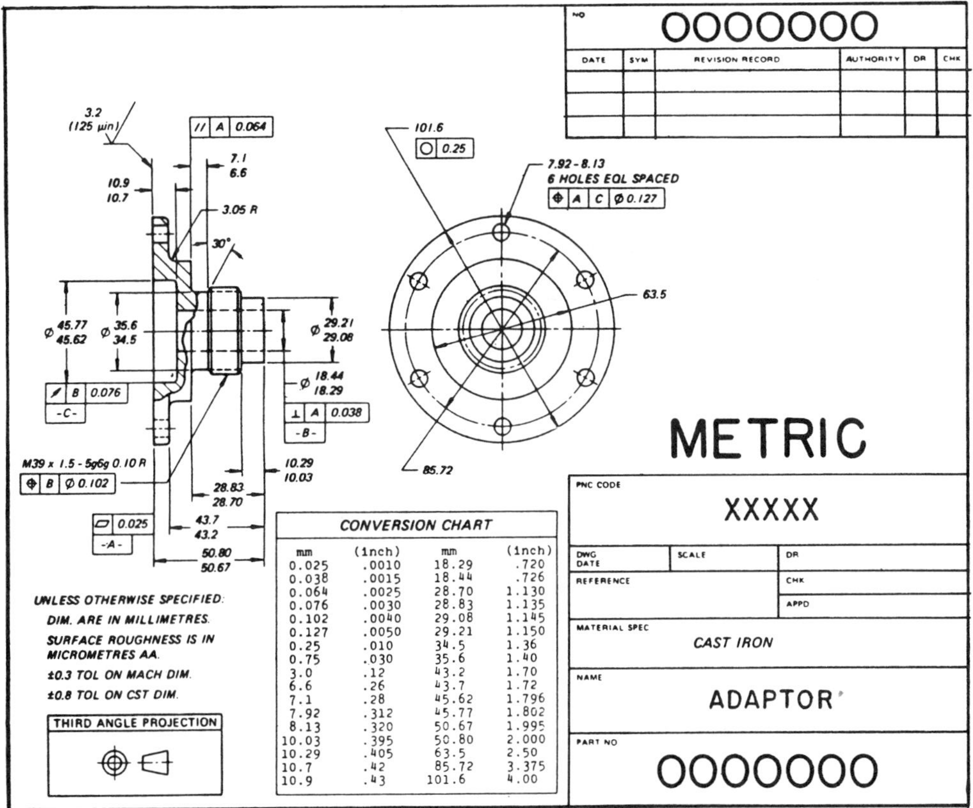

Figure 6.13 An example of a soft converted drawing from inches to millimeters.

6.5. CONVERTING A MECHANICAL DRAWING FROM INCHES TO MILLIMETERS

A belt tightener is shown in Figure 6.14 with its detail drawings appearing in Figures 6.15 – 6.20. This belt tightener is to be completely designed and drawn using hard metrics or, in other words, using hard metric conversion. Before converting this drawing to hard metrics, a tolerance analysis or tolerance stackup must be performed on all of the mating parts. The specifications related to the inch system is as follows:

a. Bushing fit in pulley: Class LN2 fit.
b. Shaft fit in bushing: Class RC5 fit.
c. Shaft fits in frame: Class RC2 fit.

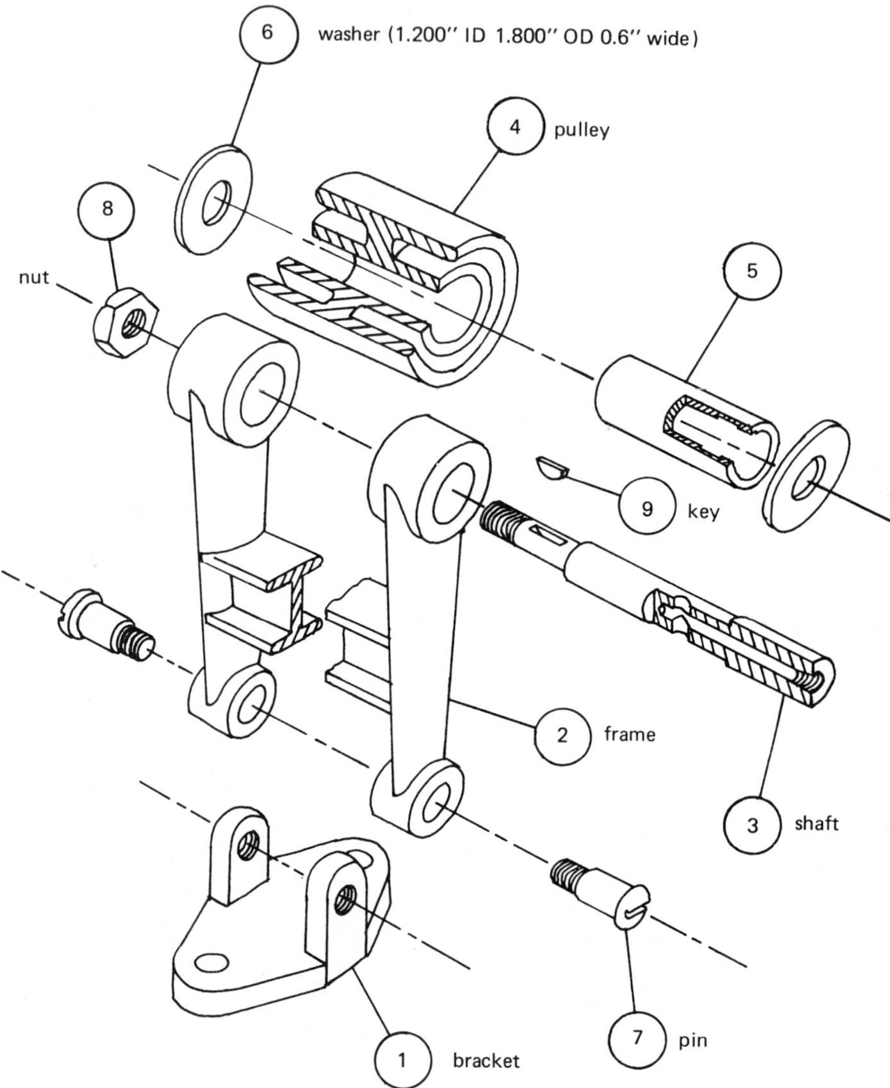

Figure 6.14 Exploded view of belt tightener.

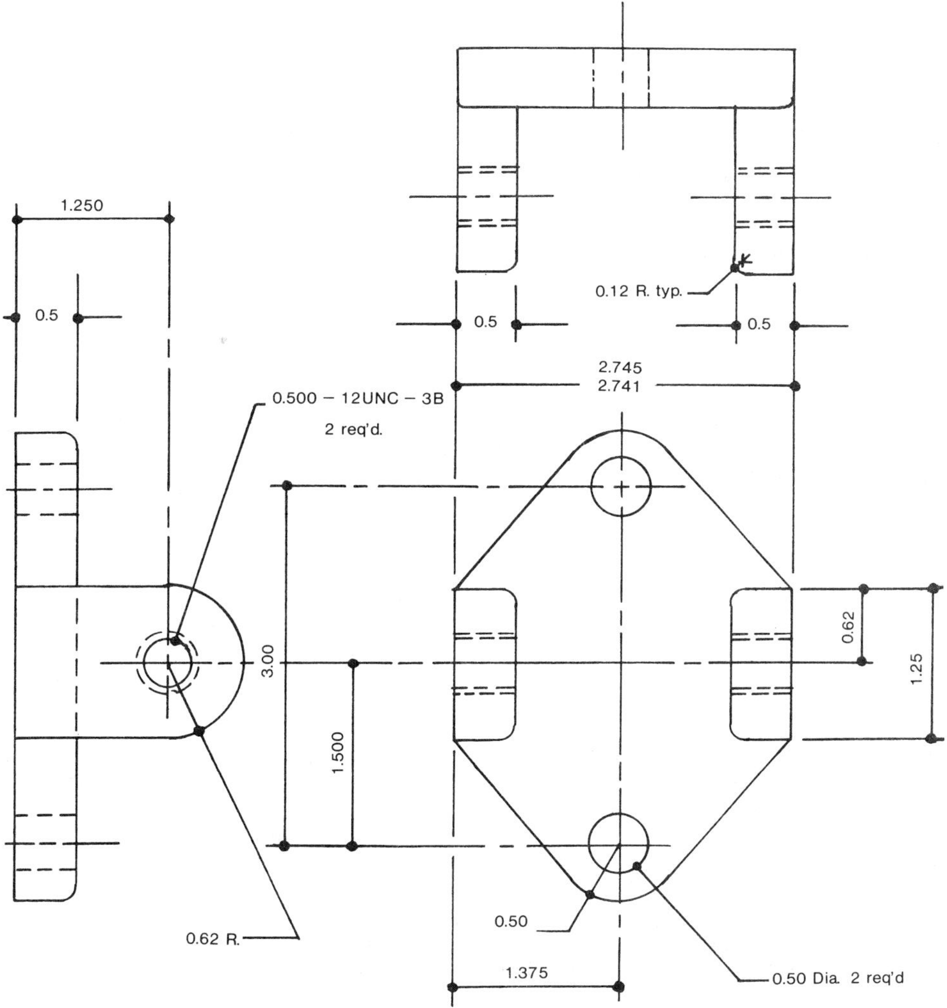

Figure 6.15 Detail drawing of base (inches).

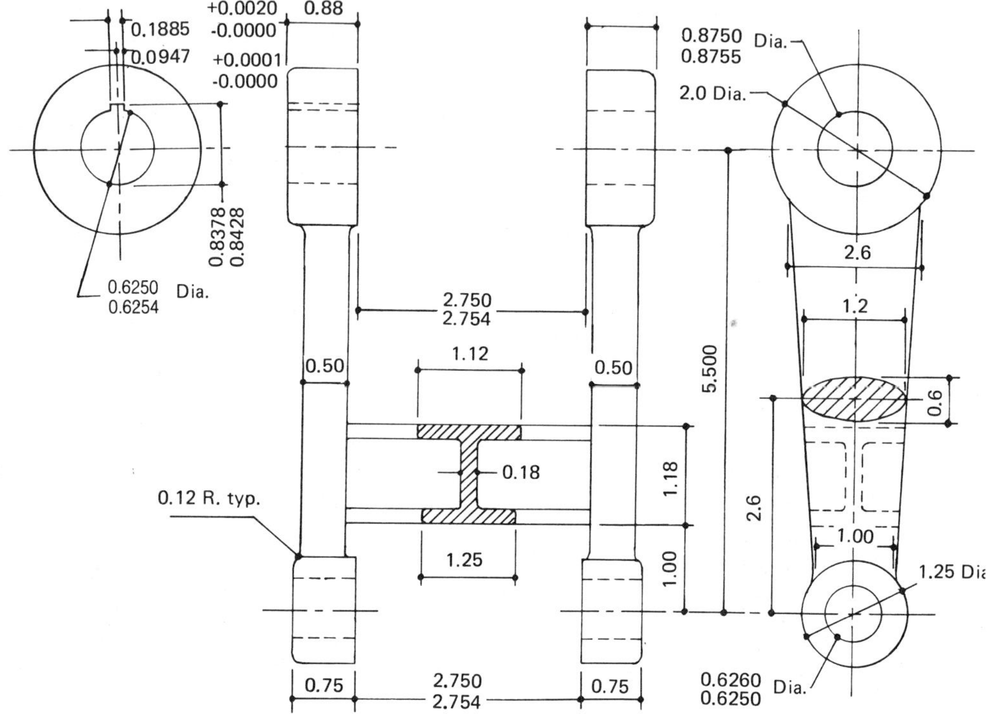

Figure 6.16 Detail drawing of frame (inches).

d. Pin fit in frame: Class RC5 fit.
e. Pulley hub length plus washers fit in frame: allowance of 0.005 inches and tolerance range of 0.004 inches.
f. Make bushing 0.010 inches shorter than pulley hub.
g. Bracket fit in frame: allowance of 0.005 inches and tolerance range of 0.004 inches.
h. All position tolerances are ϕ 0.002 inches.
i. All from tolerances are 0.001 inches.
j. Unless otherwise specified tolerances are as follows:

.x = ±0.1 inches

.xx = ±0.01 inches

.xxx = ±0.001 inches

k. All fits are based on the basic hole system.

In Chapter 1, Table 1.5 is used to find the limits related to the bushing's fit in the pulley. The basic hole system will be used for

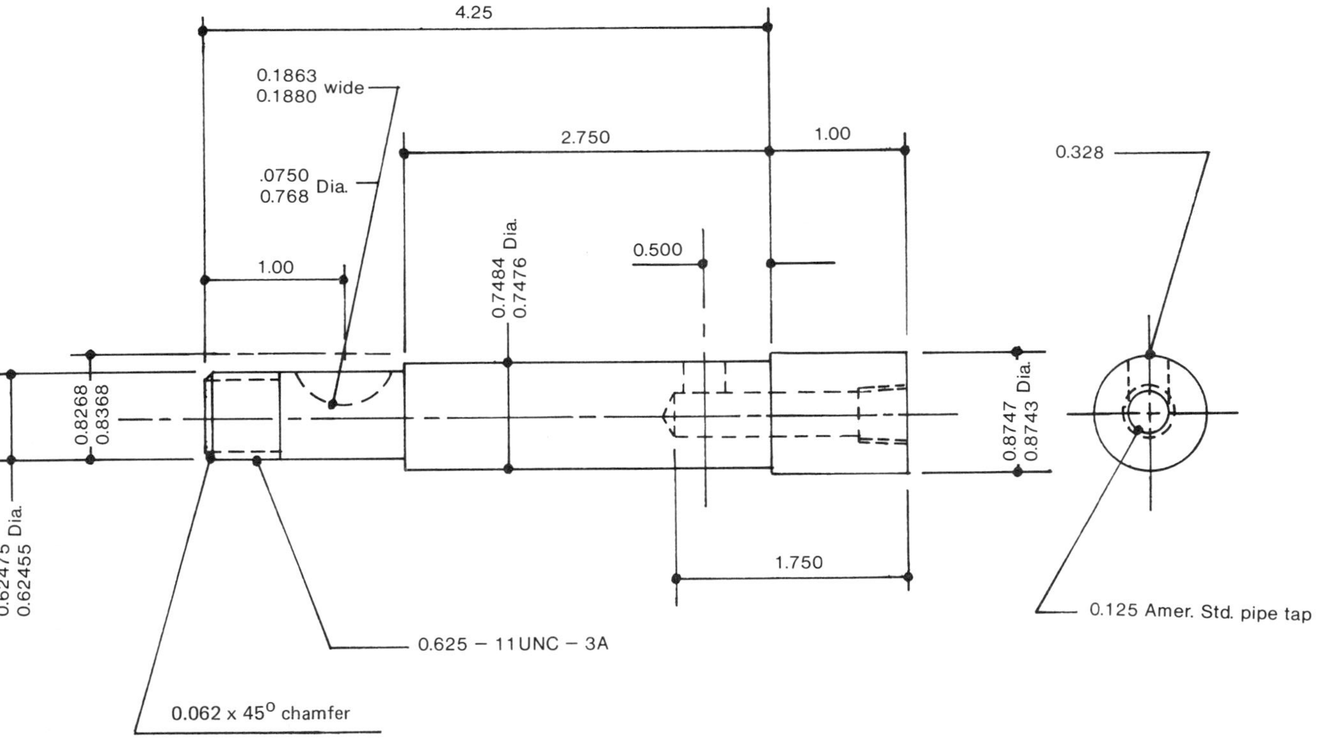

Figure 6.17 Detail drawing of shaft.

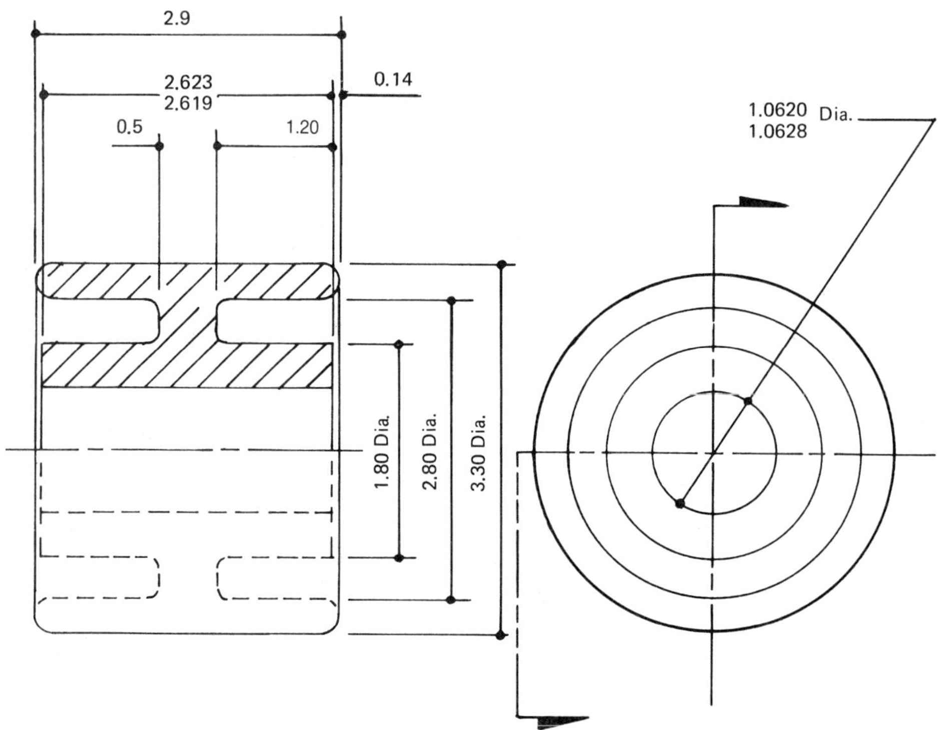

Figure 6.18 Detail drawing of pulley (inches).

calculating this fit; therefore, under the nominal size range column, the bushing's nominal diameter of 1.062 would fall under 0.71 to 1.19 which yields +0.0005/−0.0000 (H6) for the hole and +0.0010/+0.0005 (n5) for the shaft or a possible fit of −0.0000 − −0.0010 inches. The minus signs indicate an interference fit. The upper and lowers limits for the hole's diameter would be 1.0620/1.0625, respectively and for the shaft's diameter they would be 1.0625/1.0630 inches, respectively.

To change this LN2 fit to a suitable metric fit, one must refer to Table 6.8 which relates LC2 or locational interference fit to the metric fit of H7/p6. In this case, LN2 is directly related to the locational interference fit H7/p6 column of Table 6.10. Now, one can imagine why it is so important to use the standard fit tables appearing in Chapter 1 when designing and determining the proper fits using the inch system. The standard metric fit system is closely related to the standard inch fit system which makes it very easy to transfer from one

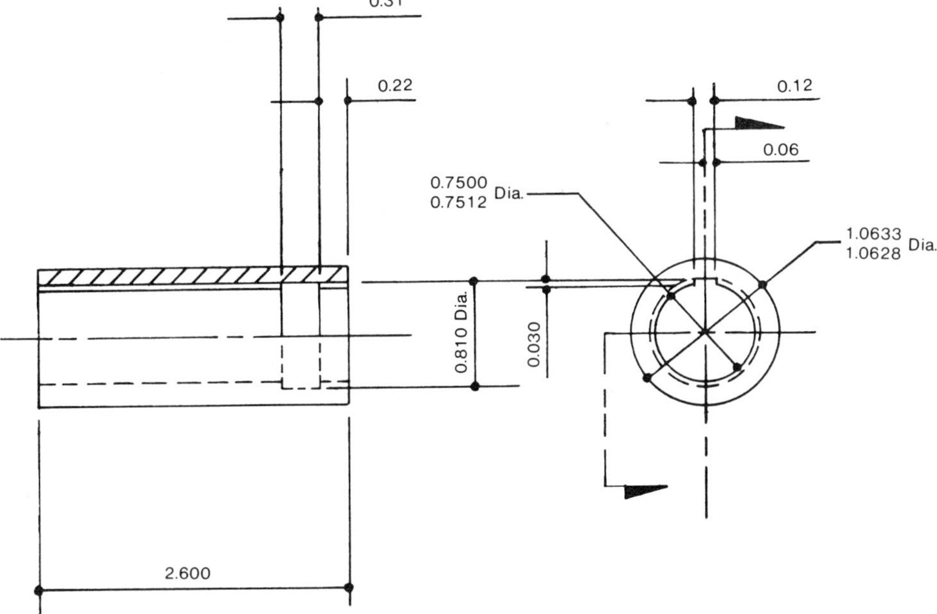

Figure 6.19 Detail drawing of bushing (inches).

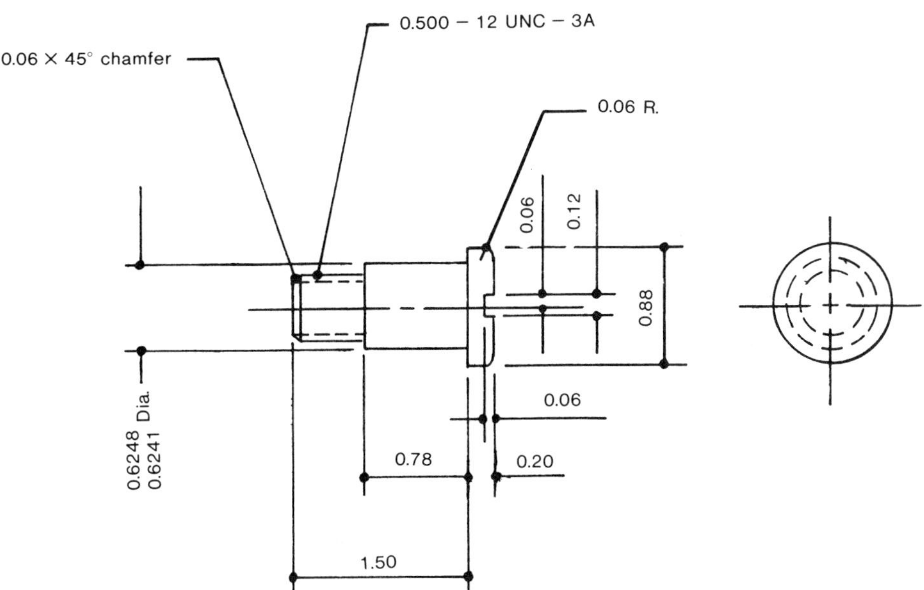

Figure 6.20 Detail drawing of pin (inches).

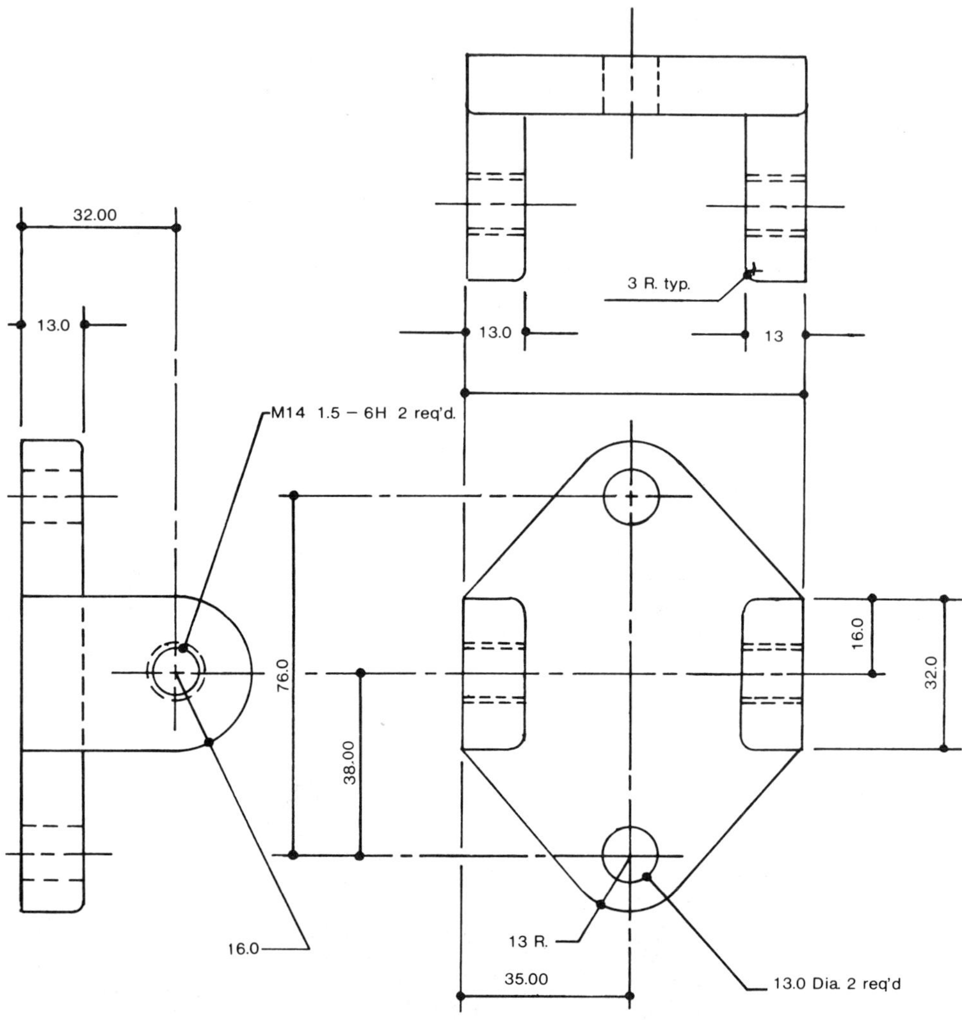

Figure 6.21 Detail drawing of base (millimeters).

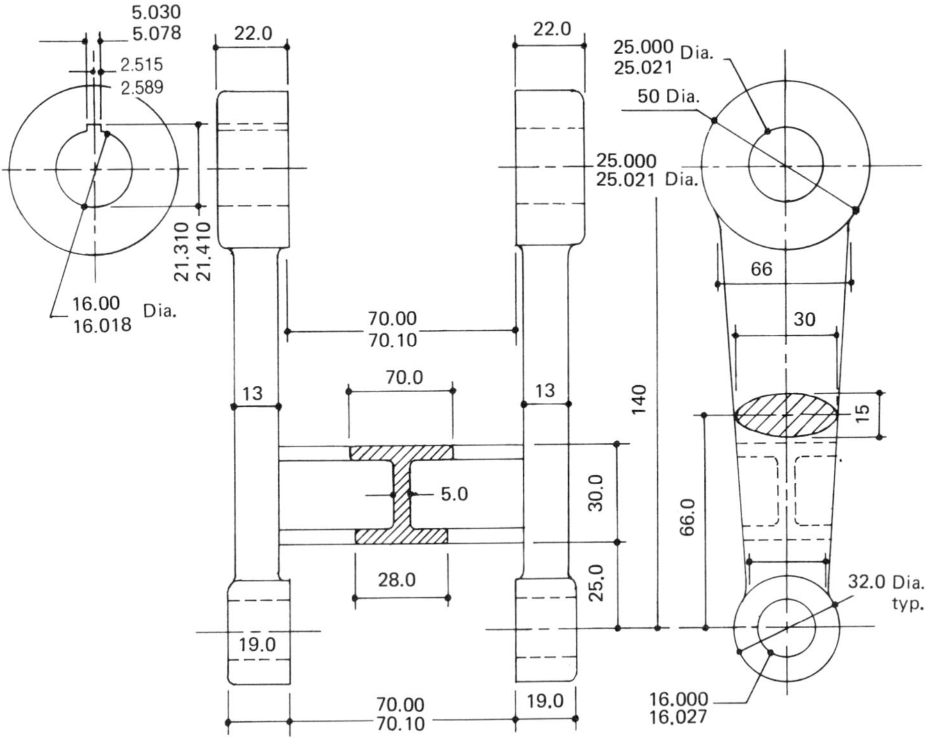

Figure 6.22 Detail drawing of frame (millimeters).

system to the other without losing the required fit, thus, retaining the integrity of the design.

The next step, is to chose a preferred metric size which would be close to 1.062 inches but given in millimeters. The closest preferred size in millimeters to 1.062 inches would be 25 mm. Locating this preferred size of 25 mm in the basic size column and moving across to the locational interference column yields 25.000/25.021 (H7) for the hole and 25.035/25.022 mm (p6) for the shaft. These values produce an interference fit of −0.001 to −0.035 mm where the minus signs indicates an interference fit.

It is observed that 0.001 mm is equal to 0.00004 inches which is considered close to the value of 0.0000 inches in the inch system and 0.035 mm is equal to 0.001378 inches which is considered close to the value of 0.0013 inches in the inch system as well. Therefore, it is concluded that the upper and lower limits of 25.000/25.021 for the hole and 25.035/25.022 mm for the shaft are very close to the equivalent of

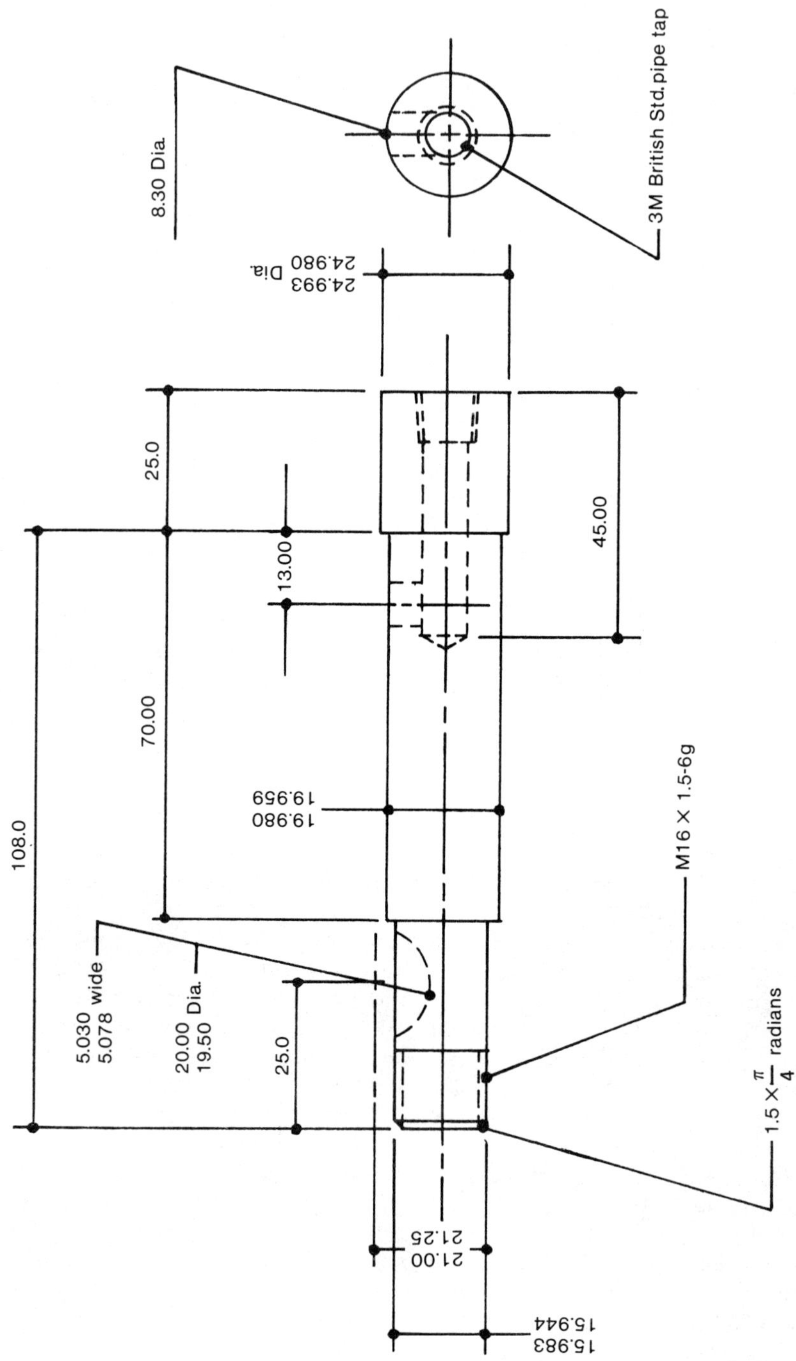

Figure 6.23 Detail drawing of shaft (millimeters).

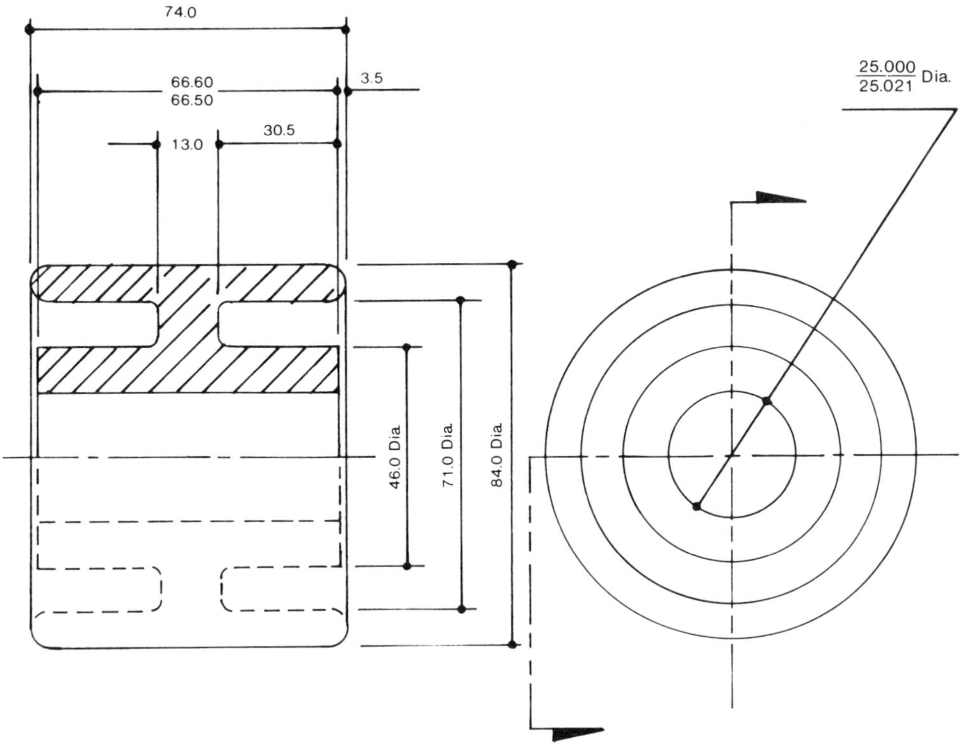

74.0
66.60
66.50
3.5
13.0
30.5
46.0 Dia.
71.0 Dia.
84.0 Dia.
25.000
25.021 Dia.

Figure 6.24 Detail drawing of pulley (millimeters).

8.0
5.5
3.0
1.5
20.000
20.033 Dia.
25.035
25.022 Dia.
20.60 Dia.
0.80
66.00

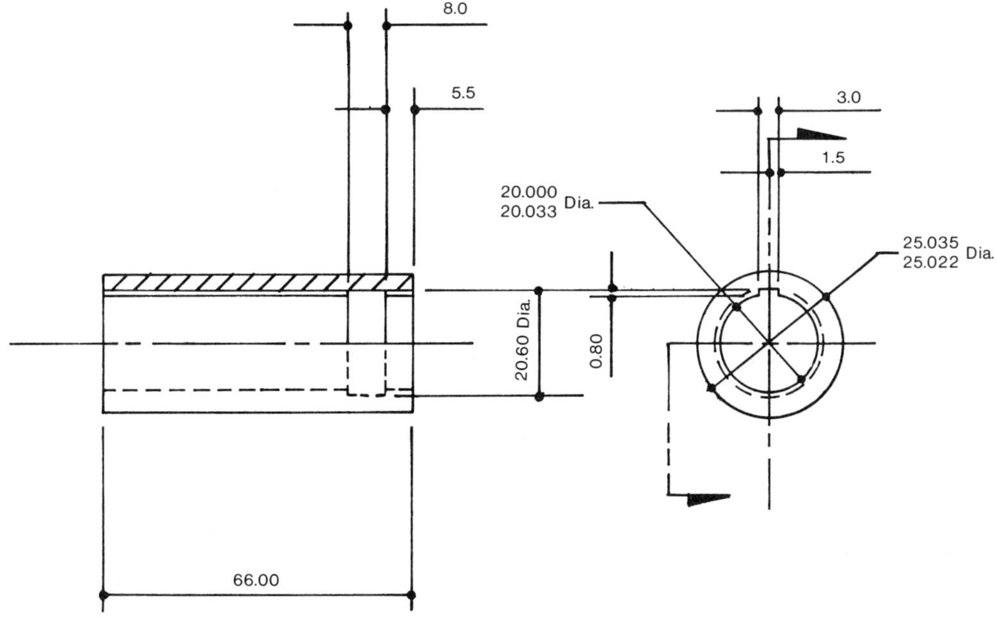

Figure 6.25 Detail drawing of bushing (millimeters).

Table 6.15 Summary of Belt Tightener's Specs in Both the Inch and Metric Systems. (From J. H. Earle, *Drafting Technology*, Addison-Wesley, Reading, Massachusets, 1982.)

System	Design Specs	Fit	Feature	Basic Size	Standard limits	Upper limits	Lower limits	Limits of (+) clearance or (−) interference.
Inch	a	LN2	Hole	1.0620 in.	+0.0008/−0.0000	1.0628	1.0620	−0.0000 TO
			Shaft	1.0620 in.	+0.0013/+0.0008	1.0633	1.0628	−0.0013
Metric	a	H7/p6	Hole	25mm		25.021	25.000	−0.001 TO
			Shaft	25mm		25.035	25.022	−0.035
Inch	b	RC5	Hole	0.7500 in.	+0.0012/+0.0000	0.7512	0.7500	+0.0016 TO
			Shaft	0.7500 in.	−0.0016/−0.0024	0.7484	0.7576	+0.0036
Metric	b	H8/f7	Hole	20 mm		20.033	20.000	+0.074 TO
			Shaft	20 mm		19.980	19.959	+0.020
Inch	c	RC2	Hole	0.6250 in.	+0.0004/−0.0000	0.6254	0.6250	+0.00025 TO
			Shaft	0.6250 in.	−0.00025/−0.00055	0.62475	0.62455	+0.00095
	c	RC2	Hole	0.8750 in.	+0.0005/+0.0000	0.8755	0.8750	+0.0003 TO
			Shaft	0.8750 in.	−0.0003/−0.0007	0.8747	0.8743	+0.0012

	c	H7/g6	Hole	16 mm		16.018	16.000	+0.035 TO
			Shaft	16 mm		15.944	15.983	+0.006
	c	H7/g6	Hole	25 mm		25.021	25.000	+0.041 TO
			Shaft	25 mm		24.993	24.980	+0.007
Inch	d	RC5	Hole	0.625 in.	+0.0010/−0.0000	0.6260	0.6250	+0.0012 TO
			Shaft	0.625 in.	−0.0012/−0.0019	0.6248	0.6241	+0.0029
Metric	d	H8/f7	Hole	16 mm		16.027	16.000	+0.061 TO
			Shaft	16 mm		15.984	15.966	+0.016
Inch	e	Allowance = 0.005 in.						
		Tolerance = 0.004 in.						
Metric	e	Allowance = 0.020 mm						
		Tolerance = 0.016 mm						
Inch	g							
		same as part e						
Metric	g							

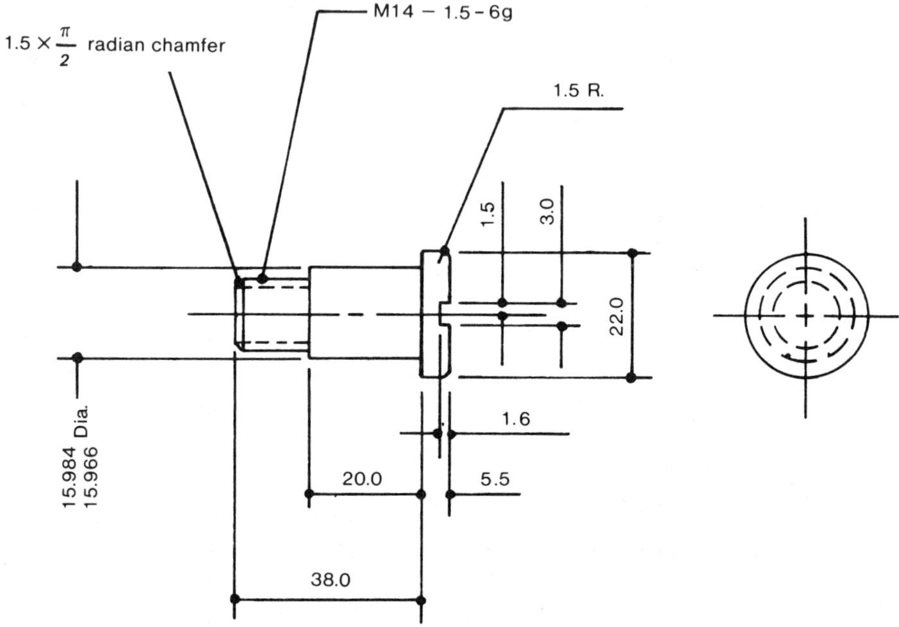

Figure 6.26 Detail drawing of pin (millimeters).

1.0620/1.0628 in. for the hole and 1.0628/1.0633 in. for the shaft related to the inch system.

This same procedure is repeated for the remainder of the belt tightener specifications and these results are summarized in Table 6.15. The remainder of the dimensions and their tolerances are converted to hard metrics resulting in the detail drawings appearing in Figures 6.21-6.26.

6.6. COMPUTER GRAPHICS APPLICATIONS

Computer graphics is widely used in the area of analyzing tolerances and their effect on the design of mating parts. Therefore, computer graphics can be easily applied to the hard conversion of inch to metric drawings. Again, it is important to emphasize the fact that when a product is being designed in the inch system, the nominal sizes and the fits of mating parts must follow the preferred sizes used in the United States and the preferred basic hole or shaft fits as shown in Chapter 1, in Tables 1.2 – 1.6. Otherwise, hard metric conversion of inch drawings become almost impossible and result in a total redesign of the product using metrics from beginning to end.

<table>
<tr><td rowspan="2">measure</td><td colspan="3">clearance</td><td rowspan="2">inch</td><td rowspan="2" colspan="2">BHS</td><td rowspan="4">/</td><td rowspan="3">unit</td><td rowspan="6">carriage return</td></tr>
<tr><td colspan="3">interference</td></tr>
<tr><td>convert</td><td colspan="3">dimension</td><td rowspan="2">millimeter</td><td rowspan="2" colspan="2">BSS</td></tr>
<tr><td>identify</td><td colspan="3">fit</td><td rowspan="3">input</td></tr>
<tr><td rowspan="2">select</td><td>0</td><td>2</td><td>4</td><td>6</td><td>8</td><td>•</td><td></td></tr>
<tr><td>1</td><td>3</td><td>5</td><td>7</td><td>9</td><td>–</td><td></td></tr>
</table>

Figure 6.27 Hard inch to millimeter conversion menu used on computer graphics system.

A typical computer graphics menu for hard metric conversion is shown in Figure 6.27. A variety of commands can be inputted into the computer via this menu. For example, if mating parts are drawn or appear on the display screen as shown in Figure 6.28, one could measure the clearance or interference between mating parts by simply inputting to the computer the commands MEASURE CLEARANCE / UNIT D_1 D_2 (CR) or MEASURE INTERFERENCE / UNIT D_1 D_2 (CR) and

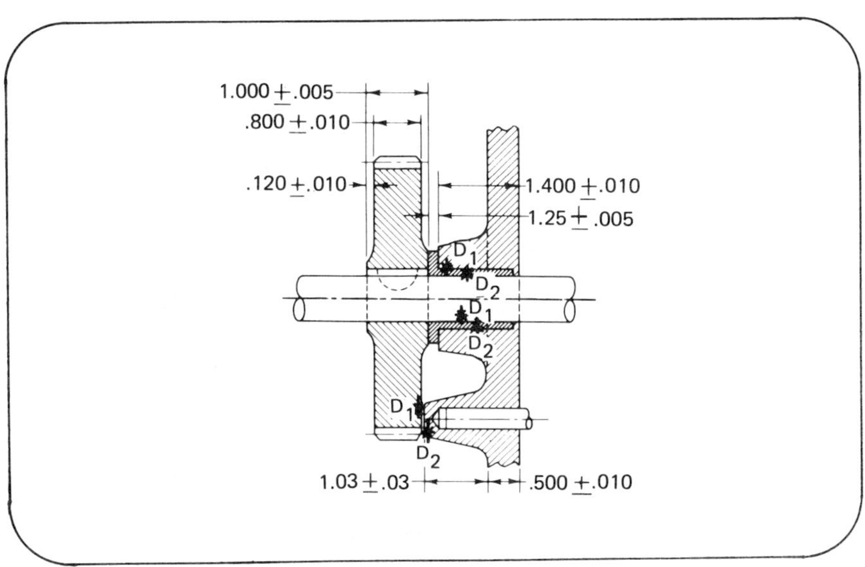

Figure 6.28 Measuring clearances or interferences on an ICGS.

identifying the mating parts on the display screen by digitizing them. This clearance or interference can be converted into millimeters by using the command CONVERT CLEARANCE MILLIMETER / UNIT D_1 D_2 (CR) or CONVERT INTERFERENCE MILLIMETER / UNIT D_1 D_2 (CR). A dimension appearing on the display screen can be converted to millimeters by using CONVERT DIMENSION MILLIMETER / INPUT 1.000 (CR) which will convert any inch value to millimeters.

A fit between two mating parts can be identified either in inches or millimeters with respect to the basic hole system or the basic shaft system by using the command IDENTIFY FIT INCH BHS / UNIT D_1 D_2 (CR) or IDENTIFY FIT MILLIMETER BSS / UNIT D_1 D_2 (CR). This fit can be converted from an inch fit to a millimeter fit by using the command CONVERT FIT MILLIMETER BHS / UNIT D_1 D_2 (CR).

or or

INCH BSS

The preferred millimeter or inch nominal values can be obtained by using the SELECT MILLIMETER / UNIT D_1 D_2 (CR) or SELECT

or

INCH

MILLIMETER / INPUT 3.52 (CR) commands. With these commands, a

or

INCH

designer can convert any inch drawing appearing on the computer graphics' display screen into an equivalent millimeter drawing and vice versa without losing any of the parts' design intent (interchangeability, serviceability, function, etc.).

To illustrate how these commands are used to convert mating parts drawn in inches to mating parts drawn in millimeters, the belt tightener shown in Figure 6.14 will be used. For example, the bushing which is to be press-fitted into the roller is shown on the display screen (Figure 6.29) of the computer graphics system. In Figure 6.30, the bushing is placed inside the roller so that the centerline of the bushing coincides with the roller's centerline. The bushing and roller are enlarged so that the interference between the bushing and the roller can be seen on the display screen.

The interference between the bushing and the roller can be measured directly in millimeters from the display screen by using the command MEASURE INTERFERENCE MILLIMETERS / UNIT D_1 D_2 (CR) even though the unit of measurement was inches when constructing the bushing and roller on the display screen. The designer can go one step further and identify the metric fit which would be equivalent to the inch fit between the bushing and the roller based on the basic hole system by using the command IDENTIFY FIT MILLIMETER BHS / D_1 D_2 (CR). The computer automatically displays the results on the display screen as shown in Figures 6.31 and 6.32.

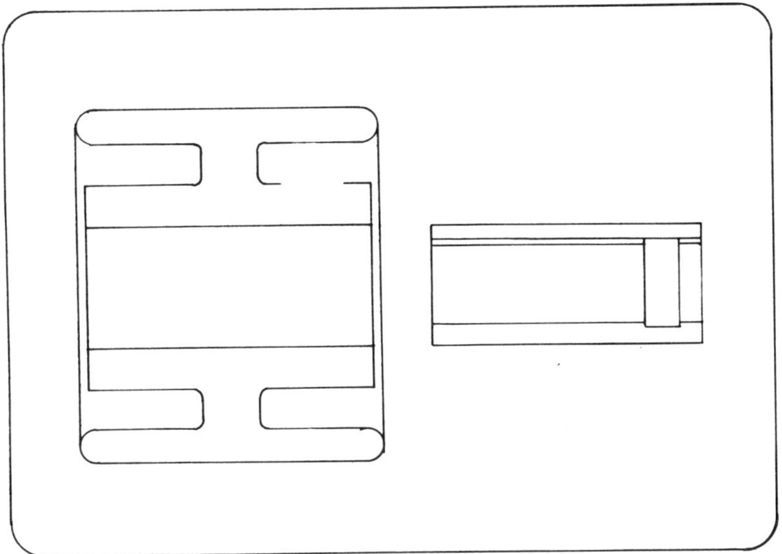

Figure 6.29 Detail drawings of pulley and bushing on computer graphics display screen.

Figure 6.30 Bushing is placed inside of pulley's hole.

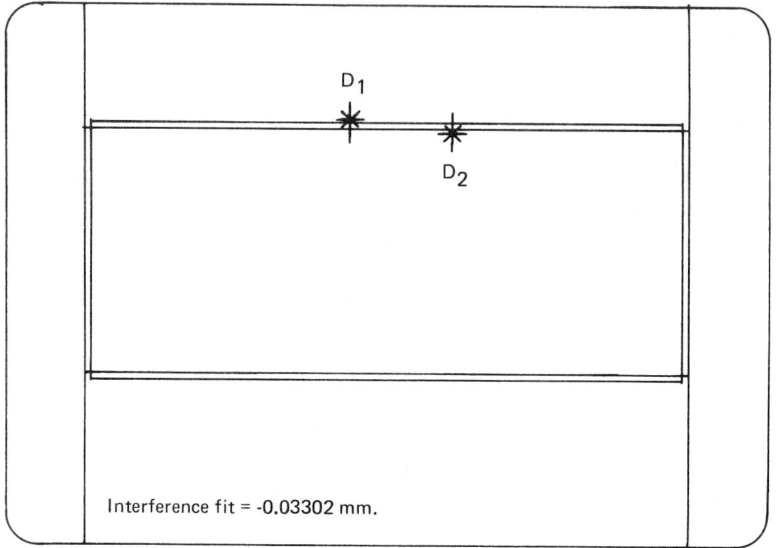

Figure 6.31 Measuring the interference in millimeters on a computer graphics display screen.

The computer has the various inch and metric tables of fits in its memory file and therefore can quickly measure a clearance or interference between two mating parts and translate that information into millimeters or inches. Also, the computer can immediately add and/or subtract the clearances and/or interference values to the maximum material conditions of the female (hole) and male (shaft) features of each part (roller and bushing) to obtain the upper and lower limits for each feature. These upper and lower limits are displayed on the computer graphics' screen with the fit information when using the IDENTIFY FIT command.

The maximum material conditions of the hole and the shaft must be used to construct the bushing and the roller in the first place. In other words, the smallest hole dimension and the largest shaft dimension must be used to construct the bushing's outside diameter and the diameter of the hole in the roller, respectively. One must remember that computers are machines and they only work with explicit values. Therefore, computers can understand inputs of one value which represent the location of points, lines, and surfaces in space instead of several values such as the case with tolerancing.

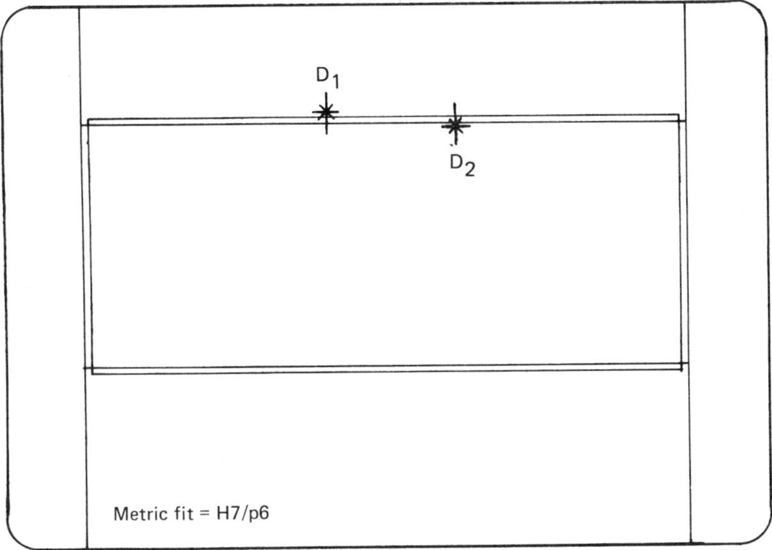

Figure 6.32 Identifying the metric fit on a computer graphics' display screen.

6.7. WORK PROBLEMS

1. Convert and make detailed working drawings of the twin cone clutch shown in Figure 6.33 from inches to millimeters using hard conversion. Use the form and position dimensioning and tolerancing system from Chapter 3 for the dimensioning scheme. Convert the American fits to metric fits by using the metric fits and preferred sizes outlined in the chapter.

2. Convert and make detailed working drawings of the lock jaw work clamp shown in Figure 6.34 from inches to millimeters using hard conversion. Use the form and position dimensioning and tolerancing system from Chapter 3 for the dimensioning scheme. Convert the American fits to metric fits by using the metric fits and preferred sizes outlined in this chapter.

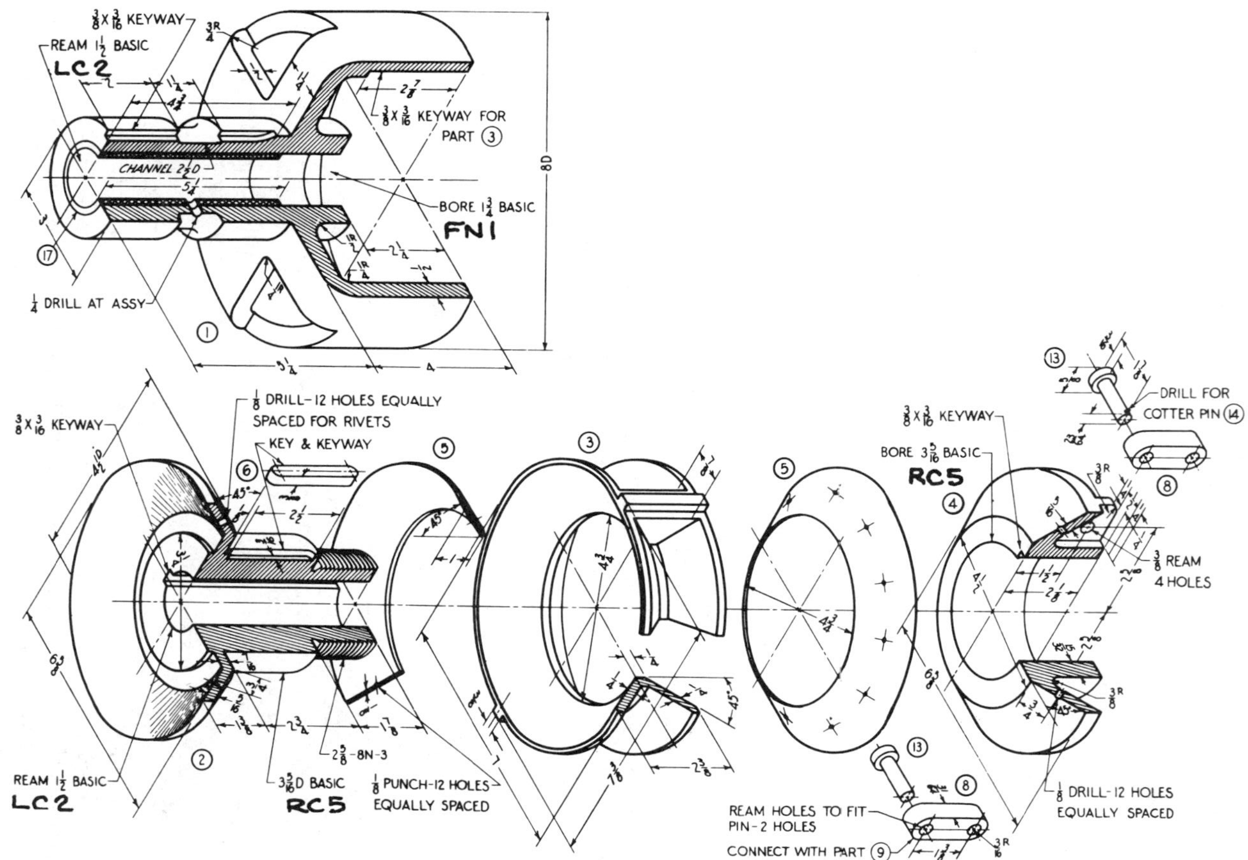
3/8 X 3/16 KEYWAY
REAM 1 1/2 BASIC
LC2
CHANNEL 2 1/2 D
3/8 X 3/16 KEYWAY FOR PART 3
BORE 1 3/4 BASIC
FN1
8D
1/4 DRILL AT ASSY
3/8 X 3/16 KEYWAY
1/8 DRILL-12 HOLES EQUALLY SPACED FOR RIVETS
KEY & KEYWAY
2 5/8-8N-3
REAM 1 1/2 BASIC
LC2
3 5/16 D BASIC
RC5
1/8 PUNCH-12 HOLES EQUALLY SPACED
3/8 X 3/16 KEYWAY
BORE 3 5/16 BASIC
RC5
DRILL FOR COTTER PIN 14
3/8 REAM 4 HOLES
REAM HOLES TO FIT PIN-2 HOLES
CONNECT WITH PART 9
1/8 DRILL-12 HOLES EQUALLY SPACED

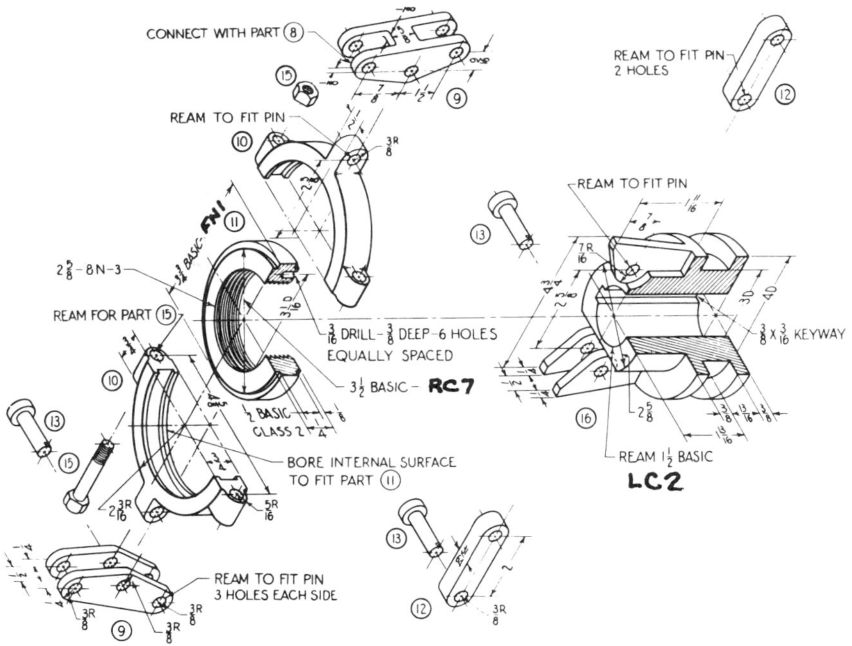

BILL OF MATERIAL

NO.	NAME	MAT'L
1	CLUTCH DRUM	C.I.
2	INSIDE FRICTION CONE	C.I.
3	FRICTION RING	C.I.
4	OUTSIDE FRICTION CONE	C.I.
5	CONE LININGS	THERM.
6	KEY	STEEL
7	LINING RIVETS	COPPER
8	SHORT TOGGLE LINK	STEEL
9	TOGGLE LEVER	C.I.
10	CLAMP COLLAR	CAST ST.
11	ADJUSTING NUT	STEEL
12	LONG TOGGLE LINK	STEEL
13	TOGGLE PINS	STEEL
14	COTTER PINS	STD.
15	5/16 BOLT & NUT	STD.
16	SHIFTING SLEEVE	C.I.
17	CLUTCH DRUM BUSH.	BRONZE

Figure 6.33 Twin cone clutch. (From H. D. Orth, R. R. Worsencroft, and H. B. Doke, *Theory and Practice of Engineering Drawing*, William C. Brown, Dubuque, Iowa, 1959.)

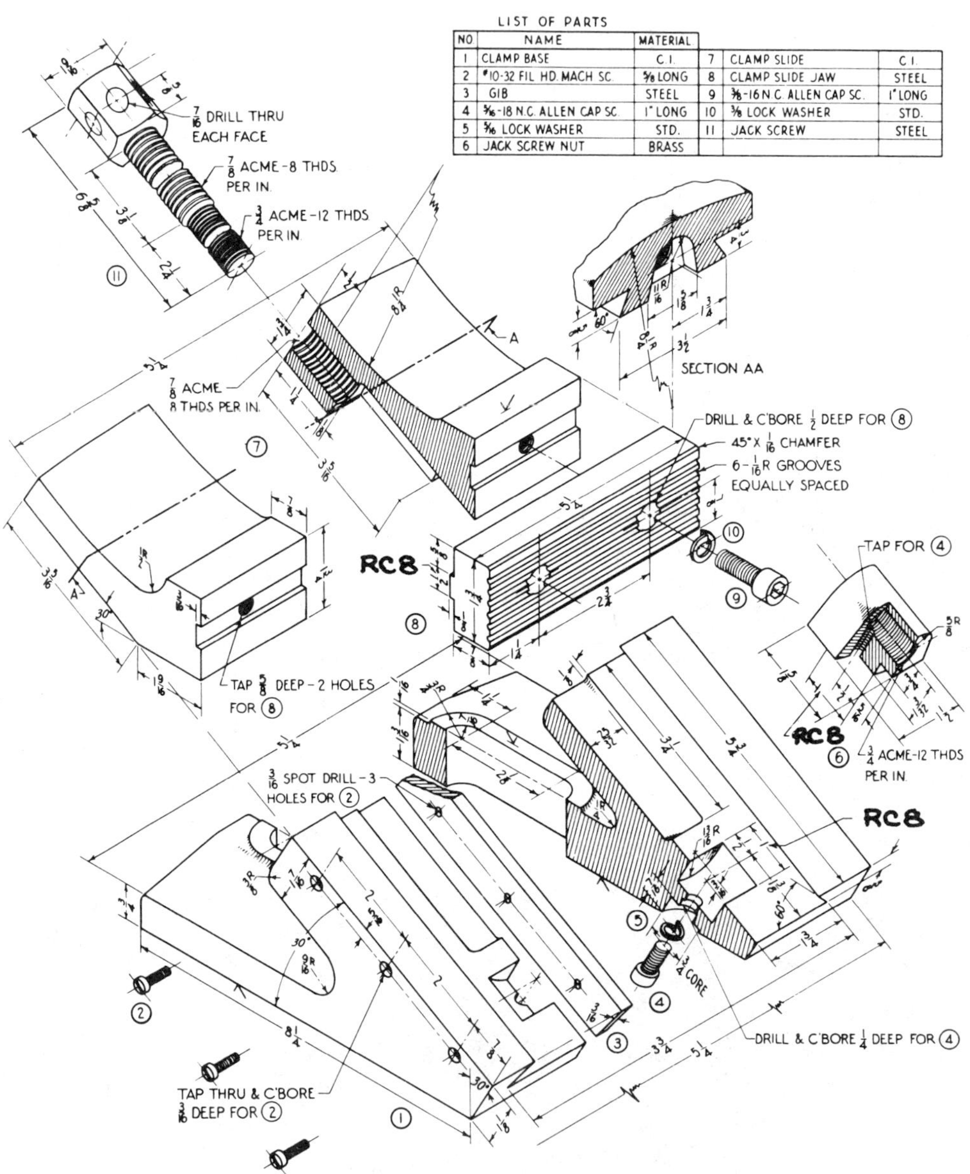

LIST OF PARTS

NO.	NAME	MATERIAL	NO.	NAME	MATERIAL
1	CLAMP BASE	C.I.	7	CLAMP SLIDE	C.I.
2	#10-32 FIL. HD. MACH. SC.	5/8 LONG	8	CLAMP SLIDE JAW	STEEL
3	GIB	STEEL	9	3/8-16 N.C. ALLEN CAP SC.	1" LONG
4	5/16-18 N.C. ALLEN CAP SC.	1" LONG	10	3/8 LOCK WASHER	STD.
5	5/16 LOCK WASHER	STD.	11	JACK SCREW	STEEL
6	JACK SCREW NUT	BRASS			

Figure 6.34 Lock-jaw work clamp. (From H. D. Orth, R. R. Worsencroft, and H. B. Doke, *Theory and Practice of Engineering Drawing*, William C. Brown, Dubuque, Iowa, 1959.)